Algebraic Combinatorics I
Association Schemes

MATHEMATICS LECTURE NOTE SERIES

Volumes of the Series published from 1961 to 1973 are not officially numbered. The parenthetical numbers shown are designed to aid librarians and bibliographers to check the completeness of their holdings.

		ISBN	
S. Lang	Algebraic Functions, 1965	0-8053-5801-3	(1)
J. Serre	Lie Algebras and Lie Groups, 1965 (3rd printing, with corrections, 1974)	0-8053-8703-X	(2)
P. J. Cohen	Set Theory and the Continuum Hypothesis, 1966 (4th printing, 1977)	0-8053-2327-9	(3)
S. Lang	Rapport sur la cohomologie des groupes, 1966	0-8053-5808-0 0-8053-5809-9	(4)
J. Serre	Algèbres de Lie semi-simples complexes, 1966	0-8053-8750-1 0-8053-8751-X	(5)
E. Artin and J. Tate	Class Field Theory, 1967 (2nd printing, 1974)	0-8053-0290-5 0-8053-0291-3	(6)
M. F. Atiyah	K-Theory, 1967	0-8053-0300-6 0-8053-0301-4	(7)
W. Feit	Characters of Finite Groups, 1967	0-8053-2434-8 0-8053-2435-6	(8)
Marvin J. Greenberg	Lectures on Algebraic Topology, 1966 (6th printing, with corrections, 1979)	0-8053-3555-4	(9)
Robin Hartshorne	Foundations of Projective Geometry, 1967 (3rd printing, 1978)	0-8053-3757-1	(10)
H. Bass	Algebraic K-Theory, 1968	0-8053-0660-9	(11)
M. Berger and M. Berger	Perspectives in Nonlinearity: An Introduction to Nonlinear Analysis, 1968	0-8053-0668-4	(12)
I. Kaplansky	Rings of Operators, 1968	0-8053-5208-2 0-8053-5209-0	(13)
I. G. MacDonald	Algebraic Geometry: Introduction to Schemes, 1968	0-8053-6690-3 0-8053-6691-1	(14)
G. W. Mackey	Induced Representation of Groups and Quantum Mechanics, 1968	0-8053-6698-9 0-8053-6699-7	(15)
R. S. Palais	Foundations of Global Non-linear Analysis, 1968	0-8053-7710-7 0-8053-7711-5	(16)
D. Passman	Permutation Groups, 1968	0-8053-7818-9 0-8053-7819-7	(17)
J. Serre	Abelian *l*-Adic Representations and Elliptic Curves, 1968	0-8053-8725-0	(18)
J. F. Adams	Lectures on Lie Groups, 1969	0-8053-0116-X	(19)
J. Barshay	Topics in Ring Theory, 1969	0-8053-0550-5	(20)
A. Borel	Linear Algebraic Groups, 1969	0-8053-1021-5	(21)
R. Bott	Lectures on K(X), 1969	0-8053-1050-9	(22)
A. Browder	Introduction to Function Algebras, 1969	0-8053-1430-X 0-8053-1431-8	(23)
G. Choquet	Lectures on Analysis (3rd printing, 1976)		
	Volume I. Integration and Topological Vector Spaces, 1969	0-8053-6955-4	(24)
	Volume II. Representation Theory, 1969	0-8053-6957-0	(25)
	Volume III. Infinite Dimensional Measures and Problem Solutions, 1969	0-8053-6959-7	(26)
E. Dyer	Cohomology Theories, 1969	0-8053-2366-X 0-8053-2367-8	(27)

MATHEMATICS LECTURE NOTE SERIES *(continued)*

		ISBN	
R. Ellis	Lectures on Topological Dynamics, 1969	0-8053-2420-8 0-8053-2421-6	(28)
J. Fogarty	Invariant Theory, 1969	0-8053-2570-0 0-8053-2571-9	(29)
William Fulton	Algebraic Curves: An Introduction to Algebraic Geometry, 1969 (5th printing, with corrections, 1978)	0-8053-3080-1 0-8053-3081-X	(30)
M. J. Greenberg	Lectures on Forms in Many Variables, 1969	0-8053-3552-8 0-8053-3553-6	(31)
R. Hermann	Fourier Analysis on Groups and Partial Wave Analysis, 1969	0-8053-3940-X 0-8053-3941-8	(32)
J. F. P. Hudson	Piecewise Linear Topology, 1969	0-8053-4551-5	(33)
K. M. Kapp and H. Schneider	Completely O-Simple Semigroups: An Abstract Treatment of the Lattice of Congruences, 1969	0-8053-5212-0 0-8053-5213-9	(34)
J. B. Keller and S. Antman, (eds.)	Bifurcation Theory and Nonlinear Eigenvalue Problems, 1969	0-8053-5240-6 0-8053-5241-4	(35)
O. Loos	Symmetric Spaces		
	Volume I. General Theory, 1969	0-8053-6620-2 0-8053-6621-0	(36)
	Volume II. Compact Spaces and Classification, 1969	0-8053-6622-9 0-8053-6623-7	(37)
H. Matsumura	Commutative Algebra, 1970 (2nd Edition—cf. Vol. 56)	0-8053-7024-2 0-8053-7025-0	(38)
A. Ogg	Modular Forms and Dirichlet Series, 1969	0-8053-7574-0 0-8053-7575-9	(39)
W. Parry	Entropy and Generators in Ergodic Theory, 1969	0-8053-7812-X 0-8053-7813-8	(40)
W. Rudin	Function Theory in Polydiscs, 1969	0-8053-8350-6 0-8053-8351-4	(41)
S. Sternberg	Celestial Mechanics Part I, 1969	0-8053-9100-2 0-8053-9101-0	(42)
S. Sternberg	Celestial Mechanics Part II, 1969	0-8053-9102-9	(43)
M. E. Sweedler	Hopf Algebras, 1969	0-8053-9254-8 0-8053-9255-6	(44)
R. Hermann	Lectures in Mathematical Physics Volume I, 1970	0-8053-3946-9 0-8053-3947-7	(45)
R. Hermann	Lie Algebras and Quantum Mechanics, 1970	0-8053-3942-6	(46)
D. L. Russell	Optimization Theory, 1970	0-8053-8364-6 0-8053-8365-4	(47)
R. K. Miller	Nonlinear Volterra Integral Equations, 1971	0-8053-7080-3 0-8053-7081-1	(48)
J. L. Challifour	Generalized Functions and Fourier Analysis, 1972	0-8053-1875-5 0-8053-1876-3	(49)
R. Hermann	Lectures in Mathematical Physics Volume II, 1972	0-8053-3952-3	(50)
I. Kra	Automorphic Forms and Kleinian Groups, 1972	0-8053-2342-2 0-8053-2343-0	(51)
G. E. Sacks	Saturated Model Theory, 1972	0-8053-8380-8 0-8053-8381-6	(52)
A. M. Garsia	Martingale Inequalities: Seminar Notes on Recent Progress, 1973	0-8053-3103-4	(53)
T. Y. Lam	The Algebraic Theory of Quadratic Forms, 1973 (2nd printing, with revisions, 1980)	0-8053-5664-3 0-8053-5666-5	(54)
George W. Mackey	Unitary Group Representations in Physics, Probability, and Number Theory, 1978	0-8053-6702-0 0-8053-6703-9	55

MATHEMATICS LECTURE NOTE SERIES *(continued)*

		ISBN	
Hideyuki Matsumura	Commutative Algebra, Second Edition, 1980	0-8053-7026-9	56
Richard Bellman	Analytic Number Theory: An Introduction, 1980	0-8053-0360-X	57
Eiichi Bannai and Tatsuro Ito	Algebraic Combinatorics I Association Schemes	0-8053-0490-8 0-8053-0491-6	58

Other volumes in preparation

Algebraic Combinatorics I
Association Schemes

Eiichi Bannai
The Ohio State University
Columbus, Ohio

Tatsuro Ito
Tsukuba University
Sakuramura, Ibaraki, Japan

1984

The Benjamin/Cummings Publishing Company, Inc.
Advanced Book Program
Menlo Park, California

London • Amsterdam • Don Mills, Ontario • Sydney • Tokyo

Library of Congress Cataloging in Publication Data

Bannai, Eiichi.
Algebraic combinatorics I.

Bibliography: p.
Includes index.
1. Combinatorial analysis. I. Itó, Tatsuró, 1948–
II. Title.
QA164.B36 1983 511′.6 83-21355
ISBN 0-8053-0490-8

American Mathematical Society (MOS) Subject Classification Scheme (1980):
05-02, 05B05, 05B25, 05B30, 05C25, 05C99, 20B25, 20C15, 33A45, 33A65, 51E99

Manufactured in the United States of America

ABCDEFGHIJ–MA–8987654

PREFACE

The purpose of these lecture notes is to give a systematic account of <u>Algebraic Combinatorics</u>. Here we mean by Algebraic Combinatorics the approach to combinatorics which was formulated in Delsarte's monumental thesis in 1973 (Delsarte (1973b)), enabling us to look at a wide range of combinatorial problems from a unified viewpoint. The origin of this approach can be of course traced back far beyond Delsarte in various branches of mathematics. One important source is found in group theory, in particular in the work of Issai Schur (cf. Schur (1933)). (This line of study is as old as any character theory of groups, including those by Frobenius, Schur, Burnside.) Another important source is found in combinatorics, in particular in the theory of experimental designs and association schemes by R. C. Bose et al. But we think it is in Delsarte's thesis that Algebraic Combinatorics is recognized as what unifies the individual approaches which are named, in each branch, algebraic graph theory, algebraic coding theory, algebraic design theory, the method of Schur rings, the method of intersection matrices in permutation group theory, and so on. It is possible to describe Algebraic Combinatorics as "a character theoretical study of combinatorial objects", or "a group theory without groups"!.

These lecture notes are based on the lectures given by the first author, Eiichi Bannai, during September 1978-December 1982 for graduate courses at The Ohio State

University. The project of publishing the lectures started while the second author, Tatsuro Ito, was visiting OSU from September 1980 to June 1981. The second author arranged them in the present form, examining all the lectures with reference to the classroom notes of Yoshimi Egawa for the lectures of the previous two years, and adding a few sections for those who are studying this area for the first time. Some developments since the time of the lectures are also incorporated in these lecture notes, and some of the conjectures mentioned in the classroom have been promoted to theorems. We tried to include the most current research in Algebraic Combinatorics as far as we could. However, since this area is progressing faster than we expected, we are afraid that we will have to update these lecture notes as soon as they are published. Nevertheless, we decided to hurry the publication in the hope that we could help those who are trying to start their own research in this area, as there seems no comprehensive treatment available on this topic at the present time other than original papers scattered in various journals and publications. We admit a lot of material is left in these lecture notes to be polished and improved both mathematically and in exposition. To help and to stimulate further research in this area are the only possible raison d'etre of these lecture notes. We should be happy if they could serve this purpose.

Part II, which is now being prepared, will be published in one or two years under the subtitle "Delsarte Theory, Codes and Designs". We should most appreciate the reader's comments, as we plan to update and improve these lecture notes to make their final version a more solid book form in some near future.

We are happy to record with gratitude that we received a lot of comments on our first manuscript from many people, and that Yoshimi Egawa made his classroom notes available, Hikoe Enomoto carefully read through our first manuscript and Dodie Shapiro typed our final manuscript. We also thank our colleagues, graduate students and secretaries of OSU who attended the lectures or helped us prepare our manuscript, in particular Joe Hemmeter for improving the English in our manuscript. The first author wishes to thank the NSF for its financial support during the preparation of these lecture notes.

Eiichi Bannai and Tatsuro Ito

TABLE OF CONTENTS

PART I. ASSOCIATION SCHEMES

NOTES FOR THE READER.

(1) The main logical dependencies between sections are as follows:

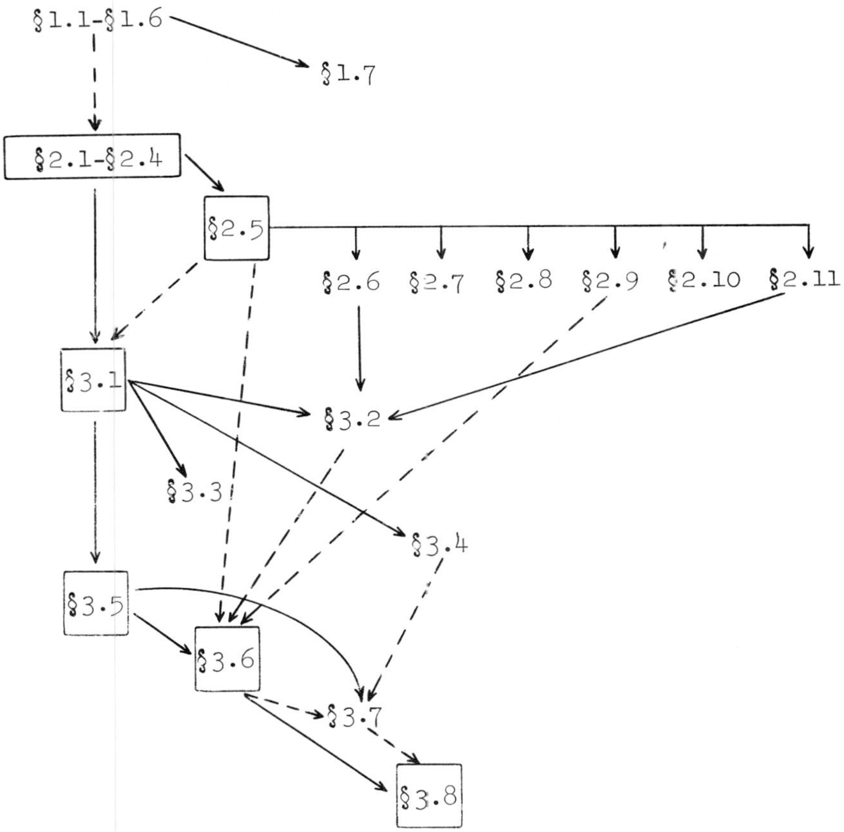

——: main dependency.

- - -: some dependency.

Some minor dependencies are not indicated here.

☐ : means key sections. Other sections could be skipped at the first reading.

(2) PROBLEM and QUESTION both mean problems open at the present time. The solution of any of them will be worth publishing as a research paper. The reader is encouraged to contact the authors about his solution. PROBLEM means it is most likely to be solved and we have some vague ideas what the solution will be, while QUESTION, though not exactly distinguished from PROBLEM, means we do not know exactly how it will be solved.

(3) For the convenience of the reader, we include a lot of literature in the references. Some of the literature is referred to in the text of Part I, some in Part II and some may not be referred to explicitly. The references are not intended to be complete, but as the authors plan to revise these lecture notes in the future, suggestions from the reader about additional references are most welcome.

INTRODUCTION

The main topic in Part I of these lecture notes is association schemes. Association schemes were, by Delsarte (1973b), recognized and fully used as the basic underlying structures of coding theory, design theory, etc. The success of Delsarte theory as well as the emergence of "Algebraic Combinatorics" can be attributed to this fact. Besides playing these roles, association schemes are important and interesting mathematical objects in their own right, having their own history of several decades. We believe that they deserve serious mathematical study themselves.

These lecture notes consist of two parts. Part I is written from this last viewpoint, and is devoted to the study of association schemes themselves as independent mathematical objects. Part II is devoted to the development of Delsarte theory, coding theory, design theory, etc., built on the framework of association schemes and their continuous analogues.

An association scheme is a finite set X and a set of relations R_i on X $(0 \leq i \leq d)$ which satisfy several axioms of regularity listed in (2.2) of section 2.2. The most important one is that the vector space $\mathfrak{A}$ spanned by the adjacency matrices A_i of R_i $(0 \leq i \leq d)$ over the complex number field becomes an algebra of dimension exactly $d+1$. Namely the product of A_i and A_j is a linear combination of A_k's: $A_i A_j = \sum_{k=0}^{d} p_{ij}^k A_k$. This means that association schemes have strong algebraic properties. If all A_i are permutation matrices, then the association scheme is nothing but a finite group. There are several other interesting ways to treat finite groups in the context of association schemes (see, for example, Example 2.1(2) of section 2.2). Association schemes are far more general structures than finite groups, but they are not arbitrarily general, as they possess the associativity and a kind of inverse element $A_{i'}$ for A_i like groups.

These are theories similar to those in group theory which work for association schemes. The most typical one is something like the character theory established for finite groups by Frobenius, Schur, Burnside, etc. In finite groups, simple groups are fundamental objects from which arbitrary finite groups are built up. In association schemes, the concept of primitive association schemes corresponds to that of simple groups, but they are so abundant that the complete classification of them seems unreasonable and unrealistic. Also the way in which association schemes are built up from primitive ones is far more complicated.

In Chapter II, we discuss these basic theories for association schemes in a general setting. In the course of the discussion, the finite group theory will often appear explicitly or implicitly for better understanding both of association schemes and finite groups. For example, the orthogonality relations for the characters of a finite group are explained beautifully from the view-point of association schemes (Hoheisel (1939)). This led Kawada (1942) to formulate the axioms of Character Algebras (C-algebras) to develop all the algebraic properties of association schemes in a beautiful way. It is interesting and remarkable that the concept of association schemes did not exist in the combinatorial context when Hoheisel and Kawada did their work.

Chapter II contains the material related to group theory such as centralizer rings of finite permutation groups (section 2.1), Schur rings (section 2.6), interpretation of the character table of a finite group from the viewpoint of association schemes (section 2.7), Norton algebras (section 2.8), primitive and imprimitive association schemes (section 2.9). Some of this material is intended to provide group theorists with background information from the combinatorial viewpoint. (The authors themselves were introduced to association schemes through the study of finite permutation groups, in particular centralizer rings.) Chapter II is intended to be a systematic introduction to the algebraic aspect of association schemes and can be used as a text or reference for association schemes.

As mentioned before, association schemes are such general structures that there is certainly a limitation to the extent of the theory as long as we are considering the general class of association schemes. Then, what classes are important association schemes? We believe that the most important ones are P-polynomial association schemes and Q-polynomial association schemes. The study of these association schemes is the main topic of Chapter III.

For the reader who meets them for the first time, we shall begin with their definitions. P-polynomial association schemes are equivalent to distance-regular graphs. They are easily understood and described in a combinatorial context as follows. A P-polynomial association scheme $(X,\{R_i\}_{0\leq i\leq d})$ is a symmetric association scheme such that a pair (x,y) of elements $x,y \in X$ is in the relation R_i if and only if x and y have distance i in the undirected graph (X,R_1), i.e., the shortest paths joining x and y have length i. Another equivalent definition is that there exist certain polynomials $v_i(x)$ of degree i such that $A_i = v_i(A_1)$ $(0 \leq i \leq d)$, where A_i is the adjacency matrix of R_i. The third equivalent condition is that the intersection matrix $B = (p^k_{1j})$ $(0 \leq j \leq d,\ 0 \leq k \leq d)$ is tridiagonal:

$$B = \begin{pmatrix} 0 & 1 & & & & 0 \\ b_0 & a_1 & c_2 & & & \\ & b_1 & a_2 & \ddots & c_{d-1} & \\ & & b_2 & \ddots & a_{d-1} & c_d \\ 0 & & & \ddots & b_{d-1} & a_d \end{pmatrix}$$

$(c_i \neq 0,\ b_i \neq 0)$.

This tridiagonal matrix is related to the polynomials $v_i(x)$ by the following 3-term recurrence

$$b_{i-1}v_{i-1}(x) + a_i v_i(x) + c_{i+1}v_{i+1}(x) = xv_i(x) .$$

As is well known in the theory of orthogonal polynomials, the existence of such a 3-term recurrence is equivalent to that of a measure on the real line on which the polynomials are orthogonal to each other with respect to the integral. (By the finiteness of the diameters d of P-polynomial association schemes, the measures are discrete point-mass measures defined only for finitely many points on the real line.) This relation between P-polynomial association schemes and orthogonal polynomials is not superficial at all. For example, the multiplicities m_i of eigenvalues of the adjacency matrix A_1 are interpretated as the Christoffer numbers in the numerical quadratures and calculated by the Christoffer-Darboux formula.

Q-polynomial association schemes were defined by Delsarte (1973b) as the duals of P-polynomial association schemes. We need some preparation to define Q-polynomial association schemes.

The adjacency (Bose-Mesner) algebra $\mathfrak{A} = \langle A_0, A_1, \ldots, A_d \rangle$ of a commutative association scheme is semi-simple and has a set of primitive idempotents $E_0, E_1, \ldots, E_d$ which are uniquely determined. The set $\{E_i\}$ is another basis of $\mathfrak{A}$. So we have

$$A_i = \sum_{j=0}^{d} p_i(j)E_j$$

and

$$E_i = \frac{1}{|X|} \sum_{j=0}^{d} q_i(j)A_j$$

for certain complex numbers $p_i(j)$, $q_i(j)$. The matrices $P = (p_i(j))$ and $Q = (q_i(j))$ are called the eigenmatrices. The eigenmatrices can be regarded as the _character tables of association schemes_. An important property of $\mathfrak{A}$ is that $\mathfrak{A}$ is closed both under the ordinary matrix multiplication and the entry-wise product $\circ$ (Hadamard product). Therefore we have

$$E_i \circ E_j = \frac{1}{|X|} \sum_{k=0}^{d} q_{ij}^k E_k$$

for certain complex numbers q_{ij}^k. The fact is that q_{ij}^k are non-negative real numbers (Krein condition).

Now Q-polynomial association schemes are defined as follows. One definition is that $E_i = v_i^*(E_1)$ for certain polynomials $v_i^*(x)$ of degree i, where the multiplication is under the Hadamard product. Another equivalent definition is that the dual intersection matrix $B = (q_{1j}^k)$ $(0 \le j \le d,\ 0 \le k \le d)$ is tridiagonal:

$$\hat{B} = \begin{pmatrix} 0 & 1 & & & & 0 \\ b_0^* & a_1^* & c_2^* & & & \\ & b_1^* & a_2^* & \ddots & c_{d-1}^* & \\ & & b_2^* & \ddots & a_{d-1}^* & c_d^* \\ 0 & & & \ddots & b_{d-1}^* & a_d^* \end{pmatrix}$$

$(c_i^* \neq 0,\ b_i^* \neq 0)$.

(a_i^*, b_i^*, c_i^* are not necessarily integers.) Unlike P-polynomial association schemes, it seems difficult to find, perhaps unlikely to exist, an equivalent combinatorial definition for Q-polynomial association schemes, but the concept of Q-polynomial association schemes is essential when we define t-designs and study them.

Chapter III is the core of Part I. The purpose of Chapter III is to study P-polynomial association schemes, in particular <u>(P and Q)-polynomial association schemes</u> with large diameters. We treat the classification of (P and Q)-polynomial association schemes as one of the central problems in Algebraic Combinatorics. (P and Q)-polynomial association schemes are an important class of association schemes in the sense that (1) they give a particularly nice framework to develop design and coding theories, (2) their "character tables" can be described in terms of Askey-Wilson orthogonal polynomials (Leonard (1982)), and (3) they are closely related to most of the finite simple groups (only excluding some Lie types of small rank and some sporadic ones). Our ambition to classify all (P and Q)-polynomial association schemes with sufficiently large diameters was set up in the very early stage of the lectures. At the beginning, we thought it might be hopelessly difficult. However, as the lectures were going on, the prospect was getting better and better, notably by the contribution of the lecture attendants (e.g., Doug Leonard, Yoshimi Egawa), and now we believe that the complete classification should not be far off, though it is still difficult. The first author, Eiichi

Bannai, wishes to acknowledge the guidance of Richard Askey on orthogonal polynomials, which helped him pursue this direction of study.

Typical and important examples of (P and Q)-polynomial association schemes are the Hamming scheme $H(d,q)$ and the Johnson scheme $J(v,d)$ (see section 3.2 for details), and they were extensively studies in connection with (ordinary) coding theory and (ordinary) design theory (cf. MacWilliams (1963), Delsarte (1973b), etc.). They are constructed from the groups S_q and S_d (the wreath product of the symmetric groups S_q and S_d) and the symmetric group S_v respectively. There are other examples of (P and Q)-polynomial association schemes which come from finite classical groups (finite simple groups of Lie type) and classical forms over a finite field $GF(q)$ (bilinear forms, alternating forms, hermitian forms, quadratic forms). Their <u>spherical functions</u> (as well as the decompositions of their permutation representations) are fully studied by Dunkl (1976 et al.) and Stanton (1980), etc. They are listed in section 3.6. This list seems to cover all (P and Q)-polynomial association schemes with large diameters except for certain relatives of them.

A program of three steps is proposed for the classification of (P and Q)-polynomial association schemes.

<u>Step 1</u>. Determination of the orthogonal polynomials which have the (P and Q)-polynomial property.

<u>Step 2</u>. Determination of the feasible parameters for (P and Q)-polynomial association schemes.

<u>Step 3</u>. Characterization of known (P and Q)-polynomial association schemes by the parameters p_{ij}^k.

Step 1 was finished by Doug Leonard in the form of Theorem 5.1 (section 3.5), concluding that such orthogonal polynomials are Asky-Wilson polynomials (balanced ${}_4\varphi_3$) or their certain limiting cases. This is the highlight of Part I. Given a (P and Q)-polynomial association scheme, let us set $u_i(x) = \frac{1}{k_i} v_i(x)$ and $u_i^*(x) = \frac{1}{m_i} v_i^*(x)$ to normalize them. Notice that the intersection matrix B (respectively dual intersection matrix $\hat{B}$) determines the 3-term recurrence of $u_i(x)$ (respectively $u_i^*(x)$)

$$c_i u_{i-1}(x) + a_i u_i(x) + b_i u_{i+1}(x) = x u_i(x) \qquad (5.1)$$

(respectively (5.2) with c_i^*, a_i^*, b_i^*),

and hence $\{u_i(x)\}$, $\{u_i^*(x)\}$ are orthogonal polynomials. The (P and Q)-polynomial property of association schemes implies

$$u_i(\theta_j) = u_j^*(\theta_i^*) \qquad (i,j = 0, 1, \ldots, d), \qquad (5.3)$$

where θ_j and θ_i^* are the eigenvalues of B and $\hat{B}$ respectively. Let us start with orthogonal polynomials $\{u_i(x)\}_{0 \le i \le d}$, $\{u_i^*(x)\}_{0 \le i \le d}$ which satisfy (5.3) with θ_j (respectively θ_i^*) determined by $\{u_i(x)\}$ (respectively $\{u_i^*(x)\}$). Such orthogonal polynomials are said to have the (P and Q)-polynomial property. Askey-Wilson polynomials (q-Racah polynomials) are balanced basic hypergeometric series ${}_4\varphi_3$, which were found to be orthogonal polynomials by Askey and Wilson (1979). They have the (P and Q)-polynomial property. Conversely, Leonard showed that orthogonal polynomials $\{u_i(x)\}$, $\{u_i^*(x)\}$ with the (P and Q)-polynomial property are balanced ${}_4\varphi_3$ or their certain limiting cases (Theorem 5.1). Askey-Wilson polynomials seem to be the widest known class of orthogonal polynomials which are expressed by using basic hypergeometric series. They are a wide and important class of orthogonal polynomials, including all the "classical" (continuous or discrete) orthogonal polynomials as their special or limiting cases. Theorem 5.1 makes it clear what is "classical" both in association schemes and orthogonal polynomials.

Step 2 is treated in section 3.7, where we show that the eigenvalues θ_i are rational integers, but much remains to be overcome to reach the final result. We give a survey of Step 3 in section 3.8 (see also section 3.6).

Step 3 is a problem which, in slightly different formulations, has a long history in many branches of combinatorics, in particular in graph theory, finite geometries and group theory. In spite of a huge amount of work done already, much remains unsolved. Recent progress can be seen in Egawa (1982) for $H(d,q)$ and Moon (1983) for $J(v,k)$. For other (P and Q)-polynomial association schemes in the list of section 3.6, the complete characterization only by the parameters p_{ij}^k seems not easy in most cases. Many researches on this characterization problem are currently going on, sometimes under additional assumptions such as local geometric structures.

Our main interest of Chapter III is in (P and Q)-polynomial association schemes. It is interesting to know how much the category of P-polynomial or Q-polynomial association schemes is wider than that of (P and Q)-polynomial association schemes. The study of known examples and continuous analogues indicates that if diameters are sufficiently large, this gap might be very small. If so, the class of (P and Q)-polynomial association schemes is really essential. We hope that in some future, this line of study could be developed beyond the classification of (P and Q)-polynomial association schemes dreamingly up to the stage where finite simple groups are classified along this line.

Part II, which is being prepared, mainly deals with Delsarte theory and coding and design theories on the framework of the (P and Q)-polynomial association schemes and their continuous analogues. We shall briefly survey them, showing some examples in depth, as it may help the reader understand the importance of P-polynomial and Q-polynomial association schemes.

(i) The original coding theory started from vector spaces over a finite field $GF(q)$ and their subspaces. Let X be a vector space of dimension d over a finite field $GF(q)$. For $x = (x_1,\ldots,x_d)$ and $y = (y_1,\ldots,y_d) \in X$, we define the distance $\partial(x,y)$ of x and y by

$$\partial(x,y) = \#\{i \mid x_i \neq y_i\}.$$

For a linear subspace Y of X, let

$$s = s(Y) = \#\{\partial(x,y) \mid x,y \in Y,\ x \neq y\} \quad \text{(degree)}$$

$$f = f(Y) = \mathrm{Min}\{\partial(x,y) \mid x,y \in Y,\ x \neq y\} \quad \text{(minimum distance)}.$$

Y is called an e-code $(e = [\frac{f-1}{2}])$, i.e., $\partial(x,y) \geq 2e + 1$ for distinct $x,y \in Y$. Intuitively, the degree s increases with the size of $|Y|$ and the minimum distance f decreases as $|Y|$ increases. One of the main concerns in coding theory is to make $|Y|$ as large as possible for a fixed e and to make e as large as possible for a fixed cardinality of Y (cf. MacWilliams and Sloane (1977)). There are well known inequalities:

$$(1) \quad |Y| \leq 1 + d(q-1) + \binom{d}{2}(q-1)^2 + \cdots + \binom{d}{s}(q-1)^s$$

$$(2) \quad |Y| \leq q^d / \{1 + d(q-1) + \binom{d}{2}(q-1)^2 + \cdots + \binom{d}{e}(q-1)^e\}$$

with $e = [\frac{f-1}{2}]$.

Let $Y^{\perp}$ be the subspace defined by

$$Y^{\perp} = \{x \in X \mid (x,y) = \sum_{i=1}^{d} x_i y_i = 0 \quad \forall y \in Y\}.$$

Let

$$t = f(Y^{\perp}) - 1 \qquad \text{(strength of } Y),$$

and

$$r = s(Y^{\perp}) \qquad \text{(external distance of } Y).$$

Y is called a t-design and coincides with an orthogonal array of strength t. Applying the inequalities (1) and (2) to $Y^\perp$, we get

$$(1)^* \quad |Y| \geq 1 + d(q-1) + \binom{d}{2}(q-1)^2 + \cdots + \binom{d}{[\frac{t}{2}]}(q-1)^{[\frac{t}{2}]}$$

$$(2)^* \quad |Y| \geq q^d/\{1 + d(q-1) + \binom{d}{2}(q-1)^2 + \cdots + \binom{d}{r}(q-1)^r\}.$$

One of the equalities in (1), $(1)^*$ implies the other, and dually one of the equalities in (2), $(2)^*$ implies the other. Such an equality-attaining subspace Y is a <u>tight</u> orthogonal array of strength t $(t = 2s)$ and a <u>perfect</u> e-code $(r = e)$. Dually, $Y^\perp$ becomes a perfect s-code and a tight orthogonal array of strength $f - 1$ $(f - 1 = 2r)$.

The whole argument is valid for the dth Cartesian power X of a q-element subset for an arbitrary integer $q \geq 2$, i.e., for the Hamming scheme $H(d,q)$, whose structure is much weaker than that of the vector space. For an arbitrary subset Y of X, there does not exist the combinatorial object $Y^\perp$ any longer, but we can still consider it algebraically. Let $\mathbb{a} = (a_0, a_1, \ldots, a_d)$ be the inner distribution of Y:

$$a_i = \frac{1}{|Y|} \#\{(x,y) \in X \times X \mid \partial(x,y) = i\}$$

$$(0 \leq i \leq d).$$

Notice that $\mathbb{a}$ is the weight distribution of Y in case Y is a subspace of a vector space X. Define the dual $\mathbb{b} = (b_0, b_1, \ldots, b_d)$ of $\mathbb{a}$ by

$$\mathbb{b} = \frac{1}{|Y|}\, \mathbb{a}\, Q,$$

where $Q = (q_i(j))$ is the second eigenmatrix of $H(d,q)$:

$$q_i(j) = \sum_{\alpha=0}^{i} (-1)^{\alpha}(q-1)^{i-\alpha}\binom{j}{\alpha}\binom{d-j}{i-\alpha} \quad \text{(Krawtchouk polynomials)}$$

$$= \binom{d}{i}(q-1)^{i}\,{}_2F_1\left(\begin{matrix}-i,\ -j\\ -d\end{matrix};\frac{q}{q-1}\right),$$

where ${}_2F_1$ is the geometric series. In the case of a subspace Y of a vector space X, this is the MacWilliams transformation and $\mathbb{b}$ is the weight distribution of $Y^{\perp}$. Using the parameters

$$s = \#\{a_i \mid a_i \neq 0,\ i = 1, 2, \ldots, d\} \quad \text{(degree)},$$

$$f = 1 + \mathrm{Max}\{h \mid a_1 = a_2 = \cdots = a_h = 0\} \quad \text{(minimum distance)},$$

$$t = \mathrm{Max}\{h \mid b_1 = b_2 = \cdots = b_h = 0\} \quad \text{(strength)},$$

$$r = \#\{b_i \mid b_i \neq 0,\ i = 1, 2, \ldots, d\} \quad \text{(external distance)},$$

the inequalities (1), (2), $(1)^*$, $(2)^*$ hold with the same equality condition as before. Y is called a t-design and coincides with an orthogonal array of strength t. Dually Y is called an e-code with $e = [\frac{f-1}{2}]$. The equality-attaining Y in (1), $(1)^*$ (respectively (2), $(2)^*$) is a tight orthogonal array of strength t (respectively a perfect e-code).

(ii) The original design theory started from the Johnson scheme $J(v,k)$ $(v/2 \geq k)$. Let X be the set of k-element subsets of a v-element set V. For $A,B \in X$, we define the distance $\partial(A,B)$ of A and B by

$$\partial(A,B) = k - |A \cap B| .$$

For a subset Y of X, the degree s and the minimum distance f are defined as in (i), and the following inequalities hold for $|Y|$:

$$(1) \quad |Y| \leq \binom{v}{s}$$

$$(2) \quad |Y| \leq \binom{v}{s} / \{1 + \binom{k}{1}\binom{v-k}{1} + \binom{k}{2}\binom{v-k}{2} + \cdots + \binom{k}{e}\binom{v-k}{e}\}$$

with $e = [\frac{f-1}{2}]$. Y is called an e-code, i.e., $\partial(A,B) \geq 2e + 1$ for distinct $A,B \in Y$.

The second eigenmatrix $Q = (q_i(j))$ of $J(v,k)$ is given by

$$\frac{q_j(i)}{m_j} = {}_3F_2\left(\begin{matrix} -i, \ -j, \ -v-1+j \\ -k, \ -v+k \end{matrix} ; 1\right) \qquad \text{(Hahn polynomials)},$$

where $m_j = \binom{v}{j} - \binom{v}{j-1}$ and ${}_3F_2$ is the hypergeometric series.

The set $Y^{\perp}$ cannot be well-defined in this case either, but the dual inner distribution $\mathbb{b}$ and hence the strength t and the external distance r can be defined algebraically as before. Then it holds that

$$(1)^* \quad |Y| \geq \binom{v}{[\frac{t}{2}]}$$

$$(2)^* \quad |Y| \geq \binom{v}{k} / \{1 + \binom{k}{1}\binom{v-k}{1} + \binom{k}{2}\binom{v-k}{2} + \cdots + \binom{k}{r}\binom{v-k}{r}\}$$

with the same equality condition as in (i). Y is called a t-design and coincides with the combinatorial t-design in the original sense. The bound-achieving Y in (1), $(1)^*$ is a <u>tight</u> t-design $(t = 2s)$ and the bound-achieving Y in (2), $(2)^*$ is a perfect e-code.

Out of the four important parameters $s, f, t, r,$ the minimum distance f and the strength t make sense only for P-polynomial and Q-polynomial association schemes respectively, because they depend on the orderings of a_i and b_i in the inner and dual inner distributions, namely, the orderings of adjacency matrices A_i and the primitive idempotents E_i, respectively. As a generalization of (i) and (ii), a subset Y of a Q-polynomial association scheme (respectively P-polynomial association scheme) is defined to be a t-design (respectively an e-code) if its strength (respectively minimum distance) is no less than t (respectively $2e + 1$). Then the following inequalities hold:

$$(1) \qquad |Y| \leq m_0 + m_1 + \dots + m_s$$

$$(1)^* \qquad |Y| \geq m_0 + m_1 + \dots + m_{[\frac{t}{2}]}$$

for a t-design Y, with one of the equalities in (1), $(1)^*$ implying the other, and

$$(2) \qquad |Y| \leq |X|/(k_0 + k_1 + \dots + k_{[\frac{f-1}{2}]})$$

$$(2)^* \qquad |Y| \geq |X|/(k_0 + k_1 + \dots + k_r)$$

for a $[\frac{f-1}{2}]$-code Y, with one of the equalities in (2), $(2)^*$ implying the other, where k_i, m_i are the valencies and multiplicities respectively.

The bound-achieving Y in (1), $(1)^*$ (respectively

(2), $(2)^*$) is called <u>tight</u> (respectively <u>perfect</u>). If a tight t-design exists in a Q-polynomial association scheme, then the polynomial $\sum_{i=0}^{[\frac{t}{2}]} v_i^*(x)$ has roots all in $\{\theta_1^*, \theta_2^*, \ldots, \theta_d^*\}$, where θ_i^* are the eigenvalues of B. Dually if a perfect e-code exists in a P-polynomial association scheme, then the polynomial $\sum_{i=0}^{e} v_i(x)$ has roots all in $\{\theta_1, \theta_2, \ldots, \theta_d\}$, where θ_i are the eigenvalues of $\hat{B}$. These are generalizations of the Lloyd theorem for perfect codes in $H(d,q)$. If the underlying association scheme has the (P and Q)-polynomial properties, then the polynomials $v_i(x)$, $v_i^*(x)$ are Askey-Wilson polynomials. Furthermore, the polynomials $\sum_{i=0}^{[\frac{t}{2}]} v_i^*(x)$ and $\sum_{i=0}^{e} v_i(x)$ are shown to be Askey-Wilson polynomials. This makes us expect that the Lloyd theorem eliminates in a unified way the possibility of the existence of tight t-designs and perfect e-codes for most of (P and Q)-polynomial association schemes. In fact, perfect e-codes $(e \geq 3)$ in $H(d,q)$ were completely classified and tight t-designs in $J(v,k)$ were known to exist for only finitely many (v,k) for a fixed t, mainly by making use of the Lloyd theorem.

The strength t is interpreted geometrically in terms of combinatorial designs and orthogonal arrays in the examples (i), (ii). But in general, there is no guarantee that a geometric interpretation of t is always possible. For the known (P and Q)-polynomial association schemes, we have a natural geometric interpretation of t in the context of semi-lattices (flag complexes) and hence there is a geometric definition of t-designs which coincides with the algebraic one in most cases. It is an open problem to define t-designs geometrically in arbitrary Q-polynomial association schemes.

(iii) This is another situation in which a similar theory holds. Let X be the d-dimensional unit sphere in $\mathbb{R}^{d+1}$. For a finite subset Y of X, define $s = s(Y)$ to be

$$\#\{\partial(x,y) \mid x,y \in Y,\ x \neq y\}$$

and $t = t(Y)$ to be the maximum integer t such that

$$\frac{1}{|X|}\int_X f(x)dw = \frac{1}{|Y|}\sum_{x \in Y} f(x)$$

for all polynomials $f(x) = f(x_1,\ldots,x_{d+1})$ of degree at most t, where $\int_X$ is the integral over X with respect to the usual measure dw and $|X|$ denotes the area of X. Y is called a spherical t-design. Then

$$(1)\qquad |Y| \leq \binom{d+s}{s} + \binom{d+s-1}{s-1}$$

$$(1)^*\qquad |Y| \geq \binom{d+[\frac{t}{2}]}{[\frac{t}{2}]} + \binom{d+[\frac{t}{2}]-1}{[\frac{t}{2}]-1}.$$

Moreover, if one of the bounds in (1), $(1)^*$ is achieved, then so is the other. Such a bound-achieving Y is called a <u>tight</u> spherical t-design $(t = 2s)$.

In the third example, we have a similar theoretical background to the previous examples. Consider the spherical representation of the orthogonal group $O(d+1)$. The ith spherical representation has degree $m_i = \binom{d+i}{i} - \binom{d+i-2}{i-2}$. The spherical functions are Gegenbauer polynomials $Q_i(x) = \frac{d+2i}{d-1} C_i^{\frac{d-1}{2}}(x)$ and they play the same role as that of $v_i^*(x)$ in Q-polynomial schemes. We have $m_i = Q_i(1)$. The inequalities (1) and $(1)^*$ can be rewritten

$$(1) \quad |Y| \leq m_0 + m_1 + \dots + m_s,$$

$$(1)^* \quad |Y| \geq m_0 + m_1 + \dots + m_{[\frac{t}{2}]} .$$

If Y attains equality in (1) and $(1)^*$, then the polynomial $\sum_{i=0}^{[\frac{t}{2}]} Q_i(x)$ has roots all in $\{(x,y) \mid x,y \in Y, x \neq y\}$, where (x,y) is the usual inner product. This fact was used to classify tight spherical t-designs for $t = 6$ and $t \geq 8$.

The same mechanism as in the example (iii) works for compact symmetric spaces of rank 1 and for some non-compact spaces such as hyperbolic spaces.

The first purpose of these lecture notes is to study the common underlying structures for the phenomena observed above and to study the character theoretical mechanisms of them. The second purpose is to study subsets Y close to the bounds (1), $(1)^*$ (or (2), $(2)^*$). This line of study includes the theories of (tight) designs, (perfect) codes, (tight) spherical designs, extremal sets, etc.

In these lecture notes, in particular in Part I, we mainly discuss the "algebraic" or "character theoretical" aspect of association schemes, and we do not discuss much of the "structure theoretical", "geometric" or "combinatorial" aspect of them. This does not mean that the structure theoretical aspect is unimportant. In finite group theory, it played a central role in the classification of finite groups. It seems natural to expect that the structure theory of association schemes will play an important role in the (future) theory of association schemes, imminently in the final stage of the (P and Q)-polynomial association scheme classification. One reason why we do not extensively discuss the structure theoretical aspect in these lecture notes is that we understand that a project of writing an extensive survey paper dealing with this aspect is going on by A. Brower, A. Cohen and A. Neumaier (Brower, Cohen and Neumaier (in preparation)). So we avoid the overlaps.

In Part II, we plan to discuss design theory, coding theory, extremal set theory. Some combinatorial or geometric discussion will be also included in connection with these theories. We also discuss finite subsets in spheres and related spaces in connection with representation of topological (classical) groups.

CHAPTER I. REPRESENTATIONS OF FINITE GROUPS

Sections 1.1 to 1.6 are devoted to an elementary and very standard introduction to ordinary representation theory of finite groups. No material is new. The reader who is familiar with this topic can skip these sections.

There are three reasons why we included these preliminary sections. (1) We want to emphasize that the representation theory of finite groups is an important main source of Algebraic Combinatorics. (2) This self-contained concise exposition on this topic is very useful for students who have no previous knowledge of representation theory of groups. (Many graduate students who are beginning to start combinatorics at OSU are usually well motivated but often lack the basic knowledge of representation theory of groups.) (3) Our treatment here follows exactly that of Chapter 6 of the textbook: Group Theory (Iwanami publisher, Tokyo, 1965) by Keizo Asano and Hirosi Nagao written in Japanese (Asano-Nagao (1965)). We believe that this text book is really excellent in its conciseness and its clearness in expositions. When the author (E. B.) lectured on this topic at OSU, he really felt sorry that this book was not available in English. So he carried this book to the class and spoke the contents of it in English. We would like to thank Professor Asano and Professor Nagao for allowing us to include here the treatment closely following that of their book, sometimes word for word.

The topic in section 1.7 (Sharp Permutation Groups) is

new and interesting. But it may be a bit isolated from the other topics in these lecture notes. We included this topic for two reasons: (1) in order to see how the previous materials in sections 1.1 to 1.6 are applied to certain problems, and (2) in order to try to broaden the realm of Algebraic Combinatorics.

1.1 REPRESENTATIONS OF GROUPS

Let $GL(n,\mathbb{C})$ be the group of all non-singular matrices of degree n over the field $\mathbb{C}$ of complex numbers. Let G be a group. A (matrix) <u>representation</u> of G is, by definition, a homomorphism of G into $GL(n,\mathbb{C})$:

$$\mathbb{A}: a \to A(a) \qquad (\forall a \in G)$$

with

$$A(ab) = A(a)A(b),$$

$$A(1) = I \text{ (the identity matrix)},$$

$$A(a^{-1}) = A(a)^{-1},$$

and n is called the <u>degree</u>. If $\mathbb{A}$ is one-to-one, the representation is called <u>faithful</u>.

EXAMPLE 1.1. The mapping which carries every element of G to $1 \in \mathbb{C}$ is a representation of degree 1. This is called the <u>identity representation</u> of G and denoted by $\mathbb{1}_G$.

EXAMPLE 1.2. Given a representation $a \to A(a)$, the mapping

$$a \to P^{-1}A(a)P \qquad (\forall a \in G)$$

becomes a representation of G for any non-singular matrix P.

Let $\mathbb{A}: a \to A(a)$ and $\mathbb{B}: a \to B(a)$ be representations of G. If there exists a non-singular matrix P such that

$$B(a) = P^{-1}A(a)P \qquad (\forall a \in G),$$

we say that $\mathbb{A}$ and $\mathbb{B}$ are <u>equivalent</u>. Equivalent representations are denoted by $\mathbb{A} \sim \mathbb{B}$. The relation $\sim$ defines <u>equivalence classes of representations</u> of G.

EXAMPLE 1.3. Let S_n be the symmetric group of degree n. For an element

$$\sigma = \begin{pmatrix} 1 & 2 & \dots & n \\ s_1 & s_2 & \dots & s_n \end{pmatrix} \in S_n,$$

let $A(\sigma)$ be the matrix whose i^{th} row is $(0,\dots,0,1,0,\dots,0)$ with 1 in the s_i-th place:

$$A(\sigma) = (\alpha_{ij}(\sigma)) \qquad (i, j = 1, 2, \dots, n)$$

with

$$\alpha_{ij}(\sigma) = \begin{cases} 1 & \text{if } j = s_i, \\ 0 & \text{otherwise.} \end{cases}$$

The mapping $\sigma \to A(\sigma)$ is a faithful representation of S_n.

EXAMPLE 1.4. Let G be a finite group consisting of elements $a_1, a_2, \dots, a_n$ and let S^G be the symmetric group on G. The mapping which carries $a \in G$ to a permutation

$$\begin{pmatrix} a_1 & a_2 & \cdots & a_n \\ a_1 a & a_2 a & \cdots & a_n a \end{pmatrix} \in S^G \tag{1.1}$$

is a one-to-one homomorphism from G to S^G. To this permutation (1.1), we attach the matrix

$$A(a) = (\alpha_{ij}(a))$$

with

$$\alpha_{ij}(a) = \begin{cases} 1 & \text{if } a_i a = a_j, \\ 0 & \text{otherwise,} \end{cases}$$

as in Example 1.3. Then the mapping $a \to A(a)$ becomes a faithful representation of G. This representation is said to be the right regular representation of G. Let $\delta(a)$ be

$$\delta(a) = \begin{cases} 1 & \text{if } a = 1, \\ 0 & \text{otherwise.} \end{cases}$$

Then

$$A(a) = \begin{pmatrix} \delta(a_1 a a_1^{-1}) & \delta(a_1 a a_2^{-1}) & \cdots & \delta(a_1 a a_n^{-1}) \\ \delta(a_2 a a_1^{-1}) & \delta(a_2 a a_2^{-1}) & \cdots & \delta(a_2 a a_n^{-1}) \\ \cdots & \cdots & \cdots & \cdots \\ \delta(a_n a a_1^{-1}) & \delta(a_n a a_2^{-1}) & \cdots & \delta(a_n a a_n^{-1}) \end{pmatrix}. \tag{1.2}$$

If $a \neq 1$, every diagonal entry is zero.

The left regular representation of G is defined similarly using the homomorphism

$$a \to \begin{pmatrix} a_1 & a_2 & \cdots & a_n \\ aa_1 & aa_2 & \cdots & aa_n \end{pmatrix}.$$

Namely,

$$A(a) = \begin{pmatrix} \delta(a_1^{-1}aa_1) & \delta(a_1^{-1}aa_2) & \cdots & \delta(a_1^{-1}aa_n) \\ \delta(a_2^{-1}aa_1) & \delta(a_2^{-1}aa_2) & \cdots & \delta(a_2^{-1}aa_n) \\ \cdots & \cdots & \cdots & \cdots \\ \delta(a_n^{-1}aa_1) & \delta(a_n^{-1}aa_2) & \cdots & \delta(a_n^{-1}aa_n) \end{pmatrix}. \qquad (1.2)'$$

Let $\varphi: a \to \varphi(a)$ be a homomorphism of G into S_n (i.e., a permutation representation of G). Expressing the permutation $\varphi(a)$ by the matrix $A(a)$ as in Example 1.3, we get a matrix representation $a \to A(a)$.

Let $\mathbb{A}: a \to A(a)$ be a representation of degree n. We say $\mathbb{A}$ is <u>reducible</u> if there exists a non-singular matrix P such that

$$P^{-1}A(a)P = \begin{pmatrix} B(a) & 0 \\ D(a) & C(a) \end{pmatrix} \qquad (\forall a \in G),$$

where $B(a)$, $C(a)$ are square matrices of degree r, s with $r \geq 1$, $s \geq 1$, $r + s = n$. Notice the representations

$$a \to A'(a) = \begin{pmatrix} B(a) & 0 \\ D(a) & C(a) \end{pmatrix}$$

and

$$a \to A''(a) = \begin{pmatrix} C(a) & D(a) \\ 0 & B(a) \end{pmatrix}$$

are equivalent, because $Q^{-1}A'(a)Q = A''(a)$ with

$$Q = \begin{pmatrix} 0 & I_r \\ I_s & 0 \end{pmatrix} \quad (I_r, I_s \text{ are the identity matrices of degree } r, s).$$

We say $\mathbb{A}$ is <u>irreducible</u> if it is not reducible. In (1.3), the mappings $a \to B(a)$ and $a \to C(a)$ become representations of degree r, s, respectively.

Given representations of G, $\mathbb{A}: a \to A(a)$ and $\mathbb{B}: a \to B(a)$ with degree n, m, respectively, the mapping

$$a \to \begin{pmatrix} A(a) & 0 \\ 0 & B(a) \end{pmatrix} \qquad (\forall a \in G)$$

becomes a representation of G with degree $n + m$. This representation is called the <u>direct sum</u> of $\mathbb{A}$ and $\mathbb{B}$ and denoted by $\mathbb{A} \oplus \mathbb{B}$.

A representation $\mathbb{A}: a \to A(a)$ of G is said to be <u>completely reducible</u> if $\mathbb{A}$ is equivalent to a direct sum of some irreducible representations, i.e., there exists a non-singular matrix P such that

$$P^{-1}A(a)P = \begin{pmatrix} F_1(a) & & & 0 \\ & F_2(a) & & \\ & & \ddots & \\ 0 & & & F_r(a) \end{pmatrix},$$

where each $\mathbb{F}_i: a \to F_i(a)$ $(i = 1, 2, \ldots, r)$ is an irreducible representation of G.

1.2 REPRESENTATIONS BY UNITARY MATRICES, AND COMPLETE REDUCIBILITY OF REPRESENTATIONS OF FINITE GROUPS

A representation $\mathbb{A}: a \to A(a)$ of G is called a <u>unitary representation</u> if $A(a)$ is a unitary matrix for all $a \in G$, i.e., ${}^t\overline{A(a)}A(a) = I$. Here ${}^t\overline{A}$ denotes the transposed of $\overline{A} = (\overline{\alpha}_{ij})$, where $A = (\alpha_{ij})$ and $\overline{\alpha}_{ij}$ is the complex conjugate of α_{ij}. In this section, we want to show that every representation of a finite group is equivalent to a unitary representation and is completely reducible.

A matrix A is said to be <u>Hermitian</u> if ${}^t\overline{A} = A$, and <u>positive definite</u> if ${}^t\overline{\mathbf{x}}A\mathbf{x} > 0$ for all nonzero column vectors $\mathbf{x}$. The following lemma is trivial.

LEMMA 2.1. For any non-singular matrix A, ${}^t\overline{A}A$ is a positive definite Hermitian matrix. The sum of positive definite Hermitian matrices is also a positive definite Hermitian matrix.

LEMMA 2.2. For any positive definite Hermitian matrix A, there exists a non-singular upper triangular matrix C such that ${}^t\overline{C}AC = I$.

<u>Proof</u>. Let $A = (\alpha_{ij})$ $(i, j = 1, 2, \ldots, n)$. Then $\alpha_{ji} = \overline{\alpha}_{ij}$ $(i, j = 1, 2, \ldots, n)$ and $\alpha_{ii} > 0$ $(i = 1, 2, \ldots, n)$. Let

$$A = \begin{pmatrix} \alpha & \mathbf{a} \\ {}^t\overline{\mathbf{a}} & B \end{pmatrix} \qquad \begin{array}{l} (\alpha = \alpha_{11} > 0, \\ \mathbf{a} = (\alpha_{12}, \alpha_{13}, \ldots, \alpha_{1n}), \\ B = (\alpha_{ij}) \quad (i, j = 2, 3, \ldots, n)). \end{array}$$

Put

$$C_1 = \begin{pmatrix} \frac{1}{\sqrt{\alpha}} & -\frac{1}{\alpha}\mathfrak{a} \\ 0 & I \end{pmatrix} .$$

Then

$${}^t\overline{C}_1 A C_1 = \begin{pmatrix} 1 & 0 \\ 0 & -\frac{1}{\alpha}\,{}^t\overline{\mathfrak{a}}\,\mathfrak{a} + B \end{pmatrix}$$

and $-\frac{1}{\alpha}\,{}^t\overline{\mathfrak{a}}\,\mathfrak{a} + B$ is a positive definite Hermitian matrix. Use induction on the degree of A. □

THEOREM 2.3. Let G be a finite group. For any representation $\mathbb{F}: a \to F(a)$ of G, there exists a non-singular upper triangular matrix C such that $C^{-1}F(a)C$ is a unitary matrix for all $a \in G$.

Proof. Set

$$A = \sum_{b \in G} {}^t\overline{F(b)}F(b) .$$

Then A is a positive definite Hermitian matrix by Lemma 2.1. So there exists a non-singular upper triangular matrix C such that

$${}^t\overline{C}AC = I$$

and so

$$A = ({}^t\overline{C})^{-1}C^{-1} .$$

Since

$$ {}^t\overline{F(a)}AF(a) = \sum_{b\in G} {}^t\overline{F(ba)}F(ba) = A, $$

we get

$$ {}^t\overline{F(a)}({}^t\overline{C})^{-1}C^{-1}F(a) = ({}^t\overline{C})^{-1}C^{-1}, $$

i.e.,

$$ {}^t\overline{(C^{-1}F(a)C)}(C^{-1}F(a)C) = I $$

and so $C^{-1}F(a)C$ is a unitary matrix. □

THEOREM 2.4. Any representation of a finite group is completely reducible.

Proof. Let $A: a \to A(a)$ be a reducible representation of a finite group G and let $A(a)$ be decomposed as

$$ A(a) = \begin{pmatrix} A_1(a) & * \\ 0 & A_2(a) \end{pmatrix}. $$

By the previous theorem, there exists a non-singular upper triangular matrix C such that $C^{-1}A(a)C$ is a unitary matrix. Set $U(a) = C^{-1}A(a)C$. Since C is upper triangular, $U(a)$ is decomposed as

$$ U(a) = \begin{pmatrix} U_1(a) & V(a) \\ 0 & U_2(a) \end{pmatrix}. $$

Since ${}^t\overline{U(a)} = U(a)^{-1} = U(a^{-1})$, we get

$$\begin{pmatrix} {}^t\overline{U_1(a)} & 0 \\ {}^t\overline{V(a)} & {}^t\overline{U_2(a)} \end{pmatrix} = \begin{pmatrix} U_1(a^{-1}) & V(a^{-1}) \\ 0 & U_2(a^{-1}) \end{pmatrix},$$

and so $V(a) = 0$. □

1.3 SCHUR'S LEMMA

LEMMA 3.1. (Schur's Lemma) Let $\mathbb{A}: a \to A(a)$ and $\mathbb{B}: a \to B(a)$ be irreducible representations of a group G with degree m and n respectively. Let P be an $m \times n$ matrix with the property that

$$A(a)P = PB(a) \qquad (\forall a \in G).$$

Then either

(i) $P = 0$

or

(ii) $m = n$ and P is non-singular.

<u>Proof</u>. Assume $P \neq 0$. We want to show (ii) holds. Suppose either $m \neq n$, or $m = n$ and P is singular. Then there exist $Q \in GL(m,\mathbb{C})$ and $R \in GL(n,\mathbb{C})$ such that

$$QPR = \begin{pmatrix} I_r & 0 \\ 0 & 0 \end{pmatrix} \qquad (I_r \text{ is the identity matrix of degree } r)$$

with $r < \max\{m,n\}$. Since $QA(a)Q^{-1} \cdot QPR = QPR \cdot R^{-1}B(a)R$, we get

$$\begin{pmatrix} A_{11} & 0 \\ A_{21} & 0 \end{pmatrix} = \begin{pmatrix} B_{11} & B_{12} \\ 0 & 0 \end{pmatrix},$$

where

$$QA(a)Q^{-1} = \begin{pmatrix} A_{11} & A_{12} \\ A_{21} & A_{22} \end{pmatrix} \qquad (A_{11} \text{ is a square matrix of degree } r)$$

and

$$R^{-1}B(a)R = \begin{pmatrix} B_{11} & B_{12} \\ B_{21} & B_{22} \end{pmatrix} \qquad (B_{11} \text{ is a square matrix of degree } r).$$

Therefore $A_{21} = 0$ if $r < m$ and $B_{12} = 0$ if $r < n$. Anyway, $\mathbb{A}$ or $\mathbb{B}$ is reducible, which is a contradiction. □

THEOREM 3.2. Let $\mathbb{A}$: $a \to A(a)$ be an irreducible representation of a group G. Let P be a matrix with the property that $A(a)P = PA(a)$ for all $a \in G$. Then $P = \lambda I$ for some $\lambda \in \mathbb{C}$.

Proof. Let λ be an eigenvalue of P. Then $\det(\lambda I - P) = 0$ and moreover

$$A(a)(\lambda I - P) = (\lambda I - P)A(a) \qquad (\forall a \in G).$$

By Schur's lemma,

$$\lambda I - P = 0. \qquad \square$$

THEOREM 3.3. Let G be an Abelian group. Then every irreducible representation of G is of degree 1.

<u>Proof</u>. Let $\mathbb{A}$: $a \to A(a)$ be an irreducible representation of G. Since $A(a)$ commutes with each $A(b)$, the previous theorem claims that $A(a) = \lambda(a)I$ for some $\lambda(a) \in \mathbb{C}$. Since $\mathbb{A}$ is irreducible, the degree of $\mathbb{A}$ must be 1. □

1.4 ORTHOGONALITY RELATIONS OF CHARACTERS

From now on, groups are always assumed to be of finite order.

CHARACTERS. For a square matrix $A = (\alpha_{ij})$ of degree n, $\mathrm{tr}A$ denotes the <u>trace</u> of A, i.e.,

$$\mathrm{tr}A = \alpha_{11} + \alpha_{22} + \dots + \alpha_{nn}.$$

Direct calculation shows the following:

LEMMA 4.1.

(1) $\mathrm{tr}(AB) = \mathrm{tr}(BA)$.

(2) $\mathrm{tr}(P^{-1}AP) = \mathrm{tr}A$ for any non-singular matrix P.

For a representation $\mathbb{A}$: $a \to A(a)$ of a group G, we set $\mathrm{tr}A(a) = \chi(a)$. Then $\chi(a)$ is a $\mathbb{C}$-valued function and called the <u>character</u> of $\mathbb{A}$. Obviously, $\chi(1)$ is equal to the degree of $\mathbb{A}$. Characters of irreducible representations are called <u>irreducible characters</u>. From Lemma 4.1(2), we have the following:

LEMMA 4.2. Equivalent representations of a group have the same character.

Since $A(x^{-1}ax) = A(x)^{-1}A(a)A(x)$, it holds that

$\chi(x^{-1}ax) = \chi(a)$.

So χ takes the same value on a conjugacy class of G. Such a function is called a <u>class function</u>.

THE FIRST ORTHOGONALITY RELATION OF CHARACTERS. Let G be a group of order g. Let $\mathbb{A}: a \to A(a) = (\alpha_{ij}(a))$ and $\mathbb{B}: a \to B(a) = (\beta_{ij}(a))$ be irreducible representations of G with degree m, n, respectively. For an arbitrary $m \times n$ matrix $U = (\gamma_{ij})$, we set

$$V = \sum_{x \in G} A(x)UB(x^{-1}).$$

Then

$$A(a)V = \sum_{x \in G} A(ax)UB(x^{-1})$$

$$= \sum_{y \in G} A(y)UB(y^{-1}a) \qquad (y = ax)$$

$$= \sum_{y \in G} A(y)UB(y^{-1})B(a).$$

Since y varies over G as x does, we get

$$A(a)V = VB(a). \tag{4.1}$$

Assume A and B are not equivalent. Then $V = 0$ by Schur's Lemma. The (i,j) entry of V is

$$\sum_{x \in G} \sum_{\mu,\nu} \alpha_{i\mu}(x)\gamma_{\mu\nu}\beta_{\nu j}(x^{-1}) = 0.$$

In particular, let $\gamma_{\mu\nu} = 1$ for a certain pair μ, ν and

$\gamma_{\rho\sigma} = 0$ otherwise to get

$$\sum_{x \in G} \alpha_{i\mu}(x)\beta_{\nu j}(x^{-1}) = 0 .$$

Next, assume $\mathbb{A} = \mathbb{B}$. Then $V = \lambda I$ for some $\lambda \in \mathbb{C}$ by Theorem 3.2. The (i,j) entry of V is

$$\sum_{x \in G} \sum_{\mu,\nu} \alpha_{i\mu}(x)\gamma_{\mu\nu}\alpha_{\nu j}(x^{-1}) = \delta_{ij}\lambda, \tag{4.2}$$

where $\delta_{ii} = 1$, $\delta_{ij} = 0$ $(i \neq j)$. Take the trace of

$$V = \sum_{x \in G} A(x)UA(x^{-1}) = \lambda I \tag{4.3}$$

to get

$$g(\gamma_{11} + \gamma_{22} + \cdots + \gamma_{nn}) = n\lambda \qquad (n \text{ is the degree of } \mathbb{A}),$$

that is,

$$\lambda = \frac{g}{n}(\gamma_{11} + \gamma_{22} + \cdots + \gamma_{nn}) . \tag{4.4}$$

Let $\gamma_{\mu\nu} = 1$ for a certain pair μ, ν and $\gamma_{\rho\sigma} = 0$ otherwise. Then

$$\sum_{x \in G} \alpha_{i\mu}(x)\alpha_{\nu j}(x^{-1}) = \delta_{ij}\delta_{\mu\nu}\frac{g}{n} .$$

So we get the following:

THEOREM 4.3. Let G be a group of order g.

(1) Let $\mathbb{A}$: $a \to A(a) = (\alpha_{ij}(a))$ be an irreducible representation of G with degree n. Then

$$\sum_{x \in G} \alpha_{i\mu}(x)\alpha_{\nu j}(x^{-1}) = \delta_{ij}\delta_{\mu\nu}\frac{g}{n}.$$

(2) Let $\mathbb{B}$: $a \to B(a) = (\beta_{ij}(a))$ be an irreducible representation which is not equivalent to $\mathbb{A}$. Then

$$\sum_{x \in G} \alpha_{i\mu}(x)\beta_{\nu j}(x^{-1}) = 0 .$$

Let χ, χ' be the characters of $\mathbb{A}, \mathbb{B}$. In the above theorem, put $\mu = i$, $\nu = j$ and take the sum over i, j to get the following:

THEOREM 4.4. (The First Orthogonality Relation of Characters) Let G be a group of order g.

(1) Let χ be an irreducible character of G. Then

$$\sum_{x \in G} \chi(x)\chi(x^{-1}) = g .$$

(2) Let χ, χ' be the characters of non-equivalent irreducible representations of G. Then

$$\sum_{x \in G} \chi(x)\chi'(x^{-1}) = 0.$$

Notice that $\chi(a^{-1}) = \overline{\chi(a)}$ for all $a \in G$, because Theorem 2.3 claims that $\mathbb{A}$ is equivalent to a unitary representation $\mathbb{U}$: $a \to U(a)$ and so

$$\chi(a^{-1}) = \mathrm{tr}U(a^{-1}) = \mathrm{tr}U(a)^{-1} = \mathrm{tr}\,{}^t\overline{U(a)} = \overline{\chi(a)}. \quad (4.5)$$

Let $\mathbb{F}_1, \mathbb{F}_2, \ldots$ be representatives of the equivalence classes of irreducible representations of a group G and let $\chi_1, \chi_2, \ldots$ be the characters of $\mathbb{F}_1, \mathbb{F}_2, \ldots$. Let $C_1, C_2, \ldots, C_k$ be the conjugacy classes of G with $|C_\alpha| = h_\alpha$ $(\alpha = 1, 2, \ldots, k)$ and let $a_1, a_2, \ldots, a_k$ be representatives of the conjugacy classes. Since characters are class functions, Theorem 4.4 is rewritten as follows:

THEOREM 4.4'.

$$\sum_{\alpha=1}^{k} h_\alpha \chi_i(a_\alpha)\overline{\chi_j(a_\alpha)} = \delta_{ij} g .$$

For $\mathbb{C}$-valued functions φ, ψ on a group G of order g, we define the <u>inner</u> <u>produce</u> $(\varphi,\psi)_G$ by

$$(\varphi,\psi)_G = \frac{1}{g} \sum_{x \in G} \varphi(x)\psi(x^{-1}) .$$

We sometimes write (φ,ψ) for $(\varphi,\psi)_G$, when it is clear we are discussing G. Obviously, the inner product is a symmetric bilinear form:

$$(\varphi,\psi) = (\psi,\varphi),$$

$$(\varphi_1 + \varphi_2,\psi) = (\varphi_1,\psi) + (\varphi_2,\psi),$$

$$(\varphi,\psi_1 + \psi_2) = (\varphi,\psi_1) + (\varphi,\psi_2),$$

$$(\lambda\varphi,\psi) = (\varphi,\lambda\psi) = \lambda(\varphi,\psi) .$$

With this notation, the first orthogonality relation of characters is expressed as follows:

THEOREM 4.4". Let $\chi_1,\chi_2,\cdots$ be the characters of non-equivalent irreducible characters of a group G. Then

$$(\chi_i,\chi_j) = \delta_{ij} .$$

MULTIPLICITIES OF IRREDUCIBLE REPRESENTATIONS. Let $\mathbb{A}$ be a representation of a group G. Since $\mathbb{A}$ is completely reducible by Theorem 2.3, $\mathbb{A}$ is equivalent to

$$\begin{pmatrix} \overbrace{\mathbb{F}_1 \ \ddots \ \mathbb{F}_1}^{m_1} & & & 0 \\ & \overbrace{\mathbb{F}_2 \ \ddots \ \mathbb{F}_2}^{m_2} & & \\ & & \ddots & \\ 0 & & & \ddots \end{pmatrix},$$

where $\mathbb{F}_1, \mathbb{F}_2, \ldots$ are non-equivalent irreducible representations. The number m_i is called the <u>multiplicity</u> of $\mathbb{F}_i$ in $\mathbb{A}$ ($m_i = 0$ if $\mathbb{F}_i$ does not appear) and we write

$$\mathbb{A} \sim m_1\mathbb{F}_1 + m_2\mathbb{F}_2 + \cdots \tag{4.6}$$

Let χ be the character of $\mathbb{A}$ and χ_i the character of $\mathbb{F}_i$ $(i = 1, 2, \ldots)$. Then

$$\chi = m_1\chi_1 + m_2\chi_2 + \cdots \tag{4.7}$$

If $m_i \neq 0$, $\mathbb{F}_i$ and χ_i are called <u>irreducible components</u> of $\mathbb{A}$ and χ respectively.

THEOREM 4.5. Let G be a group and let χ be the character of a representation of G. Let m_i be the multiplicity of an irreducible character χ_i in χ. Then

$$m_i = (\chi, \chi_i) = \frac{1}{g} \sum_{x \in G} \chi(x)\overline{\chi_i(x)}\,.$$

<u>Proof</u>. Let $\chi = \sum_j m_j\chi_j$ be the decomposition of χ into

the sum of irreducible characters with m_j the multiplicity of χ_j. Then

$$(\chi,\chi_i) = \sum_j m_j(\chi_j,\chi_i) = m_i . \qquad \square$$

THEOREM 4.6. Let $\mathbb{A}$, $\mathbb{B}$ be representations of a group G and let χ, χ' be the characters of $\mathbb{A}$, $\mathbb{B}$ respectively. Then $\mathbb{A}$ and $\mathbb{B}$ are equivalent if and only if $\chi = \chi'$.

Proof. By the previous theorem, the multiplicities of $\mathbb{F}_i$ in $\mathbb{A}$ and $\mathbb{B}$ are determined by the characters of $\mathbb{A}$ and $\mathbb{B}$. Since any representation of G is completely reducible, $\mathbb{A}$ and $\mathbb{B}$ are equivalent if and only if each irreducible representation $\mathbb{F}_i$ has the same multiplicity in $\mathbb{A}$ and $\mathbb{B}$. So $\mathbb{A} \sim \mathbb{B}$ if and only if $\chi = \chi'$. $\square$

Let π be the character of the right regular representation of a group G of order g. Notice that

$$\pi(a) = \begin{cases} g & \text{if } a = 1, \\ 0 & \text{otherwise.} \end{cases} \tag{4.8}$$

For the character χ_i of each irreducible representation $\mathbb{F}_i$, it holds that

$$(\pi,\chi_i) = \frac{1}{g} \sum_{x \in G} \pi(x)\overline{\chi_i(x)}$$

$$= \frac{1}{g} \pi(1)\chi_i(1)$$

$$= \chi_i(1) \qquad (\chi_i(1) \text{ is the degree of } \mathbb{F}_i).$$

Hence, the following holds:

THEOREM 4.7. Let π be the character of the right regular representation. Then each irreducible representation $\mathbb{F}_i$ of G has multiplicity f_i, where f_i is the degree of $\mathbb{F}_i$. So

$$\pi = \sum_i f_i \chi_i ,$$

where the summation is over all irreducible characters χ_i of G.

Notice the right and left regular representations are equivalent, since the character π' of the left regular representation satisfies (4.8) as well and so $\pi = \pi'$.

Theorem 4.7 claims that every irreducible character appears in π as a component and so G has only finitely many irreducible characters. We shall show later that the number of irreducible characters of G coincides with that of conjugacy classes of G.

THEOREM 4.8. Let $\chi_1, \chi_2, \ldots, \chi_\ell$ be all the distinct irreducible characters of a group G. Let f_i be the degree of χ_i $(i = 1, 2, \ldots, \ell)$ and let g be the order of G. Then

$$g = f_1^2 + f_2^2 + \ldots + f_\ell^2 ,$$

and for $a \neq 1$,

$$f_1\chi_1(a) + f_2\chi_2(a) + \ldots - f_\ell\chi_\ell(a) = 0 .$$

<u>Proof</u>. Evaluate $\pi = \sum_{i=1}^{\ell} f_i \chi_i$ at a, using (4.8). □

THE SECOND ORTHOGONALITY RELATION OF CHARACTERS. Let G be a group and let $C_1 = \{1\}, C_2, \ldots, C_k$ be the conjugacy classes of G. For a conjugacy class C_α, make a formal sum of the elements of C_α:

$$\mathbb{C}_\alpha = a_1 + a_2 + \dots + a_{h_\alpha} \qquad (h_\alpha = |C_\alpha|).$$

Define the product of $\mathbb{C}_\alpha$ and $\mathbb{C}_\beta$ by

$$\mathbb{C}_\alpha \mathbb{C}_\beta = \sum_{i,j} a_i b_j , \tag{4.9}$$

where $\mathbb{C}_\beta = b_1 + b_2 + \dots + b_{h_\beta}$ and the summation is over $1 \leq i \leq h_\alpha$, $1 \leq j \leq h_\beta$. For an element $c \in C_\gamma$, let t be the number of pairs $(a,b) \in C_\alpha \times C_\beta$ such that $ab = c$. Then for $c' = x^{-1}cx \in C_\gamma$, there are exactly t pairs $(a',b') \in C_\alpha \times C_\beta$ such that $a'b' = c'$, because $ab = c$ if and only if $a'b' = c'$ with $a' = x^{-1}ax$, $b' = x^{-1}bx$. So each element of C_γ appears the same number of times in the right hand side of (4.9), i e.,

$$\mathbb{C}_\alpha \mathbb{C}_\beta = \sum_{\gamma=1}^{k} t_{\alpha\beta\gamma} \mathbb{C}_\gamma . \tag{4.10}$$

The set of all elements a^{-1} with $a \in C_\alpha$ becomes a conjugacy class. We denote this conjugacy class by $C_{\alpha'}$. Then

$$t_{\alpha\beta 1} = \begin{cases} h_\alpha & \text{if } C_\beta = C_{\alpha'}, \\ 0 & \text{otherwise} . \end{cases} \tag{4.11}$$

Let $\mathbb{F}$: $a \to F(a)$ be an irreducible representation of G and let f be the degree of $\mathbb{F}$. Define $F(\mathbb{C}_\alpha)$ by

$$F(\mathbb{C}_\alpha) = \sum_{a \in C_\alpha} F(a) .$$

Then

$$F(x)^{-1} F(\mathbb{C}_\alpha) F(x) = \sum_{a \in C_\alpha} F(x^{-1}ax) = F(\mathbb{C}_\alpha) ,$$

because $x^{-1}ax$ varies over C_α as a does. So the matrix $F(\mathbb{C}_\alpha)$ commutes with every $F(x)$ and by Theorem 3.2,

$$F(\mathbb{C}_\alpha) = \omega_\alpha I . \tag{4.12}$$

Take the trace of (4.12) to get

$$h_\alpha \chi(a_\alpha) = \omega_\alpha f,$$

$$\omega_\alpha = h_\alpha \chi(a_\alpha)/f , \tag{4.13}$$

where χ is the character of $\mathbb{F}$ and $a_\alpha \in C_\alpha$. By (4.10),

$$F(\mathbb{C}_\alpha)F(\mathbb{C}_\beta) = \sum_\gamma t_{\alpha\beta\gamma} F(\mathbb{C}_\gamma),$$

$$\omega_\alpha \omega_\beta = \sum_\gamma t_{\alpha\beta\gamma} \omega_\gamma .$$

Put (4.13) into the above identity to get

$$\frac{h_\alpha \chi(a_\alpha)}{f} \frac{h_\beta \chi(a_\beta)}{f} = \sum_\gamma t_{\alpha\beta\gamma} \frac{h_\gamma \chi(a_\gamma)}{f},$$

$$\chi(a_\alpha)\chi(a_\beta) = \sum_\gamma t_{\alpha\beta\gamma} \frac{h_\gamma}{h_\alpha h_\beta} f\chi(a_\gamma) . \tag{4.14}$$

Let $\chi_1, \chi_2, \ldots, \chi_\ell$ be the distinct irreducible characters of G and let f_i be the degree of χ_i. The identity (4.14) holds for every χ_i. Take the sum of (4.14) over i to get

$$\sum_{i=1}^{\ell} \chi_i(a_\alpha)\chi_i(a_\beta) = \sum_\gamma t_{\alpha\beta\gamma} \frac{h_\gamma}{h_\alpha h_\beta} (\sum_i f_i \chi_i(a_\gamma))$$

$$= t_{\alpha\beta 1} \frac{1}{h_\alpha h_\beta} g \qquad \text{(by Theorem 4.8)}$$

$$= \begin{cases} g/h_\alpha & \text{if } C_\beta = C_{\alpha'}, \\ 0 & \text{otherwise.} \end{cases} \qquad \text{(by (4.11))}.$$

So we get

$$\sum_{i=1}^{\ell} \chi_i(a_\alpha)\chi_i(a_{\beta'}) = \delta_{\alpha\beta} \frac{g}{h_\alpha} .$$

The number g/h_α is the order of $N(a_\alpha)$, the centralizer of a_α in G. Since $\chi_i(a_{\beta'}) = \overline{\chi_i(a_\beta)}$ by (4.5), we get the following:

THEOREM 4.9. (The Second Orthogonality Relation of Characters) Let $\{\chi_i\}$ be the set of distinct irreducible characters of G and let $\{a_\alpha\}$ be a set of representatives of the conjugacy classes of G. Then

$$\sum_i \chi_i(a_\alpha)\overline{\chi_i(a_\beta)} = \delta_{\alpha\beta} n_\alpha ,$$

where n_α is the order of $N(a_\alpha)$ and the summation is over all irreducible characters χ_i of G.

THEOREM 4.10. The number of distinct irreducible characters of G is equal to that of the conjugacy classes of G.

Proof. We use the following simple fact about matrices. Let A be an $m \times n$ matrix and B an $n \times m$ matrix. If the determinant of AB, a square matrix of degree m, is not zero, then $m \leq n$.

Let $\chi_1, \chi_2, \ldots, \chi_\ell$ be the distinct irreducible characters of G and let $a_1, \ldots, a_k$ be representatives of the conjugacy classes of G. Then by Theorem 4.4',

$$\begin{pmatrix} \chi_1(a_1) & \cdots & \chi_1(a_k) \\ \cdot & \cdots & \cdot \\ \chi_\ell(a_1) & \cdots & \chi_\ell(a_k) \end{pmatrix} \begin{pmatrix} h_1\overline{\chi_1(a_1)} & \cdots & h_1\overline{\chi_\ell(a_1)} \\ \cdot & \cdots & \cdot \\ h_k\overline{\chi_1(a_k)} & \cdots & h_k\overline{\chi_\ell(a_k)} \end{pmatrix}$$

$$= \begin{pmatrix} g & & 0 \\ & \ddots & \\ 0 & & g \end{pmatrix}.$$

So $\ell \leq k$. By Theorem 4.9,

$$\begin{pmatrix} \chi_1(a_1) & \cdots & \chi_\ell(a_1) \\ \cdot & \cdots & \cdot \\ \chi_1(a_k) & \cdots & \chi_\ell(a_k) \end{pmatrix} \begin{pmatrix} \overline{\chi_1(a_1)} & \cdots & \overline{\chi_1(a_k)} \\ \cdot & \cdots & \cdot \\ \chi_\ell(a_1) & \cdots & \chi_\ell(a_k) \end{pmatrix}$$

$$= \begin{pmatrix} n_1 & & 0 \\ & \ddots & \\ 0 & & n_k \end{pmatrix}.$$

So $k \leq \ell$. Therefore, $k = \ell$. □

1.5 INDUCED REPRESENTATIONS

Let G be a group and H a subgroup of G. Let g

and h be the orders of G and H respectively. For a function φ on G, φ_H denotes the restriction of φ to H. If φ is a class function on G, then φ_H is a class function on H too. If φ is the character of a representation $\mathbb{A}$ of G, then φ_H is the character of $\mathbb{A}_H$, the restriction of $\mathbb{A}$ to H.

For a function θ on H, we define a function θ^G on G by

$$\theta^G(a) = \frac{1}{h} \sum_{x \in G} \theta(x^{-1}ax), \tag{5.1}$$

where we set $\theta(x^{-1}ax) = 0$ if $x^{-1}ax$ is not contained in H. Notice that θ^G is a class function on G even if θ is not a class function on H. If a is not conjugate to any element of H, then $\theta^G(a) = 0$.

LEMMA 5.1. Let φ be a class function on G and let θ be a class function on a subgroup H of G. Then

$$(\theta^G, \varphi)_G = (\theta, \varphi_H)_H .$$

Proof.

$$(\theta^G, \varphi)_G = \frac{1}{g} \sum_{a \in G} \theta^G(a)\varphi(a^{-1})$$

$$= \frac{1}{gh} \sum_{a \in G} \sum_{x \in G} \theta(x^{-1}ax)\varphi(a^{-1}) .$$

Only the pairs (a,x) with $x^{-1}ax \in H$ contribute to the sum. So taking the sum over (a,x) with $a = x\tilde{a}x^{-1}$ for some $\tilde{a} \in H$, we get

$$(\theta^G,\varphi)_G = \frac{1}{gh}\sum_{\tilde{a}\in H}\sum_{x\in G}\theta(\tilde{a})\varphi(x\tilde{a}^{-1}x^{-1})$$

$$= \frac{1}{h}\sum_{\tilde{a}\in H}\theta(\tilde{a})\Big(\frac{1}{g}\sum_{x\in G}\varphi(x\tilde{a}^{-1}x^{-1})\Big)$$

$$= \frac{1}{h}\sum_{\tilde{a}\in H}\theta(\tilde{a})\varphi(\tilde{a}^{-1})$$

$$= (\theta,\varphi_H)_H . \qquad \square$$

If θ is the character of a representation of H, we call θ^G an <u>induced character</u> of G and say that θ^G is induced from θ. We want to show that an induced character is actually the character of some representation of G.

Let $\{a_1,a_2,\ldots,a_r\}$ be a set of representatives of the left cosets of H in G:

$$G = Ha_1 \cup Ha_2 \cup \ldots \cup Ha_r .$$

For a representation $\mathbb{A}\colon \tilde{a} \to A(\tilde{a})$ $(\tilde{a} \in H)$ of H, define a matrix $A^G(a)$ by

$$A^G(a) = \begin{pmatrix} A(a_1aa_1^{-1}) & A(a_1aa_2^{-1}) & \ldots & A(a_1aa_r^{-1}) \\ A(a_2aa_1^{-1}) & A(a_2aa_2^{-1}) & \ldots & A(a_2aa_r^{-1}) \\ \cdots & \cdots & \cdots & \cdots \\ A(a_raa_1^{-1}) & A(a_raa_2^{-1}) & \ldots & A(a_raa_r^{-1}) \end{pmatrix}, \qquad (5.2)$$

where we set $A(x) = 0$ if x is not contained in H. This is a generalization of the right regular representation of G. We want to show that

$$\mathbb{A}^G\colon a \to A^G(a) \qquad (\forall a \in G)$$

becomes a representation of G and has degree nr, where $r = |G : H|$ and n is the degree of $\mathbb{A}$. For a fixed a_k^{-1} and $x \in G$, the set $\{a_i x a_k^{-1} \mid i = 1, 2, \ldots, r\}$ is a set of representatives of the left cosets of H and so there is just one non-zero matrix among $A(a_i a a_k^{-1})$, $i = 1, 2, \ldots, r$. Similarly the set $\{a_i x a_k^{-1} \mid k = 1, 2, \ldots, r\}$ is a set of representatives of the right cosets of H and so there is just one non-zero matrix among $A(a_i a a_k^{-1})$, $k = 1, 2, \ldots, r$. Let C_{ik} be the (i,k) block of $A^G(a)A^G(b)$. Then

$$C_{ik} = \sum_{j=1}^{r} A(a_i a a_j^{-1})A(a_j b a_k^{-1}) .$$

We want to show $C_{ik} = A(a_i ab a_k^{-1})$. There is just one $s \in \{1, 2, \ldots, r\}$ such that $a_i a a_s^{-1} \in H$ and there is just one $t \in \{1, 2, \ldots, r\}$ such that $a_t b a_k^{-1} \in H$. If $s = t$, then $C_{ik} = A(a_i a a_t^{-1})A(a_t b a_k^{-1}) = A(a_i ab a_k^{-1})$. If $s \neq t$, then $C_{ik} = 0$ and $A(a_i ab a_k^{-1}) = 0$, because $a_i ab a_k^{-1} = a_i a a_t^{-1} \cdot a_t b a_k^{-1} \notin H$. Anyway $C_{ik} = A(a_i ab a_k^{-1})$ and hence $A^G(a)A^G(b) = A^G(ab)$. Since $A^G(a)A^G(a^{-1}) = A^G(1) = I$, $A^G(a)$ is non-singular. So A^G is a representation of G.

Let θ be the character of $\mathbb{A}$ and let χ be the character of $\mathbb{A}^G$. Then

$$\chi(a) = \sum_{i=1}^{r} \theta(a_i a a_i^{-1}) = \sum_{i=1}^{r} \frac{1}{h} \sum_{\tilde{b} \in H} \theta(\tilde{b} a_i a a_i^{-1} \tilde{b}^{-1})$$

$$= \frac{1}{h} \sum_{x \in G} \theta(x a x^{-1}) \qquad (x = \tilde{b} a_i)$$

$$= \theta^G(a) .$$

So we get $\chi = \theta^G$. We call $\mathbb{A}^G$ an <u>induced</u> <u>representation</u> of G and say that $\mathbb{A}^G$ is induced from $\mathbb{A}$. To summarize the above:

THEOREM 5.2. Let G be a group and H a subgroup of G. Let $\mathbb{A}$ be a representation of H with degree n and let θ be the character of $\mathbb{A}$. Then the induced representation $\mathbb{A}^G$ has degree nr with $r = |G:H|$ and the character of $\mathbb{A}^G$ is

$$\theta^G(a) = \frac{1}{h} \sum_{x \in G} \theta(x^{-1}ax) \qquad (h = |H|).$$

THEOREM 5.3. (Frobenius Reciprocity Theorem) Let H be a subgroup of G. Let $\chi_1, \chi_2, \ldots, \chi_r$ be the irreducible characters of G and let $\theta_1, \theta_2, \ldots, \theta_s$ be the irreducible characters of H. Then

$$(\chi_i)_H = \sum_j r_{ij}\theta_j$$

if and only if

$$\theta_j^G = \sum_i r_{ij}\chi_i .$$

In other words, given irreducible representations $\mathbb{A}$ of G and $\mathbb{B}$ of H, $\mathbb{B}$ is an irreducible component of $\mathbb{A}_H$ with multiplicity r if and only if $\mathbb{A}$ is an irreducible component of $\mathbb{B}^G$ with multiplicity r.

<u>Proof</u>. Let $(\chi_i)_H = \sum_\ell r_{i\ell}\theta_\ell$ and $\theta_j^G = \sum_k s_{kj}\chi_k$. By Lemma 5.1,

$$\begin{aligned} r_{ij} &= ((\chi_i)_H, \theta_j)_H \\ &= (\chi_i, \theta_j^G)_G \\ &= s_{ij}. \qquad \square \end{aligned}$$

EXAMPLE 5.4. Let H be a subgroup of G and let 1_H be the identity representation of H. Then the induced representation $(1_H)^G$ is the permutation representation of G acting on the left cosets $H\backslash G$ from the right.

To see this fact, let $a_1, a_2, \ldots, a_r$ be representatives of $H\backslash G$: $G = Ha_1 \cup Ha_2 \cup \ldots \cup Ha_r$. Then $(1_H)^G$ carries $a \in G$ to the matrix $(\delta(a_i a a_j^{-1}))$, where $\delta(x) = 1$ if $x \in H$ and $\delta(x) = 0$ if $x \notin H$. As seen in Example 1.4, the matrix $(\delta(a_i a a_j^{-1}))$ corresponds to the permutation

$$\begin{pmatrix} Ha_1 & Ha_2 & \ldots & Ha_r \\ Ha_1 a & Ha_2 a & \ldots & Ha_r a \end{pmatrix} .$$

EXAMPLE 5.5. Let G act transitively on a set Ω. Then the permutation representation contains the identity representation 1_G as an irreducible component with multiplicity 1.

Proof. Let H be the stabilizer of $\alpha \in \Omega$ in G:

$$H = \{a \in G \mid \alpha^a = \alpha\} .$$

Then there is a one-to-one correspondence between Ω and the left cosets $H\backslash G$ in such a way that the correspondence is preserved by the action of G. So the permutation representation is $(1_H)^G$ by Example 5.4. Let m be the multiplicity of 1_G in $(1_H)^G$. Then m is the multiplicity of 1_H in $(1_G)_H = 1_H$ by the Frobenius reciprocity theorem. So $m = 1$. □

EXAMPLE 5.6. Let G be a permutation group on Ω. Suppose Ω decomposes into exactly k orbits under the action of G. Then

$$k = (\theta, 1_G)_G ,$$

where θ is the permutation character of G.

Proof. Let $\Omega_1, \ldots, \Omega_k$ be the orbits of G. Then G acts transitively on each Ω_i and we may write the permutation representation $\mathbb{A}$ of G as

$$\mathbb{A} = \begin{pmatrix} \mathbb{A}_1 & & 0 \\ & \ddots & \\ 0 & & \mathbb{A}_k \end{pmatrix} ,$$

where $A_i(a)$ is the matrix describing the action of a on Ω_i as in Example 1.3. Let θ_i be the character of $\mathbb{A}_i$. Then $(\theta_i, 1_G) = 1$ by Example 5.5. Since $\theta = \sum_{i=1}^{k} \theta_i$, $(\theta, 1_G) = k$. $\square$

1.6 PRODUCT OF REPRESENTATIONS

Let A, B be square matrices of degree n, m respectively and let $A = (\alpha_{ik})$. We define the Kronecker product $A \otimes B$ of A and B by

$$A \otimes B = \begin{pmatrix} \alpha_{11}B & \cdots & \alpha_{1n}B \\ \cdots & \cdots & \cdots \\ \alpha_{n1}B & \cdots & \alpha_{nn}B \end{pmatrix} .$$

$A \otimes B$ is a square matrix of degree mn. Direct calculation shows the following:

LEMMA 6.1.

(1) $\mathrm{tr}(A \otimes B) = (\mathrm{tr}A)(\mathrm{tr}B)$.

(2) Let A, A' be of degree n and B, B' of degree m. Then $(A \otimes B)(A' \otimes B') = (AA') \otimes (BB')$.

Let $\mathbb{A}_1: a \to A_1(a)$ and $\mathbb{A}_2: a \to A_2(a)$ be representations of a group G. By Lemma 6.1(2), the mapping

$$a \to A_1(a) \otimes A_2(a)$$

becomes a representation of G. This representation is called the <u>product</u> of $\mathbb{A}_1$ and $\mathbb{A}_2$ and denoted by $\mathbb{A}_1 \otimes \mathbb{A}_2$. Let χ_1, χ_2, χ be the characters of $\mathbb{A}_1, \mathbb{A}_2, \mathbb{A}_1 \otimes \mathbb{A}_2$ respectively. By Lemma 6.1(1),

$$\chi(a) = \chi_1(a)\chi_2(a) \qquad (\forall a \in G)$$

$$\chi = \chi_1\chi_2 .$$

Let $\mathbb{F}_1, \mathbb{F}_2, \ldots, \mathbb{F}_r$ be the irreducible representations of G and let χ_i be the character of $\mathbb{F}_i$. The mapping $a \to \overline{F_i(a)}$ is also an irreducible representation and its character is $\overline{\chi_i}$, where $\overline{\chi_i}(a) = \overline{\chi_i(a)}$. Let $\overline{\chi_i} = \chi_{i'}$.

THEOREM 6.2.

$$\chi_i\chi_j = \sum_\nu r_{ij\nu}\chi_\nu$$

if and only if

$$\chi_{i'}\chi_\nu = \sum_j r_{ij\nu}\chi_j .$$

Proof.

$$(\chi_i\chi_j,\chi_\nu) = \frac{1}{g}\sum_{a\in G}\overline{\chi_i(a)\chi_j(a)}\chi_\nu(a)$$

$$= \frac{1}{g}\sum_{a\in G}\chi_{i'}(a)\chi_\nu(a)\overline{\chi_j(a)}$$

$$= (\chi_{i'}\chi_\nu,\chi_j)\ .$$

So the multiplicity of χ_ν in $\chi_i\chi_j$ is equal to that of χ_j in $\chi_{i'}\chi_\nu$. □

THEOREM 6.3. Let $\mathbb{A}$ be a faithful representation of a group G and let χ be the character of $\mathbb{A}$. Let m be the number of distinct values of χ on G: $m = \#\{\chi(a) \mid a \in G\}$. Then any irreducible representation of G appears in

$$\mathbb{A}^r = \overbrace{\mathbb{A}\otimes\ldots\otimes\mathbb{A}}^{r}$$

for some $r \in \{0,1,2,\ldots,m-1\}$, where $\mathbb{A}^0 = \mathbb{1}_G$.

Proof. Suppose an irreducible representation $\mathbb{F}_i$ does not appear in $\mathbb{A}^r$ $(r = 0, 1, \ldots, m-1)$. Let χ, χ_i be the characters of $\mathbb{A}$, $\mathbb{F}_i$ respectively. Then

$$(\chi^r,\chi_i) = \frac{1}{g}\sum_a \chi(a)^r\,\overline{\chi_i(a)} = 0 \tag{6.1}$$

for $r = 0,1,\ldots,m-1$. Let $\lambda_1,\lambda_2,\ldots,\lambda_m$ be the distinct values of χ on G. Let $M_\alpha = \{a \in G \mid \chi(a) = \lambda_\alpha\}$ and $\Phi_\alpha = \sum_{a\in M_\alpha}\overline{\chi_i(a)}$. By (6.1),

$$\sum_{\alpha=1}^{m}\lambda_\alpha^r\Phi_\alpha = 0 \tag{6.2}$$

for $r = 0, 1, \ldots, m-1$. Regard (6.2) as a system of linear equations for $\Phi_1, \ldots, \Phi_m$. Since $\det(\lambda_\alpha^r) = \prod_{i>j} (\lambda_i - \lambda_j) \neq 0$, the system of linear equations has the solution $\Phi_1 = \ldots = \Phi_m = 0$.

Let n be the degree of $\mathbb{A}$, i.e., $n = \chi(1)$. We may assume $\lambda_1 = n$. We want to show $M_1 = \{1\}$. Suppose $a \in M_1$, i.e., $\chi(a) = n$. Let H be the cyclic group generated by a. By Theorem 3.3, A_H is equivalent to a direct sum of representations of degree 1. So for some non-singular matrix P,

$$P^{-1}A(a)P = \begin{pmatrix} \epsilon_1 & & & 0 \\ & \epsilon_2 & & \\ & & \ddots & \\ 0 & & & \epsilon_n \end{pmatrix}. \tag{6.3}$$

Let h be the order of a. Then $\epsilon_i^h = 1$. Taking the trace of (6.3), $\chi(a) = \epsilon_1 + \epsilon_2 + \ldots + \epsilon_n = n$. This implies $\epsilon_1 = \epsilon_2 = \ldots = \epsilon_n = 1$. So $A(a) = I$. Since $\mathbb{A}$ is faithful, we get $a = 1$. Therefore $M_1 = \{1\}$ and $\Phi_1 = \chi_i(1) \neq 0$, which is a contradiction.

CHARACTER TABLES. Let G be a group and let $C_1 = \{1\}$, $C_2, \ldots, C_k$ be the conjugacy classes of G. Let $\chi_1 = 1_G$, $\chi_2, \ldots, \chi_k$ be the irreducible characters of G and let $a_1 = 1, a_2, \ldots, a_k$ be representatives of the conjugacy classes of G. Notice that the number of irreducible characters is the same as that of conjugacy classes by Theorem 4.10. We arrange the values $\{\chi_i(a_j)\}$ so as to form the character table of G, in which the rows are indexed by the distinct irreducible characters, starting with 1_G, and the columns by the conjugacy classes of G, starting with the class $\{1\}$.

	C_1	$\cdots$	C_k
χ_1	$\chi_1(a_1)$	$\cdots$	$\chi_1(a_k)$
$\vdots$	$\cdots$	$\cdots$	$\cdots$
χ_k	$\chi_k(a_1)$	$\cdots$	$\chi_k(a_k)$

The different rows of the character table are orthogonal to each other in the sense of Theorem 4.4', and by Theorem 4.9, the columns are orthogonal to each other in the usual sense of vectors in a complex unitary space.

EXAMPLE 6.4. CHARACTER TABLE OF S_4. The conjugacy classes of S_4, the symmetric group of degree 4, are $C_1 = \{(1)\}$, $C_2 = \{2\text{-cycles}\}$, $C_3 = \{3\text{-cycles}\}$, $C_4 = \{4\text{-cycles}\}$, $C_5 = \{(12)(34),(13)(24),(14)(23)\}$.

	C_1	C_2	C_3	C_4	C_5
χ_1	1	1	1	1	1
χ_2	1	-1	1	-1	1
χ_3	2	0	-1	0	2
χ_4	3	1	0	-1	-1
χ_5	3	-1	0	1	-1

EXERCISE FOR STUDENTS. Make the character tables of S_n, A_n for $n \leq 5$. It may be helpful to consider induced characters, permutation characters or product of characters.

1.7 SHARP PERMUTATION GROUPS

Let S_n be the symmetric group on $\Omega = \{1,2,\ldots,n\}$ and let θ be the permutation character of S_n. Then $\theta(x)$ is the number of points fixed by x:

$$\theta(x) = \#\{\alpha \in \Omega \mid \alpha^x = \alpha\}.$$

For $x,y \in S_n$, set

$$\partial(x,y) = n - \theta(x^{-1}y)$$

and call $\partial(x,y)$ the distance of x and y. In fact, it is easy to check that (i) $\partial(x,y) = 0$ if and only if $x = y$, (ii) $\partial(x,y) = \partial(y,x)$, (iii) $\partial(x,z) \leq \partial(x,y) + \partial(y,z)$.

For a subset X of S_n, set

$$D(X) = \{\partial(x,y) \mid x,y \in X,\ x \neq y\}.$$

QUESTION. (1) Given $D(X)$, what is the largest possible $|X|$?

In any metric space, this kind of question arises and in general it is hard to answer this question. But in our case, we have the following theorem, provided that X is a subgroup of S_n.

THEOREM 7.1. (Kiyota (1979)) Let G be a faithful permutation group on $\Omega = \{1,2,\ldots,n\}$ and let θ be the permutation character. Let $\{\ell_1,\ldots,\ell_r\} = \{\theta(x) \mid x \in G, x \neq 1\}$ with $\ell_1 < \ell_2 < \ldots < \ell_r$. Then the order of G divides the product of $(n - \ell_i)$ $(i = 1, 2, \ldots, r)$. In particular,

$$|G| \leq \prod_{a_i \in D(G)} a_i = \prod_{i=1}^{r} (n - \ell_i).$$

Proof. A generalized character χ of G is, by definition, a linear combination of characters of G with integer coefficients: $\chi = \sum_i n_i \chi_i$ with $n_i \in \mathbb{Z}$. For example, $\theta - \ell_i 1_G$ is a generalized character. Since a product of characters is also a character, a product of generalized characters is also a generalized character. In particular, the product of $\theta - \ell_i 1_G$ $(i = 1, 2, \ldots, r)$ is a generalized character and we denote this by $\hat{\theta}$. For $x \neq 1$, $(\theta - \ell_i 1_G)(x) = \theta(x) - \ell_i = 0$ for some i. So we get

$$\hat{\theta}(x) = \begin{cases} \prod_{i=1}^{r} (n - \ell_i) & \text{if } x = 1, \\ 0 & \text{otherwise.} \end{cases} \tag{7.1}$$

Let m be the multiplicity of 1_G in $\hat{\theta}$. Then

$$\begin{aligned} m &= (1_G, \hat{\theta})_G \\ &= \frac{1}{g} \sum_{x \in G} 1_G(x)\hat{\theta}(x) \qquad (g = |G|) \\ &= \frac{1}{g} \prod_{i=1}^{r} (n - \ell_i) \qquad (\text{by } (7.1)). \end{aligned}$$

Therefore g divides $\prod_{i=1}^{r} (n - \ell_i)$. In fact, $\hat{\theta}$ is $\frac{1}{g} \prod_i (n - \ell_i)$ times the character of the right regular representation of G by (4.8). □

To study the upper bound of $|X|$ for fixed $L = \{\theta(x^{-1}y) \mid x,y \in X, x \neq y\}$ was suggested by Deza (cf. Deza-Frankl (1977)). If X is not a group, then a similar plausible assertion that $|X| \leq \prod_{i=1}^{r} (n - \ell_i)$ is not always true. (See, for example, Fuji-Hara and Vanstone (1981).) It is an open problem to get the exact bound of $|X|$ for an arbitrary subset X in terms of L.

QUESTION. (2) Given $L = \{\ell_1, \ell_2, \dots, \ell_r\}$, let X be a subset of S_n such that $L = \{\theta(x^{-1}y) \mid x,y \in X,\ x \neq y\}$. Then under what condition can one get

$$|X| \leq \prod_{i=1}^{r} (n - \ell_i)?$$

G is said to be a <u>sharp</u> <u>(permutation)</u> <u>group</u> (or an L-sharp group) if G attains the equality in Theorem 7.1. If G is L-sharp, then $\hat{\theta} = \prod_{i=1}^{r} (\theta - \ell_i 1_G)$ is the character of the right regular representation of G.

LEMMA 7.2. Let G be a $\{0, \ell_2, \dots, \ell_r\}$-sharp group on $\Omega = \{1,2,\dots,n\}$ with $0 < \ell_2 < \dots < \ell_r$. For $\alpha \in \Omega$, let G_α be the stabilizer of α. Then G is transitive on Ω and G_α is an $\{\ell_2 - 1, \ell_3 - 1, \dots, \ell_r - 1\}$-sharp group on $\Omega - \{\alpha\}$.

<u>Proof</u>. Let Ω' be the orbit of G containing α. Then $|G| = |\Omega'| \cdot |G_\alpha|$. So

$$|G_\alpha| = |G|/|\Omega'|$$

$$= \frac{n}{|\Omega'|} \prod_{i=2}^{r} (n - \ell_i)$$

$$\geq \prod_{i=2}^{r} (n - \ell_i).$$

By Theorem 7.1, $|G_\alpha| \leq \prod_{i=2}^{r} (n - \ell_i)$. So,

$$|G_\alpha| = \prod_{i=2}^{r} (n - \ell_i) \quad \text{and} \quad n = |\Omega'|. \qquad \square$$

A permutation group G on Ω is said to be exactly t-transitive (or sharply t-transitive) if for any t-tuples $(\alpha_1, \ldots, \alpha_t)$, $(\beta_1, \ldots, \beta_t)$ of distinct elements of Ω, there exist a unique element $a \in X$ such that $\alpha_i^a = \beta_i$ $(i = 1, 2, \ldots, t)$. G is exactly t-transitive if and only if G is transitive on Ω and G_α is exactly $(t-1)$-transitive on $\Omega - \{\alpha\}$. It is easy to see that exactly t-transitive groups are $\{0, 1, \ldots, t-1\}$-sharp, and conversely, $\{0, 1, \ldots, t-1\}$-sharp groups are exactly t-transitive by Lemma 7.2. For $t \geq 2$, such groups were classified by Zassenhaus in the 1930's (the result for $t \geq 4$ is due to Jordan in the 1870's).

Let us set

$$F(G) = \{\alpha \in \Omega \mid \alpha^a = \alpha \quad \text{for all} \quad a \in G\}.$$

Let G be a $\{0, \ell_2, \ell_3, \ldots, \ell_r\}$-sharp group on Ω. Suppose $|F(G_\alpha)| = \ell_2$. Then G_α is a $\{0, \ell_3 - \ell_2, \ldots, \ell_r - \ell_2\}$-sharp group on $\Omega - F(G_\alpha)$ by Lemma 7.2. A $\{0, \ell_2, \ell_3, \ldots, \ell_r\}$-sharp group G on Ω is, by inductive definition, of standard type if $|F(G_\alpha)| = \ell_2$ and G_α is of standard type on $\Omega - F(G_\alpha)$. L-sharp groups of standard type are a generalization of exactly t-transitive groups.

PROBLEM. (1) Classify L-sharp groups of standard type.

L-sharp groups are transitive if $0 \in L$ (by Lemma 7.2), but this is no longer true if $0 \notin L$. Let G be L-sharp on Ω with $L = \{\ell_1, \ell_2, \ldots, \ell_r\}$, $\ell_1 < \ell_2 < \ldots < \ell_r$. then

$$(\theta, 1_G)_G > \ell_1, \tag{7.2}$$

because $(\theta, 1_G) = \frac{1}{|G|} \sum_{a \in G} \theta(a)$ and $\theta(a) \geq \ell_1$ for $a \neq 1$,

$\theta(1) = |\Omega|$. So G has more than ℓ_1 orbits by Example 5.6. Nevertheless, most of the orbits of G are of length 1 in some cases.

THEOREM. Let G be an $\{\ell, \ell+1, \ldots, \ell+r-1\}$-sharp group on Ω with $r \geq 2$. Then $|F(G)| = \ell$ and G is exactly r-transitive on $\Omega - F(G)$.

For the proof, see Ito and Kiyota (1981).

THEOREM 7.3. Let G be an $\{\ell, \ell+s\}$-sharp group on Ω. Then there exists a function $f(s)$, which depends only on s, such that $|F(G)| \geq \ell - f(s)$. In fact, we can choose $f(s) = (s-1)s' - s'^2 + 1$ with $s' = \max\{1, [\frac{s-1}{2}]\}$.

Proof. Let θ be the permutation character of G. Let

$$\theta = \sum_i a_i \chi_i,$$

where $\chi_0 = 1_G$, $\chi_1, \chi_2, \ldots$ are the irreducible characters of G. By Example 5.6, a_0 is the number of orbits of G on Ω. Set $a_0 = k$. Let $\Omega_1, \Omega_2, \ldots, \Omega_k$ be the orbits of G and let θ_i be the permutation character of G on Ω_i. Then $\theta = \theta_1 + \theta_2 + \ldots + \theta_k$ and each θ_i contains 1_G with multiplicity one. If $|\Omega_i| \neq 1$, then θ_i contains some non-trivial irreducible character of G. So we get

$$\sum_{i \neq 0} a_i \geq a_0 - |F(G)|. \tag{7.3}$$

By the proof of Theorem 7.1, $\hat{\theta} = (\theta - \ell 1_G)(\theta - (\ell+s)1_G)$ is the character of the right regular representation of G. Hence

$$(\hat{\theta}, 1_G) = 1.$$

Since $(\chi\overline{\chi}', 1_G) = (\chi, \chi')$ in general, it follows that

$$(\theta - \ell 1_G, \theta - (\ell + s)1_G) = 1,$$

$$\sum_i a_i^2 - (2\ell + s)a_0 + \ell(\ell + s) = 1,$$

and so

$$\sum_{i \neq 0} a_i^2 = 1 - (a_0 - \ell)(a_0 - \ell - s). \tag{7.4}$$

Since $\sum_{i \neq 0} a_i^2 \geq 1$, $(a_0 - \ell)(a_0 - \ell - s) \leq 0$. By (7.2), $a_0 > \ell$. So we have $\ell < a_0 \leq \ell + s$. Thus

$$|F(G)| \geq a_0 - \sum_{i \neq 0} a_i \qquad \text{(by (7.3))}$$

$$\geq a_0 - \sum_{i \neq 0} a_i^2$$

$$= a_0 - 1 + (a_0 - \ell)(a_0 - \ell - s) \qquad \text{(by (7.4))}.$$

An elementary calculation shows that

$$\text{Min}\{a_0 - 1 + (a_0 - \ell)(a_0 - \ell - s) \mid \ell < a_0 \leq \ell + s,$$

$$a_0 \in \mathbb{Z}\} = \ell + (1 - s)s' + s'^2 - 1$$

with $s' = \max\{1, [\frac{s - 1}{2}]\}$. □

Put $s = 2, 3$ in Theorem 7.3. Then $|F(G)| \geq \ell - 1$ for $s = 2$ and $|F(G)| \geq \ell - 2$ for $s = 3$. L-sharp groups with $L = \{\ell, \ell + 2\}$, $\{\ell, \ell + 3\}$ are classified (Ito and Kiyota (1981)).

PROBLEM. (2) Is there a function $f(s,t)$ such that $|F(G)| \geq \ell - f(s,t)$ for any $\{\ell, \ell + s, \ell + t\}$-sharp group G?

Let G be an $\{\ell\}$-group (i.e., each non-identity element of G fixes exactly ℓ points) with $|F(G)| \neq \ell$. The set $\mathfrak{F}$ of pointwise stabilizers of ℓ-element subsets of Ω satisfies the following property: (i) for any non-identity $a \in G$, there exists a unique subgroup $H \in \mathfrak{F}$ such that $a \in H$, (ii) for any $H \in \mathfrak{F}$ and $a \in G$, $a^{-1}Ha$ belongs to $\mathfrak{F}$ and, (iii) $\mathfrak{F} \neq \{G\}$. Such a set $\mathfrak{F}$ of subgroups is said to be a G-invariant proper partition. Conversely, it is an easy exercise to show that a group G with G-invariant proper partition has a permutation representation as an $\{\ell\}$-group with $|F(G)| \neq \ell$. Non-solvable groups G with G-invariant proper partition are classified (Suzuki (1961)).

I.8 NOTES TO CHAPTER I

For further reading on representation theory of finite groups, the reader is referred to any advanced textbook on group theory. The following books seem to be standard, though each book has its own purpose, level, and way of treatment, and they are quite different from each other: Curtis and Reiner (1962), Dornhoff (1971, 1972), Feit (1967, 1982), Gorenstein (1968), Hall (1959), Huppert (1967), Issacs (1976), Serre (1971).

We did not discuss modular representation theory of finite groups at all, but to relate modular representation theory to Algebraic Combinatorics seems to be a very hopeful and rewarding research direction in the future. This line of study has not yet started. We think that Scott (1971, 1977) and Ott (1981) may be interesting to read for this purpose. In particular, we think that Ott (1981) stands on the same basis as ours in its attitude toward combinatorics.

Topics in section 1.7 are interesting in the following sense: (i) The framework of the theory is not covered by the subsequent discussions of association schemes. There

is an underlying structure of association schemes, but that has quite a different property from P-polynomial or Q-polynomial schemes, etc. However, this structure is closely related to groups, enabling the representation of groups to be a possible key method for developing the theory in this area. Thus, we suggest that the theory should be handled along the line of Delsarte's philosophy, and we should broaden the realm of Algebraic Combinatorics.
(ii) Although this is a problem of the finite group theory, it is unlikely to be solved as a direct consequence of the classification of finite simple groups. Perhaps we cannot use the classification directly to determine L-sharp groups with $L = \{\ell_1, \ldots, \ell_r\}$, $\ell_1 < \ldots < \ell_r$ and ℓ_1 large.

For further reading about section 1.7, see Ito and Kiyota (1981), Babai, Cameron, Deza and Singhi (1981), Cameron and Deza (1979), Deza and Frankl (1977). So far, this theory is not yet developed to a very satisfactory stage. It seems that some of the most important properties are left to be found in this area.

Sharply t-transitive (or exactly t-transitive) groups with $t \geq 2$ are classified and there exist no non-trivial such groups for $t \geq 6$. A <u>subset</u> X of S^{Ω}, the symmetric group on Ω, is said to be a <u>sharply t-transitive subset</u> if for any t-tuples $(\alpha_1, \ldots, \alpha_t)$ and $(\beta_1, \ldots, \beta_t)$ of distinct elements of Ω, there exists a unique element $a \in X$ such that $\alpha_i^a = \beta_i$ $(i = 1, \ldots, t)$. It would be an interesting problem whether there exist sharply t-transitive subsets for large t (cf. Lorimer (1973)).

We will return to these subjects in Part II in connection with extremal set theory.

CHAPTER II. ASSOCIATION SCHEMES

The purpose of this chapter is to give a systematic introduction to the theory of association schemes.

The concept of association schemes is very fundamental, perhaps the most important, in Algebraic Combinatorics. Association schemes are the frameworks on which coding theory, design theory and other theories are developed in a unified and satisfactory way. This line of study was first started consciously and systematically by Delsarte (1973b), and the success of Delsarte can be attributed to his recognition and full use of association schemes as the basic underlying structures of coding theory, design theory, etc.

Besides being the basic frameworks for such theories, association schemes have their own rich mathematical content and deserve serious mathematical study in their own right. Part I of these lecture notes is written from this last viewpoint, leaving the Delsarte theory build on the framework of association schemes to be the main topic of Part II.

There are many mathematical objects whose essence is that of association schemes and many different names are given to essentially the same mathematical concept: Adjacency algebra, Bose-Mesner algebra, centralizer ring, Hecke ring, Schur ring, character algebra, hypergroup, probabilistic group, etc. Correspondingly, there are many different approaches to the study of association schemes. In these lecture notes, we understand association schemes as purely combinatorial objects, that is we do not make further assumptions such as the action of groups.

Yet there are several approaches to association schemes. In this chapter, we shall describe many of these approaches in a systematic and encyclopedic way and explain the relations between them. So there are some overlaps and repetitions in our treatment. Also this chapter will concern only general classes of association schemes. The study of association schemes with certain more specific important properties is the topic of Chapter III.

We choose a standard but rather unsophisticated approach to discuss objects in sections 2.2-2.4. We use the combinatorial structures of association schemes to derive all the algebraic properties of association schemes. The treatment of section 2.5 is quite different. We axiomatize the algebraic aspect of association schemes in several terms and show that all their algebraic properties are derived from the axioms without assuming the existence of actual combinatorial objects. Therefore, from a logical viewpoint, all calculations in sections 2.2-2.4 are superseded by and contained in those in section 2.5. The calculations in section 2.5 are ingenious and more difficult, while those in sections 2.2-2.4 are natural and easy to understand. So, for the reader who is not familiar with association schemes, we recommend that they read sections 2.1 to 2.5 in numerical order. For the reader who is already quite familiar with association schemes, section 2.5 will still be interesting to read. It should be emphasized that the approach in section 2.5 is very important, and is useful for the understanding of the duality in association schemes, and is sometimes used in later sections (in particular, it is very useful when we study Q-polynomial association schemes). An interesting feature of Algebraic Combinatorics is that some concepts are <u>fake</u> in combinatorial context, but are <u>real</u> only in algebraic context, while the study of such fake structures sometimes really benefits the whole understanding of the subject.

The content of each section will be given in the following. Note that sections 2.6-2.11 are not all strongly connected to each other and can be read separately. Also it is possible to skip these sections at the first reading, go to Chapter III and then come back later, although some parts of these sections are necessary in some later sections of Chapter III.

In section 2.1 the centralizer rings of permutation groups are discussed, following Wielandt (1964). This is an important source of Algebraic Combinatorics from group theory. Association schemes are defined in section 2.2. Actually, there are three different levels of definitions of association schemes, each of which is reasonable in its own setting. Namely, (a) symmetric association schemes (Bose-Mesner), (b) commutative association schemes (Delsarte), and (c) association schemes which are not necessarily commutative, i.e., coherent configurations (D. G. Higman). These three levels of definitions are discussed and compared in connection with some examples which come from groups (as in section 2.1). Sections 2.2 to 2.4 are devoted to a very standard introduction to the theory of association schemes. Many good expository articles on these topics are already in the literature. See, for example, Delsarte (1973b), Higman (1975, 1976), Sloane (1975), etc.

In section 2.5, we study the algebraic structure of the algebra $\mathfrak{A}$ and the duality more precisely than in previous sections. As for the duality, we found that an old paper by Kawada (1942) did solve this problem in a most general and desirable form for commutative C-algebras. Here, C-algebra is a term introduced by Kawada, and it is the concept which is obtained by algebraically and abstractly distilling the concept of association schemes. When the paper was written, no concept (or definition) of the association scheme existed. Kawada proved a duality for commutative C-algebras, motivated abstractly by the famous Tannaka duality for noncompact topological groups. This duality is essentially the duality of commutative association schemes discussed in the previous sections, and to our surprise, most of the calculations were done in the old paper of Kawada, which was not much noticed before. A generalization of this duality, i.e., a duality for noncommutative association schemes will also be mentioned in the later section 2.10.

The success of Delsarte's theory in Delsarte (1973b) was that he defined the dual concept only algebraically.

The algebraic dual structure of an association scheme is combinatorially "fake" in general, nonetheless important. If a permutation group G contains a regular abelian subgroup H, then the dual structure is combinatorially "real". This is the content in which S-rings (Schur rings) were extensively (and successfully) studied in permutation group theory. See Schur (1933), Wielandt (1964), Tamaschke (1970), etc. This topic is briefly reviewed in section 2.6.

In section 2.7, the character table of a finite group is interpreted in the framework of association schemes. An interesting consequence is an interpretation of a part of one of McKay's interesting observations concerning the character tables of finite groups. This direction of research supports our belief that Algebraic Combinatorics is a natural generalization of the character theory of finite groups.

The contents of sections 2.8 and 2.9 are mainly based on Cameron-Goethals-Seidel (1978b).

In section 2.11, spherical functions of multiplicity-free permutation representations will be discussed. This is another source of Algebraic Combinatorics from representation theory of topological groups. This source is, when ignoring topology, equivalent to an approach in permutation group theory, but the way of development and the basic philosophical emphasis are not exactly the same.

2.1 CENTRALIZER RINGS OF PERMUTATION REPRESENTATIONS

Let G be a transitive permutation group on $\Omega = \{1,2,\ldots,n\}$ and let θ be the permutation character. Let G act on $\Omega \times \Omega$ in such a way that $(\alpha,\beta)^a = (\alpha^a,\beta^a)$ for $\alpha,\beta \in \Omega$, $a \in G$. Let $\Lambda_0,\Lambda_1,\ldots,\Lambda_d$ be the orbits of G on $\Omega \times \Omega$, where $\Lambda_0 = \{(\alpha,\alpha) \mid \alpha \in \Omega\}$. The Λ_i's are called the orbitals. For $\alpha \in \Omega$, we denote the set

$\{\beta \in \Omega \mid (\alpha,\beta) \in \Lambda_i\}$ by $\Lambda_i(\alpha)$. Then $\Lambda_0(\alpha) = \{\alpha\}$, $\Lambda_1(\alpha),\ldots,\Lambda_d(\alpha)$ are the orbits of G_α on Ω, where G_α is the stabilizer of $\alpha \in \Omega$ in G, and it holds that $\Lambda_i(\alpha)^a = \Lambda_i(\alpha^a)$ for all $a \in G$, $0 \le i \le d$. We set $k_i = |\Lambda_i(\alpha)|$. (The numbers k_i are independent of the choice of α.) Let ${}^t\Lambda_i = \{(\alpha,\beta) \mid (\beta,\alpha) \in \Lambda_i\}$. Since ${}^t\Lambda_i$ is also an orbit of G, there exists some j such that ${}^t\Lambda_i = \Lambda_j$. We denote this index j by i', and $\Lambda_{i'}$ is called the <u>paired orbital</u> of Λ_i. Clearly $k_i = k_{i'}$ and it holds that $\beta \in \Lambda_i(\alpha)$ if and only if $\alpha \in \Lambda_{i'}(\beta)$.

LEMMA 1.1. Let $d + 1$ be the number of orbits of G on $\Omega \times \Omega$. Then $(\theta,\theta)_G = d + 1$.

<u>Proof</u>. Since $\theta(a)$ is the number of points in Ω fixed by a $(a \in G)$, $\theta(a)\theta(a)$ is the number of points in $\Omega \times \Omega$ fixed by a. This means that θ^2 is the permutation character of G on $\Omega \times \Omega$. By Example 5.6 in Chapter I, we get $d + 1 = (\theta^2, 1_G)_G = (\theta,\theta)_G$. □

For an arbitrary matrix A, we denote the (α,β) entry of A by $A_{\alpha\beta}$ (or by $A_\beta(\alpha)$, Delsarte's notation). With each of the orbitals Λ_i, we associate a matrix A_i of size n with

$$(A_i)_{\alpha\beta} = \begin{cases} 1 & \text{if } (\alpha,\beta) \in \Lambda_i, \\ 0 & \text{otherwise,} \end{cases}$$

where the rows and columns of A_i are indexed by the elements of Ω. In particular, A_0 is the identity matrix I. We call A_i the <u>adjacency matrix</u> of G on Ω with respect to Λ_i. It is clear that the following lemma holds.

LEMMA 1.2. (i) Let J be the n by n matrix whose entries are all 1. Then

$$\sum_{i=0}^{d} A_i = J.$$

In particular, $A_0, A_1, \ldots, A_d$ are linearly independent over $\mathbb{C}$.

(ii) ${}^tA_i = A_{i'}$.

(iii) Every column or row of A_i contains exactly k_i ones and

$$\mathrm{tr}(A_{j'}A_i) = nk_i\delta_{ij} \qquad (i, j = 0, 1, \ldots, d).$$

Let $\mathbb{A}: a \to A(a)$ be the permutation representation of G on Ω, where $A(a)$ is the matrix describing the action of a on Ω, i.e., $(A(a))_{\alpha\beta} = \delta_{\alpha^a\beta}$. For an arbitrary matrix $X = (X_{\alpha\beta})$,

$$(A(a)XA(a)^{-1})_{\alpha\beta} = \sum_{\mu,\nu} \delta_{\alpha^a\mu} X_{\mu\nu} \delta_{\nu^{a^{-1}}\beta}$$

$$= X_{\alpha^a\beta^a}.$$

This means that $X = (X_{\alpha\beta})$ commutes with all $A(a)$ $(a \in G)$ if and only if $X_{\alpha\beta} = X_{\alpha^a\beta^a}$ for all $a \in G$, $\alpha,\beta \in \Omega$. In particular, each A_i commutes with all $A(a)$ $(a \in G)$. Since $\sum_i A_i = J$ (Lemma 1.2(i)), every matrix which commutes with all $A(a)$ $(a \in G)$ is a $\mathbb{C}$-linear combination of $A_0, A_1, \ldots, A_d$. Thus we have the following theorem.

THEOREM 1.3. Let G be a transitive permutation group on $\Omega = \{1,2,\ldots,n\}$ and let $\mathbb{A}: a \to A(a)$ be the permutation representation. Let $\mathfrak{A}$ be the set of all the $n \times n$ matrices over $\mathbb{C}$ which commute with all $A(a)$ $(a \in G)$.

Then $\mathfrak{A}$ becomes an algebra over $\mathbb{C}$, and $\mathfrak{A}$ is spanned by $A_0, A_1, \ldots, A_d$ over $\mathbb{C}$ as a linear space, where $A_0, A_1, \ldots, A_d$ are the adjacency matrices of G on Ω. In particular, $\dim_{\mathbb{C}} \mathfrak{A} = d + 1$.

We call $\mathfrak{A}$ the <u>centralizer</u> <u>ring</u> (algebra) of $\mathbb{A}$ or the <u>Hecke</u> <u>ring</u> (algebra) of $\mathbb{A}$.

Let $\mathbb{F}_0 = 1_G, \mathbb{F}_1, \ldots, \mathbb{F}_r$ be the distinct irreducible representations of G over $\mathbb{C}$ appearing in $\mathbb{A}$. Let f_i be the degree of $\mathbb{F}_i$ and e_i the multiplicity of $\mathbb{F}_i$ in $\mathbb{A}$. Then by the complete reducibility of representations of G, there exists a non-singular matrix U (we can choose U as an unitary matrix) such that

$$U^{-1}A(a)U = \mathrm{diag}(\underbrace{F_0(a),}_{e_0}\underbrace{F_1(a),\ldots,F_1(a)}_{e_1},\ldots, \qquad (1.1)$$

$$\underbrace{F_r(a),\ldots,F_r(a)}_{e_r}) \qquad (\forall a \in G)$$

Hence $e_0 = 1$, because G is transitive on Ω. By Schur's lemma (section 1.3 in Chapter I), the matrices which commute with all $U^{-1}A(a)U$ are characterized by the form

$$\mathrm{diag}(M_0, M_1 \otimes I_{f_1}, \ldots, M_r \otimes I_{f_r}), \qquad (1.2)$$

where M_i is an arbitrary $e_i \times e_i$ matrix, I_{f_i} is the $f_i \times f_i$ identity matrix and $M_i \otimes I_{f_i}$ denotes the Kronecker product of M_i and I_{f_i}. So $U^{-1}\mathfrak{A}U$ consists of all the matrices of the form (1.2). Thus we get the following theorem.

THEOREM 1.4. Let G be a transitive permutation group on Ω and let $\mathbb{A}$ be the permutation representation. Let $e_0 = 1, e_1, \ldots, e_r$ be the multiplicities of the irreducible components $\mathbb{F}_0 = 1_G, \mathbb{F}_1, \ldots, \mathbb{F}_r$ of $\mathbb{A}$. Let $\mathfrak{A}$ be the centralizer ring of $\mathbb{A}$. Then

$$\text{(i)} \quad \dim_{\mathbb{C}} \mathfrak{A} = \sum_{i=0}^{r} e_i^2 = d + 1,$$

where $d + 1$ is the number of orbits of G_α on Ω $(\alpha \in \Omega)$.

(ii) (Multiplicity Free Condition) $\mathfrak{A}$ is commutative if and only if $e_i = 1$ for all $i = 0, 1, \ldots, r$.

LEMMA 1.5. If $i' = i$ for all $i = 0, 1, \ldots, d$, then $\mathfrak{A}$ is commutative. The condition $i' = i$ holds if and only if for any $(\alpha, \beta) \in \Lambda_i$, there exists an element $a \in G$ such that $\alpha^a = \beta$ and $\beta^a = \alpha$.

Proof. First we remark that each element of $\mathfrak{A}$ is a symmetric matrix if $i' = i$, because ${}^tA_i = A_{i'} = A_i$ holds for the basis $A_0, A_1, \ldots, A_d$ of $\mathfrak{A}$. In particular, A_iA_j is symmetirc for all i, j. So

$$A_jA_i = {}^tA_j\,{}^tA_i = {}^t(A_iA_j) = A_iA_j,$$

and hence $\mathfrak{A}$ is commutative.

The latter half of the claim is trivial by the definition of A_i. □

REMARK. The converse of the first assertion of Lemma 1.5 is not always true. See Examples 2.1(2) in the next section, in which the right translations and inner automorphisms act on a group. Another counterexample is the Paley tournament $(X, \{R_i\}_{0 \leq i \leq 2})$, where $X = GF(q)$ $(q \equiv 3 \bmod 4)$ and $R_1 = \{(x,y) \mid x - y$ is a non-zero square$\}$.

Let $C_0 = \{1\}, C_1, \ldots, C_k$ be the conjugacy classes of G. We make a formal sum of the elements of C_i:

$$\mathbb{C}_i = a_1 + a_2 + \cdots + a_{h_i} \qquad (h_i = |C_i|).$$

For the permutation representation $\mathbb{A}: a \to A(a)$ of G, define $A(\mathbb{C}_i)$ by

$$A(\mathbb{C}_i) = A(a_1) + A(a_2) + \cdots + A(a_{h_i}).$$

We call $A(\mathbb{C}_i)$ the ith <u>class matrix</u>.

LEMMA 1.6. Each class matrix $A(\mathbb{C}_i)$ is an element of the centralizer ring $\mathfrak{A}$. The class matrices commute with each other.

<u>Proof</u>. Since $b^{-1}ab$ ranges over C_i as a does, we have

$$\begin{aligned} A(b)^{-1}A(\mathbb{C}_i)A(b) &= \sum_{a \in C_i} A(b^{-1}ab) \\ &= A(\mathbb{C}_i) \end{aligned}$$

for all $b \in G$. In particular, $A(\mathbb{C}_i)A(\mathbb{C}_j) = A(\mathbb{C}_j)A(\mathbb{C}_i)$. □

THEOREM 1.7. (Wielandt (1964, Theorem 29.8).) Let G be a transitive permutation group on Ω. Then the centralizer ring $\mathfrak{A}$ is commutative if and only if $\mathfrak{A}$ is spanned by the class matrices as a linear space.

<u>Proof</u>. By Lemma 1.6, $\mathfrak{A}$ is commutative if it is spanned by the class matrices.

Conversely, assume $\mathfrak{A}$ is commutative and let $A(\mathbb{C}_0), A(\mathbb{C}_1), \ldots, A(\mathbb{C}_k)$ be the class matrices. By Theorem 1.4, all the $e_i = 1$ and $U^{-1}\mathfrak{A}U$ consists of all the matrices of the form $\operatorname{diag}(\lambda_0, \lambda_1 I_{f_1}, \ldots, \lambda_r I_{f_r})$ $(\lambda_i \in \mathbb{C})$.

In particular, $\dim_{\mathbb{C}} \mathfrak{A} = r + 1$.

Let $\mathbb{F}_0 = 1_G, \mathbb{F}_1, \ldots, \mathbb{F}_r$ be the irreducible components of the permutation representation of G on Ω, and let χ_i be the character of $\mathbb{F}_i$. Let $\{a_0 = 1, a_1, \ldots, a_k\}$ be a set of representatives of the conjugacy classes $C_0, C_1, \ldots, C_k$ of G. Then by Theorem 4.4' in Chapter I, the row vectors $(\chi_i(a_0), \chi_i(a_1), \ldots, \chi_i(a_k))$ $(i = 0, 1, \ldots, r)$ are linearly independent. So with appropriate renumbering,

$$\det(\chi_i(a_j))_{i,j=0,1,\ldots,r} \neq 0.$$

By (4.12), (4.13) in Chapter I, $F_i(\mathbb{C}_j) = \frac{h_j}{f_i} \chi_i(a_j) I$, where $h_j = |C_j|$ and f_i is the degree of $\mathbb{F}_i$. Therefore, in (1.1),

$$U^{-1}A(\mathbb{C}_j)U = \operatorname{diag}(F_0(\mathbb{C}_j), \ldots, F_r(\mathbb{C}_j))$$
$$= \operatorname{diag}\left(\frac{h_j}{f_0} \chi_0(a_j) I_{f_0}, \ldots, \frac{h_j}{f_r} \chi_r(a_j) I_{f_r}\right).$$

Since the matrix

$$\left(\frac{h_j}{f_i} \chi_i(a_j)\right)_{i,j=0,1,\ldots,r}$$

$$= \operatorname{diag}\left(\frac{1}{f_0}, \ldots, \frac{1}{f_r}\right)(\chi_i(a_j))\operatorname{diag}(h_0, \ldots, h_r)$$

is non-singular, $U^{-1}A(\mathbb{C}_j)U$ $(j = 0, 1, \ldots, r)$ are linearly independent. Since $\dim_C \mathfrak{A} = r + 1$, $U^{-1}\mathfrak{A}\, U$ is spanned by $U^{-1}A(\mathbb{C}_j)U$ $(j = 0, 1, \ldots, r)$. $\square$

2.2 ASSOCIATION SCHEMES

Let X be a set of cardinality n and let R_i $(i = 0, 1, \ldots, d)$ be subsets of $X \times X$ with the property that

(i) $R_0 = \{(x,x) \mid x \in X\}$.

(ii) $X \times X = R_0 \cup \ldots \cup R_d$, $R_i \cap R_j = \emptyset$ if $i \neq j$.

(iii) ${}^tR_i = R_{i'}$ for some $i' \in \{0,1,\ldots,d\}$, where ${}^tR_i = \{(x,y) \mid (y,x) \in R_i\}$.

(iv) For $i,j,k \in \{0,1,\ldots,d\}$, the number of $z \in X$ such that $(x,z) \in R_i$ and $(z,y) \in R_j$ is constant whenever $(x,y) \in R_k$. This constant is denoted by p_{ij}^k.

(v) $p_{ij}^k = p_{ji}^k$ for all i, j, k.

(2.1)

Such a configuration $\mathfrak{X} = (X, \{R_i\}_{0 \leq i \leq d})$ is called a <u>commutative association scheme of class d on X</u>. The non-negative integers p_{ij}^k are called the <u>intersection numbers</u> of $\mathfrak{X}$. A configuration $\mathfrak{X} = (X, \{R_i\}_{0 \leq i \leq d})$ with only the properties (i), (ii), (iii), (iv) is called a <u>(non-commutative) association scheme</u> (or a <u>homogeneous coherent configuration</u> after D. G. Higman (1975)). An association scheme with the additional property

(vi) $i' = i$

is called <u>symmetric</u> or of <u>Bose-Mesner type</u>.

The <u>i^{th} adjacency matrix</u> A_i of $\mathfrak{X}$ is defined to be the matrix of degree $|X|$ whose rows and columns are indexed by the elements of X and whose (x,y) entries are

$$(A_i)_{xy} = \begin{cases} 1 & \text{if } (x,y) \in R_i, \\ 0 & \text{otherwise.} \end{cases}$$

A_i is a $(0,1)$ matrix. It is easy to see that the conditions (i),...,(v) are equivalent to the following (i)',...,(v)':

(i)' $A_0 = I$, the identity matrix.

(ii)' $A_0 + A_1 + \ldots + A_d = J$, the matrix whose entries are all 1.

(iii)' ${}^tA_i = A_{i'}$ for some $i' \in \{0,1,\ldots,d\}$.

(iv)' $A_iA_j = \sum_{k=0}^{d} p_{ij}^k A_k$ for all i, j.

(v)' $A_iA_j = A_jA_i$ for all i, j.

(2.1)'

Here ${}^tA_i$ is the transposed of A_i. $\mathfrak{X}$ is symmetric if and only if

(vi)' ${}^tA_i = A_i$

for all i.

EXAMPLES 2.1. (1) Let G be a transitive permutation group on Ω ($|\Omega| = n$) and let $\Lambda_0, \Lambda_1, \ldots, \Lambda_d$ be the orbits of G on $\Omega \times \Omega$. Set $X = \Omega$, $R_i = \Lambda_i$ ($i = 0, 1, \ldots, d$), and let A_i be the ith adjacency matrix. By Lemma 1.2, A_i's satisfy (i)', (ii)', (iii)'. By Theorem 1.3, the product A_iA_j is a linear combination of A_k's and so (iv)' holds. Hence $\mathfrak{X} = (\Omega, \{\Lambda_i\}_{0 \le i \le d})$ is a (non-commutative) association scheme. $\mathfrak{X}$ is a commutative association scheme if and only if the centralizer ring $\mathfrak{A}$ is commutative or equivalently the permutation character is multiplicity free (Theorem 1.4). $\mathfrak{X}$ is symmetric if and only if for any $(x,y) \in \Lambda_i$ $(i = 0, 1, \ldots, d)$, there exists an element $a \in G$ such that $x^a = y$ and $y^a = x$ (Lemma 1.5), i.e., all Λ_i are self-paired. Such a group is called <u>generously</u> <u>transitive</u>. (This terminology is due to Neumann (1975).)

(2) Let G be a finite group and let $C_0 = \{1\}, C_1, \ldots, C_d$ be the conjugacy classes of G. Set $X = G$ and define the ith relation R_i by $(x,y) \in R_i$ if and only if $yx^{-1} \in C_i$. Then (i), (ii), (iii) are trivial with $C_{i'} = \{a \in G \mid a^{-1} \in C_i\}$. By (4.10) of Chapter I, $\mathbf{C}_i\mathbf{C}_j = \sum_k t_{ijk}\mathbf{C}_k$, where $\mathbf{C}_\alpha = \sum_{a \in C_\alpha} a$. Check (iv) holds with $p_{ij}^k = t_{ijk}$. Since $\mathbf{C}_i x = x(x^{-1}\mathbf{C}_i x) = x\mathbf{C}_i$ for all $x \in G$, $\mathbf{C}_i\mathbf{C}_j = \mathbf{C}_j\mathbf{C}_i$ and so (v) holds. Therefore $\mathfrak{X} = (G, \{R_i\}_{0 \le i \le d})$ is a commutative association scheme of class d. Note that this association scheme is symmetric if and only if x and x^{-1} are conjugate in G for all $x \in G$.

(3) Let F be a set of cardinality q $(q \ge 2)$ and $X = \underbrace{F \times F \times \cdots \times F}_{n}$. Define R_i by $(x,y) \in R_i$ if and only if x and y differ at exactly i coordinates, i.e., $\#\{j \mid x_j \ne y_j\} = i$, where $x = (x_1, x_2, \ldots, x_n)$ and $y = (y_1, y_2, \ldots, y_n)$. Since the wreath product of the symmetric groups, S_q wr S_n of order $(q!)^n n!$, acts transitively on each relation R_i, it is easy to see $\mathfrak{X} = (X, \{R_i\}_{0 \le i \le n})$ is a symmetric association scheme of class n. (For the wreath product, see Hall (1959), page 81.) $\mathfrak{X}$ is called the <u>Hamming</u> <u>scheme</u> and denoted by $H(n,q)$.

(4) Let V be a set of cardinality v and let k

be a positive integer $(k \leq \frac{v}{2})$. Let X be the set of k-element subsets of V $(|X| = \binom{v}{k})$. Define R_i by $(x,y) \in R_i$ $(x,y \in X)$ if and only if $|x \cap y| = k - i$. Since the symmetric group S_v acts transitively on each relation R_i, it is easy to see $\mathcal{X} = (X, \{R_i\}_{0 \leq i \leq k})$ is a symmetric association scheme of class k. $\mathcal{X}$ is called the Johnson scheme and denoted by $J(v,k)$.

In what follows, we always assume $\mathcal{X} = (X, \{R_i\})$ is a commutative association scheme of class d unless otherwise stated. Set

$$k_i = p^0_{ii'} . \tag{2.2}$$

Then k_i is the number of elements $y \in X$ such that $(x,y) \in R_i$ for a fixed $x \in X$. The positive integer k_i is called the valency of R_i. It is clear that

$$\begin{aligned} k_0 &= 1, \\ k_i &= k_{i'}, \\ |X| &= k_0 + k_1 + \dots + k_d . \end{aligned} \tag{2.3}$$

PROPOSITION 2.2.

(i) $p^k_{0j} = \delta_{jk}$,

(ii) $p^k_{i0} = \delta_{ik}$,

(iii) $p^0_{ij} = k_i \delta_{ij'}$,

(iv) $p^k_{ij} = p^{k'}_{i'j'}$,

(v) $\sum_{j=0}^{d} p^k_{ij} = k_i$,

(vi) $k_\gamma p^\gamma_{\alpha\beta} = k_\beta p^\beta_{\alpha'\gamma} = k_\alpha p^\alpha_{\gamma\beta'}$,

(vii) $\sum_{\alpha=0}^{d} p^\alpha_{ij} p^\ell_{k\alpha} = \sum_{\beta=0}^{d} p^\beta_{ki} p^\ell_{\beta j}$.

<u>Proof</u>. The identities (i), (ii), (iii), (iv) follow immediately from the definition of p^k_{ij}.

(v) For a fixed pair $(x,y) \in R_k$, count the number of points $z \in X$ such that $(x,z) \in R_i$.

(vi) Count the number of triangles (x,y,z) such that $(x,y) \in R_\gamma$, $(x,z) \in R_\alpha$, $(z,y) \in R_\beta$.

(vii) For a fixed pair $(x,y) \in R_\ell$, count the number of pairs (z,w) such that $(x,z) \in R_k$, $(z,w) \in R_i$, $(w,y) \in R_j$.

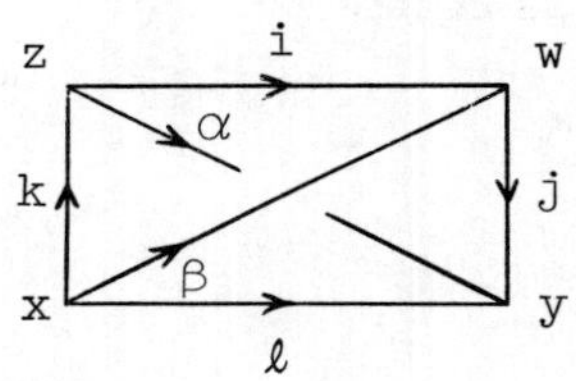

□

Let $\mathfrak{A}$ be the subalgebra of $M_n(\mathbb{C})$ spanned by the adjacency matrices $A_0, A_1, \ldots, A_d$, where $M_n(\mathbb{C})$ is the full matrix algebra of degree $n = |X|$ over $\mathbb{C}$. $\mathfrak{A}$ is a commutative algebra of degree $d+1$ by (iv)', (v)' of (2.1)'. $\mathfrak{A}$ is called the <u>adjacency algebra</u> of $\mathfrak{X}$ or the <u>Bose-Mesner algebra</u> of $\mathfrak{X}$. In Example 2.1(1), $\mathfrak{A}$ is nothing but the centralizer ring or the Hecke ring of G on Ω.

Regard the left multiplication of $\mathfrak{A}$ as linear transformations of $\mathfrak{A}$ and express them in matrix form with respect to the basis $\{A_0, A_1, \ldots, A_d\}$ of $\mathfrak{A}$. Then we get an algebra homomorphism from $\mathfrak{A}$ to $M_{d+1}(\mathbb{C})$.

This is called the <u>left regular representation</u> of $\mathfrak{A}$ with respect to the basis $\{A_0, A_1, \ldots, A_d\}$. It is faithful, because $\mathfrak{A}$ has the identity element. By (iv)' of (2.1)', the image of A_i is ${}^tB_i$, where B_i is the matrix of degree $d+1$ whose (j,k) entry is p_{ij}^k. So the correspondence $A_i \to {}^tB_i$ is an isomorphism of $\mathfrak{A}$. Since the anti-isomorphism $B_i \to {}^tB_i$ becomes an isomorphism by the commutativity of $\mathfrak{A}$, the correspondence $A_i \to B_i$ gives an isomorphism of $\mathfrak{A}$. So we get the following theorem.

THEOREM 2.3. Let $\mathfrak{X} = (X, \{R_i\}_{0 \le i \le d})$ be a commutative association scheme of class d. Let $A_0, A_1, \ldots, A_d$ be the adjacency matrices and let p_{ij}^k be the intersection numbers. Set $B_i = (p_{ij}^k)$ and let $\mathfrak{B}$ be the subalgebra of $M_{d+1}(\mathbb{C})$ spanned by $B_0, B_1, \ldots, B_d$. Then the adjacency algebra $\mathfrak{A}$ of $\mathfrak{X}$ is isomorphic to $\mathfrak{B}$ by the correspondence of A_i to B_i. In particular, A_i and B_i have the same minimal polynomial.

$B_i = (p_{ij}^k)$ is called the ith <u>intersection matrix</u> of $\mathfrak{X}$ and $\mathfrak{B}$ is called the <u>intersection algebra</u> of $\mathfrak{X}$.

REMARK. Let $\mathfrak{X} = (X, \{R_i\}_{0 \le i \le d})$ be a commutative association scheme and let p_{ij}^k be the intersection numbers. Define new relations $\tilde{R}_k$ by $\tilde{R}_i = R_i \cup R_{i'}$. Then $\tilde{\mathfrak{X}} = (X, \{\tilde{R}_i\})$ becomes a symmetric association scheme. Let $\tilde{p}_{ij}^k$ be the intersection numbers of $\tilde{\mathfrak{X}}$. Then

$$\tilde{p}_{ij}^k = p_{ij}^k + \delta_{ii'} p_{i'j}^k + \delta_{jj'} p_{ij'}^k + \delta_{ii'}\delta_{jj'} p_{i'j'}^k .$$

$\tilde{\mathfrak{X}}$ is called the <u>symmetrization</u> of $\mathfrak{X}$.

QUESTION. When can a symmetric association scheme be the symmetrization of a certain commutative but non-symmetric association scheme?

2.3 ADJACENCY ALGEBRAS (BOSE-MESNER ALGEBRAS), ORTHOGONALITY RELATIONS OF EIGENMATRICES, AND KREIN PARAMETERS

Let $\mathfrak{X} = (X, \{R_i\}_{0 \leq i \leq d})$ be a commutative association scheme. Set $X = \{x_1, x_2, \ldots, x_n\}$ and let $e_x = (0,\ldots,0,1,0,\ldots,0) \in \mathbb{C}^n$ with 1 in the ith place for $x = x_i \in X$. Let V be the Hermitian space with an orthonormal basis $\{e_x \mid x \in X\}$. The adjacency matrices $A_0, A_1, \ldots, A_d$ are pairwise commutative normal matrices by (2.1)'(iii)', (v)'. So, by a well known theorem in linear algebra that pairwise commutative normal matrices are diagonalized simultaneously by a unitary matrix, we have

$$V = V_0 + V_1 + \ldots + V_r \quad \text{(a direct sum)}, \tag{3.1}$$

where each V_i is a common eigenspace of $A_0, A_1, \ldots, A_d$. Here we choose V_i to be maximal, i.e., if $i \neq j$, then there exists A_k such that the eigenvalue of A_k on V_i is different from that on V_j. So for $i \neq j$, V_i and V_j are orthogonal to each other. Since the eigenspace of $J = \sum_{i=0}^{d} A_i$ belonging to the eigenvalue n is the subspace spanned by $(1,1,\ldots,1)$, the one-dimensional subspace $\langle(1,1,\ldots,1)\rangle$ is exactly V_i for some i. We may choose $i = 0$:

$$V_0 = \langle(1,1,\ldots,1)\rangle . \tag{3.2}$$

Let E_i be the orthogonal projection $V \to V_i$ expressed in a matrix form with respect to the basis $\{e_x | x \in X\}$. Then,

$$E_0 + E_1 + \dots + E_r = I \quad \text{and} \quad E_i E_j = \delta_{ij} E_i , \tag{3.3}$$

$$E_0 = \frac{1}{|X|} J \qquad (\tfrac{1}{|X|} J \text{ is the orthogonal projection } V \to V_0), \tag{3.4}$$

and there exists a unitary matrix U such that

$$U^{-1} E_i U = \operatorname{diag}(0,\dots,0,\underbrace{1,\dots,1}_{\dim V_i},0,\dots,0) . \tag{3.5}$$

Let $p_i(j)$ be the eigenvalue of A_i on V_j. Then

$$A_i = \sum_{j=0}^{r} p_i(j) E_j , \tag{3.6}$$

and $U^{-1}A_iU = \operatorname{diag}(p_i(0), p_i(1), \dots, p_i(2), \dots, p_i(r), \dots)$. Since A_i is a normal matrix, the eigenvalue of ${}^t\overline{A_i}$ on V_j is $\overline{p_i(j)}$, where $\overline{p_i(j)}$ is the complex conjugate of $p_i(j)$. Since ${}^t\overline{A_i} = A_{i'}$, we have

$$p_{i'}(j) = \overline{p_i(j)} . \tag{3.7}$$

THEOREM 3.1. Let $\mathcal{X} = (X, \{R_i\}_{0 \le i \le d})$ be a commutative association scheme ($|X| = n$). Let $V = V_0 + \dots + V_r$ be the decomposition of the n-dimensional Hermitian space V into the direct sum of maximal common eigenspaces of the adjacency matrices A_i. Let E_i be the orthogonal projection matrix $V \to V_i$ with respect to the orthonormal basis $\{(0,\dots,0,1,0,\dots,0)\}$ with 1 in the jth place.

Then each E_i is a linear combination of $A_0, A_1, \ldots, A_d$. In particular, $r = d$ and $E_0, E_1, \ldots, E_d$ are primitive idempotents of the adjacency algebra $\mathfrak{A}$.

Proof. It suffices to show that a linear combination of $A_0, A_1, \ldots, A_d$ has a nonzero eigenvalue on V_i and vanishes on V_j $(j \neq i)$. For each j $(j \neq i)$, there exists some k such that $p_k(i) \neq p_k(j)$ by the maximality of V_i. Set $F_j = A_k - p_k(j)I$. Then F_j has a nonzero eigenvalue $p_k(i) - p_k(j)$ on V_i and vanishes on V_j. So $\prod_{j \neq i} F_j$ has a nonzero eigenvalue on V_i and vanishes on V_j $(j \neq i)$. □

Let

$$E_i = \frac{1}{|X|} \sum_{j=0}^{d} q_i(j) A_j \tag{3.8}$$

and set

$$\begin{aligned} P &= (p_i(j)) \qquad \text{(the } (j,i) \text{ entry is } p_i(j)) \\ Q &= (q_i(j)) \qquad \text{(the } (j,i) \text{ entry is } q_i(j)). \end{aligned} \tag{3.9}$$

The matrices P and Q of degree $d + 1$ are called the first and second eigenmatrices of $\mathfrak{X}$ respectively. By (3.6) and (3.8),

$$PQ = QP = nI\,. \tag{3.10}$$

For matrices A, B of the same size, $A \circ B$ is, by definition, the matrix obtained by the entry-wise product of A and B, i.e., $(A \circ B)_{ij} = A_{ij}B_{ij}$. We call $A \circ B$ the Hadamard product of A and B. Notice that

$$\begin{aligned} &A_0 + A_1 + \cdots + A_d = J \quad \text{and} \\ &A_i \circ A_j = \delta_{ij} A_i\,, \end{aligned} \tag{3.11}$$

and compare this with (3.3).

For a matric C, $\tau(C)$ denotes the sum of all entries C. Direct calculation shows the following:

LEMMA 3.2. For any square matrices A, B of the same size, $\mathrm{tr}(A \cdot {}^tB) = \mathrm{tr}({}^tA \cdot B) = \tau(A \circ B)$.

Let $m_i = \dim V_i = \mathrm{rank}\, E_i$. The positive integers m_i are called the <u>multiplicities</u> of $\mathfrak{X}$.

PROPOSITION 3.3.

(i) $$\mathrm{tr}A_i = \begin{cases} |X| & \text{if } i = 0, \\ 0 & \text{otherwise.} \end{cases}$$

(ii) $\tau(A_i) = |X|k_i$.

(iii) $\mathrm{tr}E_i = m_i$.

(iv) $$\tau(E_i) = \begin{cases} |X| & \text{if } i = 0, \\ 0 & \text{otherwise.} \end{cases}$$

<u>Proof</u>.

(i) Trivial.

(ii) Since k_i is the number of elements y such that $(x,y) \in R_i$ for any fixed $x \in X$, A_i has a constant row sum k_i and so (ii) holds.

(iii) Take the trace of (3.5).

(iv) By (3.4), $\tau(E_0) = |X|$. Suppose $i \neq 0$. Apply tr to $E_0E_i = 0$ to get

$$\begin{aligned} 0 &= \mathrm{tr}(E_0 E_i) \\ &= \tau({}^tE_0 \circ E_i) \qquad \text{(by Lemma 3.2)} \\ &= \tau(\frac{1}{|X|} J \circ E_i) \qquad \text{(by (3.4))} \\ &= \frac{1}{|X|}\tau(E_i), \end{aligned}$$

because J is the identity with respect to the Hadamard product. □

PROPOSITION 3.4. For each i $(i = 0, 1, \ldots, d)$, it holds that

(i) $p_0(i) = 1$,

(ii) $p_i(0) = k_i$,

(iii) $q_0(i) = 1$,

(iv) $q_i(0) = m_i$.

Proof.

(i) Compare the coefficients of E_i in $A_0 = \sum_i p_0(i)E_i$ and $A_0 = I = E_0 + E_1 + \ldots + E_d$.

(ii) Apply τ to $A_i = \sum_j p_i(j)E_j$ and use Proposition 3.3.

(iii) Compare the coefficients of A_i in $E_0 = \frac{1}{|X|} J = \frac{1}{|X|}(A_0 + A_1 + \ldots + A_d)$.

(iv) Apply tr to $E_i = \frac{1}{|X|} \sum_j q_i(j)A_j$ and use Proposition 3.3. □

THEOREM 3.5. Let $P = (p_i(j))$ and $Q = (q_i(j))$ be the first and second eigenmatrices of a commutative association scheme $\mathfrak{X}$ respectively. Then the following holds.

(i) $q_j(i)/m_j = \overline{p_i(j)}/k_i$.

(ii) (The First Orthogonality Relation)

$$\sum_{\nu=0}^{d} \frac{1}{k_\nu} p_\nu(i)\overline{p_\nu(j)} = \frac{|X|}{m_i} \delta_{ij}.$$

(iii) (The Second Orthogonality Relation)

$$\sum_{\nu=0}^{d} m_\nu p_i(\nu)\overline{p_j(\nu)} = |X| k_i \delta_{ij}.$$

Proof.

(i) Multiply $E_j = \frac{1}{|X|} \sum_{\nu=0}^{d} q_j(\nu) A_\nu$ by A_i under Hadamard product. By (3.11), $E_j \circ A_i = \frac{1}{|X|} q_j(i) A_i$. Applying τ, we get

$$\begin{aligned}
\tau(E_j \circ A_i) &= tr(E_j \cdot {}^tA_i) && \text{(by Lemma 3.2)} \\
&= tr(E_j A_{i'}) && \text{(by (2.1)'(iii)')} \\
&= tr(p_{i'}(j)E_j) && \text{(by (3.6))} \\
&= p_{i'}(j) m_j && \text{(by Proposition 3.3)} \\
&= m_j \overline{p_i(j)} && \text{(by (3.7))}
\end{aligned}$$

and

$$\tau\left(\frac{1}{|X|} q_j(i) A_i\right) = q_j(i) k_i \qquad \text{(by Proposition 3.3)}.$$

(ii) Write the relation $PQ = |X|I$ entry-wise, using (i).

(iii) Write the relation $QP = |X|I$ entry-wise using (i). □

$\mathfrak{A}$ is closed under the Hadamard product by (3.11). This commutative algebra with respect to the Hadamard product is denoted by $\hat{\mathfrak{A}}$ and $A_0, A_1, \ldots, A_d$ are the primitive idempotents. Obviously $E_i \circ E_j$ is a linear combination of $E_0, E_1, \ldots, E_d$ by Theorem 3.1:

$$E_i \circ E_j = \frac{1}{|X|} \sum_{k=0}^{d} q_{ij}^k E_k \tag{3.12}$$

for some complex numbers q_{ij}^i. (Later q_{ij}^k are shown to be non-negative real numbers.) The coefficients q_{ij}^k are called the <u>Krein parameters</u>. $\hat{\mathfrak{A}}$ together with the basis $E_0, E_1, \ldots, E_d$ is a dual of $\mathfrak{A}$ together with the basis $A_0, A_1, \ldots, A_d$ in the sense that

(i) $|X|E_0 = J$, the identity of $\hat{\mathfrak{A}}$.

(ii) $E_0 + E_1 + \ldots + E_d = I$, the identity of $\mathfrak{A}$.

(iii) ${}^tE_i = E_{\hat{i}}$ for some $\hat{i} \in \{0,1,\ldots,d\}$. (3.13)

(iv) $E_i \circ E_j = \frac{1}{|X|} \sum_{k=0}^{d} q_{ij}^k E_k$.

(v) $E_i \circ E_j = E_j \circ E_i$.

Compare (3.13) with (2.1)'. The property (3.13)(iii) follows from the uniqueness of primitive idempotents of $\mathfrak{A}$, which is easily proved from the simultaneous diagonalization of the A_i's, and the fact that ${}^tE_0, {}^tE_1, \ldots, {}^tE_d$ satisfy (3.3), i.e., they are the primitive idempotents of $\mathfrak{A}$. Since $E_{\hat{i}} = {}^tE_i = \overline{E}_i$, taking the complex conjugate of (3.8), we get

$$q_{\hat{i}}(j) = \overline{q_i(j)}. \tag{3.14}$$

Compare this with (3.7).

THEOREM 3.6.

(i) $$q_{ij}^{k} = \frac{|X|}{m_k}\tau(E_i \circ E_j \circ E_{\hat{k}})$$

$$= \frac{m_i m_j}{|X|}\sum_{\nu=0}^{d}\frac{1}{k_\nu^2}p_\nu(i)p_\nu(j)\overline{p_\nu(k)}.$$

(ii) $$p_{ij}^{k} = \frac{1}{|X|k_k}\operatorname{tr}(A_iA_jA_{k'})$$

$$= \frac{k_i k_j}{|X|}\sum_{\nu=0}^{d}\frac{1}{m_\nu^2}q_\nu(i)q_\nu(j)\overline{q_\nu(k)}.$$

Proof. Take the trace of the identity

$$\frac{1}{|X|}q_{ij}^{k}E_k = (E_i \circ E_j)E_k.$$

Then

$$\frac{1}{|X|}q_{ij}^{k}m_k = \operatorname{tr}((E_i \circ E_j)E_k) \quad \text{(by Proposition 3.3)} \tag{3.15}$$

$$= \tau(E_i \circ E_j \circ {}^{t}E_k) \quad \text{(by Lemma 3.2)}$$

$$= \tau(E_i \circ E_j \circ E_{\hat{k}}) \quad \text{(by (3.13)(iii))} \tag{3.16}$$

$$= \tau\Big(\frac{1}{|X|^3}\sum_{\nu} q_i(\nu)q_j(\nu)q_k(\nu)A_\nu\Big)$$

(by (3.8))

$$= \frac{1}{|X|^2}\sum_{\nu} q_i(\nu)q_j(\nu)\overline{q_k(\nu)}k_\nu$$

(by Proposition 3.3)

$$= \frac{m_i m_j m_k}{|X|^2} \sum_\nu \frac{1}{k_\nu^2} \overline{p_\nu(i)}\, \overline{p_\nu(j)}\, p_\nu(k) \qquad (3.17)$$

(by Theorem 3.5(i)).

Since $\overline{p_{\nu'}(i)} = p_\nu(i)$ and $k_{\nu'} = k_\nu$, replacing ν in (3.17) by ν', we get

$$\frac{1}{|X|} q_{ij}^k m_k = \frac{m_i m_j m_k}{|X|^2} \sum_\nu \frac{1}{k_\nu^2} p_\nu(i) p_\nu(j) \overline{p_\nu(k)}. \qquad (3.18)$$

By (3.16) and (3.18), we get the assertion (i). By (3.17) and (3.18), q_{ij}^k's are real numbers.

(ii) Apply τ to

$$p_{ij}^k A_k = (A_i A_j) \circ A_k.$$

Then

$$\begin{aligned} p_{ij}^k |X| k_k &= \tau((A_i A_j) \circ A_k) \\ &= \operatorname{tr}(A_i A_j A_{k'}) \\ &= \sum_\nu p_i(\nu) p_j(\nu) \overline{p_k(\nu)} m_\nu \\ &= k_i k_j k_k \sum_\nu \frac{1}{m_\nu^2} \overline{q_\nu(i)}\, \overline{q_\nu(j)}\, q_\nu(k). \end{aligned}$$

Since p_{ij}^k's are real numbers, we get the desired result. □

The following relations between q_{ij}^k are the dual of the relations between p_{ij}^k in Proposition 2.2.

PROPOSITION 3.7.

(i) $q^k_{0j} = \delta_{jk}$,

(ii) $q^k_{i0} = \delta_{ik}$,

(iii) $q^0_{ij} = m_i \delta_{i\hat{j}}$,

(iv) $q^k_{ij} = q^{\hat{k}}_{\hat{i}\hat{j}}$,

(v) $\sum_{j=0}^{d} q^k_{ij} = m_i$,

(vi) $m_k q^k_{ij} = m_j q^j_{\hat{i}k} = m_i q^i_{k\hat{j}}$,

(vii) $\sum_{\alpha=0}^{d} q^\alpha_{ij} q^\ell_{k\alpha} = \sum_{\beta=0}^{d} q^\beta_{ki} q^\ell_{\beta j}$.

<u>Proof</u>. (i) Compare the coefficients of E_k in $E_0 \circ E_j = \frac{1}{|X|} \sum_k q^k_{0j} E_k$ with those in

$$E_0 \circ E_j = \frac{1}{|X|} J \circ E_j$$

$$= \frac{1}{|X|} E_j .$$

(ii) By (3.13)(v), $q^k_{i0} = q^k_{0i}$. Use (i) to get $q^k_{i0} = \delta_{ik}$.

(iii) By Theorem 3.6(i),

$$q^0_{ij} = |X| \tau (E_i \circ E_j \circ E_0)$$

$$= \tau (E_i \circ E_j) \qquad \text{(by (3.4))}$$

$$= \operatorname{tr}(E_i E_{\hat{j}}) \qquad \text{(by Lemma 3.2)}$$

$$= \operatorname{tr}(\delta_{i\hat{j}} E_i)$$

$$= m_i \delta_{i\hat{j}} \qquad \text{(by Proposition 3.3).}$$

(iv) Take the transpose of $E_i \circ E_j = \frac{1}{|X|} \sum_{k=0}^{d} q_{ij}^k E_k$.

(v) Take the sum of $E_i \circ E_j = \frac{1}{|X|} \sum_{k=0}^{d} q_{ij}^k E_k$ over j. Then the left-hand side is

$$\sum_j (E_i \circ E_j) = E_i \circ (\sum_j E_j)$$

$$= (\frac{1}{|X|} \sum_\nu q_i(\nu) A_\nu) \circ A_0$$

$$= \frac{1}{|X|} q_i(0) A_0$$

$$= \frac{1}{|X|} m_i \sum_k E_k \qquad \text{(by Proposition 3.4(iv)),}$$

while the right-hand side is $\frac{1}{|X|} \sum_k (\sum_j q_{ij}^k) E_k$. Compare the coefficients of E_k to get the desired result.

(vi) By the proof of Theorem 3.6, $\tau(E_i \circ E_j \circ E_{\hat{k}})$ are real numbers. So taking the complex conjugate, we get

$$\tau(E_i \circ E_j \circ E_{\hat{k}}) = \tau(E_{\hat{i}} \circ E_k \circ E_{\hat{j}}) = \tau(E_k \circ E_{\hat{j}} \circ E_{\hat{i}}).$$

Use Theorem 3.6(i) and rewrite the above identities in terms of Krein parameters.

(vii) By the associativity of the Hadamard product, $E_k \circ (E_i \circ E_j) = (E_k \circ E_i) \circ E_j$. Now

$$E_k \circ (E_i \circ E_j) = E_k \circ (\frac{1}{|X|} \sum_\alpha q_{ij}^\alpha E_\alpha)$$

$$= \frac{1}{|X|^2} \sum_\ell (\sum_\alpha q_{ij}^\alpha q_{k\alpha}^\ell) E_\ell$$

and

$$(E_k \circ E_i) \circ E_j = \left(\frac{1}{|X|} \sum_\beta q_{ki}^\beta E_\beta\right) \circ E_j$$
$$= \frac{1}{|X|^2} \sum_\ell \left(\sum_\beta q_{ki}^\beta q_{\beta j}^\ell\right) E_\ell .$$

So we get the desired result. □

The Krein parameters q_{ij}^k are non-negative real numbers. This fact was found by Dunkl-Scott (cf. Scott (1973)) in Krein's paper (1950, English translation 1963) on harmonic analysis and is now called the Krein condition.

THEOREM 3.8. (The Krein Condition.) The Krein parameters q_{ij}^k of a commutative association scheme $\mathfrak{X} = (X, \{R_i\}_{0 \le i \le d})$ are non-negative real numbers.

We need a lemma to prove this theorem.

LEMMA 3.9. Let A, B be positive definite (respectively non-negative definite) Hermitian matrices. Then the Hadamard product $A \circ B$ is also a positive definite (respectively non-negative definite) Hermitian matrix.

Proof. The Kronecker product $A \otimes B$ is a positive definite (respectively non-negative definite) Hermitian matrix, which is an easy exercise of linear algebra. $A \circ B$ is a principal minor of $A \otimes B$, hence a positive definite (respectively non-negative definite) Hermitian matrix. □

Proof of Theorem 3.8. Each E_i is a non-negative definite Hermitian matrix. By Lemma 3.9, $E_i \circ E_j$ is a non-negative definite Hermitian matrix. Since $E_i \circ E_j = \frac{1}{|X|} \sum_{k=0}^{d} q_{ij}^k E_k$, we obtain that $\frac{1}{|X|} q_{ij}^k$ are eigenvalues of $E_i \circ E_j$ and so are non-negative real numbers. □

There are many cases in which all the q_{ij}^k's are rational numbers.

QUESTION. When is q_{ij}^k irrational?

The following is one of the few examples we know in which some q_{ij}^k are irrational. Let Γ be the generalized 6-gon with 126 vertices. Namely, let V be a Hermitian space of dim 3 over $GF(3^2)$. There are 63 non-isotropic 1-dimensional subspaces of V. Call them points and denote the set of points by X_1. There are 63 unordered triples $\{x_1, x_2, x_3\}$ such that x_i $(i = 1, 2, 3)$ are non-isotropic 1-dimensional subspaces of V and x_i and x_j $(i \neq j)$ are orthogonal. Call them lines and denote the set of lines by X_2. Define the adjacency between points and lines by inclusion. Then we get the generalized 6-gon $\Gamma = X_1 \cup X_2$. The intersection matrix of Γ (Chapter III, section 3.1) is

$$B = \begin{pmatrix} 0 & 1 & \cdot & \cdot & \cdot & \cdot & 0 \\ 3 & 0 & \cdot & & & & \cdot \\ \cdot & 2 & 0 & \cdot & & & \cdot \\ \cdot & & \cdot & 0 & \cdot & & \cdot \\ \cdot & & & \cdot & 0 & 1 & \cdot \\ \cdot & & & & \cdot & 0 & 3 \\ 0 & \cdot & \cdot & \cdot & \cdot & 2 & 0 \end{pmatrix} \quad \text{(of size 7)},$$

and the minimal polynomial of B is $x(x^2 - 9)(x^2 - 6)(x^2 - 2)$. The first eigenmatrix is

$$P = \begin{pmatrix} 1 & 3 & 6 & 12 & 24 & 48 & 32 \\ 1 & \sqrt{6} & 3 & \sqrt{6} & 0 & -2\sqrt{6} & -4 \\ 1 & \sqrt{2} & -1 & -3\sqrt{2} & -4 & 2\sqrt{2} & 4 \\ 1 & 0 & -3 & 0 & 6 & 0 & -4 \\ 1 & -\sqrt{2} & -1 & 3\sqrt{2} & -4 & -2\sqrt{2} & 4 \\ 1 & -\sqrt{6} & 3 & -\sqrt{6} & 0 & 2\sqrt{6} & -4 \\ 1 & -3 & 6 & -12 & 24 & -48 & 32 \end{pmatrix}$$

Then $\frac{|X|}{m_1 m_2} q_{12}^4 = \frac{147 - 49\sqrt{6}}{144}$, which is irrational. (Use Theorem 3.6(i) to check this.)

2.4 FORMULAE FOR m_i

Let $\mathfrak{X} = (X, \{R_i\}_{0 \leq i \leq d})$ be a commutative association scheme of class d. In this section, we discuss Biggs' formula for m_i, Frame's theorem, and the absolute bound condition.

The following theorem is essentially due to Biggs (1974) and tells us how to calculate m_i from the information of the intersection matrices B_i. We shall say that a vector $\mathbf{x}$ is <u>standard</u> when the first entry of $\mathbf{x}$ is 1.

THEOREM 4.1. Let $\mathfrak{X} = (X, \{R_i\}_{0 \leq i \leq d})$ be a commutative association scheme of class d. Let $P = (p_i(j))$ and $Q = (q_i(j))$ be the first and second eigenmatrices of $\mathfrak{X}$, and let k_i and m_i be the valencies and multiplicities of $\mathfrak{X}$. Let $\mathbf{v}_i = (p_0(i), p_1(i), \ldots, p_d(i))$ and $\mathbf{u}_i =$

$(\overline{p_0(i)}/k_0, \overline{p_1(i)}/k_1, \ldots, \overline{p_d(i)}/k_d)$ be the row vectors obtained by standardizing the ith row of P and the ith column of Q respectively. Then the following holds:

$$\text{(i)} \quad m_i = \frac{|X|}{\sum_{\nu=0}^{d} \frac{1}{k_\nu} |p_\nu(i)|^2}$$

$$= \frac{|X|}{(\mathfrak{u}_i, \mathfrak{v}_j)} .$$

(ii) $\mathfrak{u}_i$ is characterized as the unique standardized common left eigenvector $\mathfrak{u}$ of the intersection matrices B_j such that $\mathfrak{u}B_j = p_j(i)\mathfrak{u}$ $(j = 0, 1, \ldots, d)$. ${}^t\mathfrak{v}_i$ is characterized as the unique standardized common right eigenvector ${}^t\mathfrak{v}$ of the intersection matrices B_j such that $B_j{}^t\mathfrak{v} = p_j(i){}^t\mathfrak{v}$ $(j = 0, 1, \ldots, d)$.

Proof. (i) This follows immediately from the first orthogonality relation.

(ii) Consider the left regular representation of the adjacency algebra $\mathfrak{A}$. If we choose a basis $\{A_0, A_1, \ldots, A_d\}$ of $\mathfrak{A}$, the image of A_j is ${}^tB_j$. If we choose a basis $\{E_0, E_1, \ldots, E_d\}$ of $\mathfrak{A}$, the image of A_j is $\mathrm{diag}(p_j(0), p_j(1), \ldots, p_j(d))$. Since $(E_0, E_1, \ldots, E_d)P = (A_0, A_1, \ldots, A_d)$, we have

$${}^tB_j = P^{-1}\mathrm{diag}(p_j(0), p_j(1), \ldots, p_j(d))P$$

and so

$$B_j{}^tP = {}^tP\,\mathrm{diag}(p_j(0), p_j(1), \ldots, p_j(d)). \tag{4.1}$$

This means the ith column vector ${}^t\mathfrak{v}_i$ of ${}^tP$ is a right eigenvector of B_j belonging to the eigenvalue $p_j(i)$.

Multiplying (4.1) by ${}^tQ$ from the right and left, we get

$$ {}^tQB_j = \mathrm{diag}(p_j(0), p_j(1), \ldots, p_j(d))\, {}^tQ. \qquad (4.2)$$

This means the ith row vector $(q_i(0), q_i(1), \ldots, q_i(d))$ of ${}^tQ$ is a left eigenvector of B_j belonging to the eigenvalue $p_j(i)$. By Theorem 3.5, $\frac{1}{m_i}(q_i(0), q_i(1), \ldots, q_i(d)) = \mathfrak{u}_i$.

Let $\mathfrak{u}$ be a common left eigenvector of B_j such that $\mathfrak{u}B_j = p_j(i)\mathfrak{u}$ for all j. We can write $\mathfrak{u} = \sum_k \lambda_k \mathfrak{u}_k$, because $\mathfrak{u}_k$ $(0 \le k \le d)$ are linearly independent. Applying B_j, we get $p_j(i)\sum_k \lambda_k \mathfrak{u}_k = \sum_k \lambda_k p_j(k)\mathfrak{u}_k$ and so $\lambda_k p_j(i) = \lambda_k p_j(k)$ for all j. If $\lambda_k \neq 0$, then $p_j(i) = p_j(k)$ for all j, which implies $i = k$ by $\det P \neq 0$. So $\mathfrak{u}$ is a scalar multiple of $\mathfrak{u}_i$. The characterization of ${}^t\mathfrak{v}_i$ is proved similarly. □

In case the intersection algebra $\mathfrak{B}$ is generated by a certain B_j, $\mathfrak{u}_i$ and $\mathfrak{v}_i$ $(i = 0, 1, \ldots, d)$ are determined by the single matrix B_j. This situation happens in P-polynomial schemes (Chapter III, section 3.1). An important consequence of Theorem 4.1 is that the quantity $|X|/(\mathfrak{u}_i, \mathfrak{v}_i)$ must be an integer, and this integrality condition is often useful to show the non-existence of certain association schemes.

The following is another integrality condition due to Frame (1941). (Theorem 4.2(iii) is due to Wielandt.) Our proof follows Wielandt (1964).

THEOREM 4.2. Let $\mathfrak{X} = (X, \{R_i\}_{0 \le i \le d})$ be a commutative association scheme of class d. Set $|X| = n$, and let

k_i, m_i be the valencies and multiplicities of $\mathfrak{X}$. Then the following assertions hold.

(i) The rational number

$$q = n^{d-1} \prod_{\nu=0}^{d} \frac{k_\nu}{m_\nu}$$

is an integer.

(ii) If all the entries $p_i(j)$ of the first eigenmatrix P are rational, then q is a square. The hypothesis is always fulfilled if the multiplicities m_i $(i \neq 0)$ are all different.

(iii) For any prime p and any positive integer ℓ,

$$\#\{m_i \mid m_i \text{ is divisible by } p^\ell\} \leq$$

$$\#\{nk_i \mid nk_i \text{ is divisible by } p^\ell\}.$$

Proof. (i) By writing the second orthogonality relation in the matrix form, we get

$$n \operatorname{diag}(k_0, k_1, \ldots, k_d) = {}^t\overline{P} \operatorname{diag}(m_0, m_1, \ldots, m_d) P. \quad (4.3)$$

Take the determinant of (4.3) to get

$$(\det P)(\overline{\det P}) = n^{d+1} \prod_{\nu=0}^{d} \frac{k_\nu}{m_\nu}. \quad (4.4)$$

Since $nE_0 = \sum_\nu A_\nu = \sum_i (\sum_\nu p_\nu(i)) E_i$, we have

$$\sum_{\nu=0}^{d} p_\nu(i) = \begin{cases} n & \text{if } i = 0, \\ 0 & \text{otherwise.} \end{cases}$$

So, by adding the ith columns $(i = 1, 2, \ldots, d)$ to the

0th one,

$$\det P = \begin{vmatrix} n & p_1(0) & \cdots & p_d(0) \\ 0 & p_1(1) & \cdots & p_d(1) \\ \vdots & \vdots & & \vdots \\ 0 & p_1(d) & \cdots & p_d(d) \end{vmatrix} \tag{4.5}$$

$$= n \begin{vmatrix} p_1(1) & \cdots & p_d(1) \\ \vdots & & \vdots \\ p_1(d) & \cdots & p_d(d) \end{vmatrix}$$

Since $p_i(j)$'s are eigenvalues of A_i, they are all algebraic integers. So $(\det P)/n$ is an algebraic integer by (4.5). Thus the rational number $q = n^{d-1} \prod_\nu \frac{k_\nu}{m_\nu}$ is an algebraic integer by (4.4) and hence a rational integer.

(ii) Suppose $p_i(j)$'s are all rational. Since they are algebraic integers, they are rational integers. By (4.5), $(\det P)/n$ is a rational integer and hence q is a square by (4.4).

Suppose m_i's are all different. Let K be the field generated by $p_i(j)$ $(i, j = 0, 1, \ldots, d)$ over the rational field $\mathbb{Q}$. For an arbitrary element σ of the Galois group of K over $\mathbb{Q}$, let E_i^σ be the matrix obtained by applying σ to E_i entry-wise. Then $E_0^\sigma, E_1^\sigma, \ldots, E_d^\sigma$ satisfy (3.3), i.e., they are the primitive idempotents of $\mathfrak{A}$. By the uniqueness of primitive idempotents, σ permutes $E_0, E_1, \ldots, E_d$ and so $E_i^\sigma = E_{i^\sigma}$ for some $i^\sigma \in \{0,1,\ldots,d\}$. Then $m_i = m_{i^\sigma}$, because $m_i = \text{rank } E_i$. But m_i are all different by the assumption and hence $i = i^\sigma$ for all i. Apply σ to $A_i = \sum_j p_i(j)E_j$. Then, since $A_i^\sigma = A_i$ and $E_j^\sigma = E_j$, we get $p_i(j)^\sigma = p_i(j)$ for all i, j, which implies $p_i(j)$'s are all rational.

(iii) We use the following theorem.

THEOREM*. Let A and B be square matrices of the same degree whose coefficients are algebraic integers. Then the ith elementary divisor of A divides that of AB.

For the proof of Theorem*, the reader may consult a book which treats the elementary divisor theory over Dedekind domains, e.g., Curtis and Reiner (1962, section 22). (Notice that elementary divisors are no longer numbers but ideals in general.) Applying this theorem to (4.3), the ith elementary divisor of $\text{diag}(m_0, m_1, \ldots, m_d)$ divides that of $\text{diag}(nk_0, nk_1, \ldots, nk_d)$. Thus the assertion (iii) holds. □

THEOREM 4.3. If all the multiplicities m_i $(i \neq 0)$ of a commutative association scheme coincide with each other, then so do all the valencies k_i $(i \neq 0)$.

<u>Proof</u>. Set $m_i = m$ $(i = 1, 2, \ldots, d)$. By Theorem 4.2, $q = \dfrac{n^{d-1}}{m^d} \prod_{\nu=1}^{d} k_\nu$ is an integer. Since n and m are relatively prime by $n = 1 + md$, m^d divides $\prod_{\nu=1}^{d} k_\nu$. But the geometric mean of the k_ν's is at most equal to the arithmetic mean, thus

$$\left(\prod_{\nu=1}^{d} k_\nu\right)^{1/d} \le \frac{1}{d}\sum_{\nu=1}^{d} k_\nu$$

$$= \frac{n-1}{d}$$

$$= m .$$

Therefore $m^d = \prod_{\nu=1}^{d} k_\nu$ and this occurs only if $k_\nu = m$ $(\nu = 1, \ldots, d)$. □

From Theorem 4.3, we have the following theorem immediately.

THEOREM 4.4. If the valencies k_i $(i = 1, 2)$ of a commutative association scheme of class 2 are different, then so are the multiplicities m_i $(i = 1, 2)$.

REMARKS. (1) The assumption of commutativity in Theorem 4.4 is redundant, because association schemes of class at most 4 are always commutative.

(2) Theorem 4.4 is proved by direct calculation as follows. Let

$$B = \begin{pmatrix} 0 & 1 & 0 \\ k & \lambda & \mu \\ 0 & k-\lambda-1 & k-\mu \end{pmatrix}$$

be the intersection matrix of an association scheme of class 2. Then

$$q = (\lambda - \mu)^2 + 4(k - \mu),$$

$$\begin{Bmatrix} \theta_1 \\ \theta_2 \end{Bmatrix} = \frac{\lambda - \mu \pm \sqrt{q}}{2},$$

$$\begin{Bmatrix} m_1 \\ m_2 \end{Bmatrix} = \frac{1}{2}\{n - 1 \mp \frac{2k + (n - 1)(\lambda - \mu)}{\sqrt{q}}\},$$

$$k_1 = k,$$

$$k_2 = \frac{k(k - \lambda - 1)}{\mu}.$$

Thus if $m_1 = m_2$, then $2k + (n - 1)(\lambda - \mu) = 0$. This implies $0 < \mu - \lambda = \frac{2k}{n - 1} < 2$ and so $\mu - \lambda = 1$, $k = \frac{n - 1}{2}$.

(3) By Theorem 4.4 and Theorem 4.2(ii), if a commutative association scheme of class 2 has valencies k_i $(i \neq 0)$ all different, then q is a square. But there are many examples of commutative association schemes of class greater than 2 in which the valencies k_i $(i \neq 0)$ are all different and q is not a square. For example, consider the smallest Janko group of order 175,560 acting transitively on 266 letters. The permutation character is multiplicity free and the one-point stabilizer has 5 orbits of length $k_0 = 1$, $k_1 = 11$, $k_2 = 110$, $k_3 = 132$, $k_4 = 12$. Example 2.1(1) is commutative in this case and has an intersection matrix

$$B_1 = \begin{pmatrix} 0 & 1 & & & \\ 11 & 0 & 1 & & \\ & 10 & 4 & 5 & \\ & & 6 & 5 & 11 \\ & & & 1 & 0 \end{pmatrix}.$$

The multiplicities are $m_0 = 1$, $m_1 = 76$, $m_2 = 77$, $m_3 = 56$, $m_4 = 56$. Thus $q = 3^2 \cdot 5 \cdot 11^2 \cdot 19^2$ is not a square, though the k_i's are all different.

We shall discuss the absolute bound condition for primitive commutative association schemes (Neumaier (1981b)). Let $\mathfrak{X} = (X, \{R_i\}_{0 \leq i \leq d})$ be a commutative association scheme of class d. To begin with, we discuss the primitivity and imprimitivity of commutative association schemes briefly (the details will be given in section 2.9). Let $\Gamma_i = (X, R_i)$ be the graph whose vertex and edge sets are X and R_i respectively. A <u>path</u> (in Γ_i) of length s from x to y is, by definition, a sequence of vertices $x_0 = x, x_1, \ldots, x_{s-1}, x_s = y$ such that $(x_\nu, x_{\nu+1}) \in R_i$ for $\nu = 0, 1, \ldots, s-1$. $\mathfrak{X}$ is said to be <u>primitive</u> if all the Γ_i are connected $(i = 1, 2, \ldots, d)$, i.e., for any distinct $x, y \in X$, there exists a path from x to y in Γ_i.

PROPOSITION 4.5. For a fixed i $(i = 1, 2, \ldots, d)$, define a relation $\sim$ on X by $x \sim y$ if and only if there exists a path from x to y in Γ_i. Then $\sim$ is an equivalence relation which is a union of some association relations R_j.

<u>Proof</u>. Notice that there exists a path of length s from x to y in Γ_i if and only if the (x,y) entry of A_i^s is nonzero. Since A_i^s is a linear combination of $A_0, A_1, \ldots, A_d$, the relation $\sim$ is a union of R_j whose adjacency matrix A_j has a nonzero coefficient in A_i^s for some s.

The transitivity of $\sim$ is trivial. Since $|X|$ is finite, there is a loop in Γ_i, i.e., a sequence $x_0 = x, x_1, \ldots, x_{s-1}, x_s = x$ $(s \geq 2)$ such that each $(x_\nu, x_{\nu+1}) \in \Gamma_i$ $(\nu = 0, 1, \ldots, s-1)$. This means A_0 appears in the linear combination expression of A_i^s with respect to the basis $A_0, A_1, \ldots, A_d$. So the relation $\sim$ contains R_0. Since $p_{ij}^0 = k_i \delta_{ij'}$, the appearance of A_0 in A_i^s implies that $A_{i'}$ appears in the linear combination expression of A_i^{s-1} with respect to the basis $A_0, A_1, \ldots, A_d$. So the relation $\sim$ contains $R_{i'}$. □

The following theorem is due to Cameron, Goethals and Seidel (1978b).

THEOREM 4.6. Let $\mathfrak{X} = (X, \{R_i\}_{0 \le i \le d})$ be a commutative association scheme of class d, $A_0, A_1, \ldots, A_d$ the adjacency matrices of $\mathfrak{X}$ and $E_0, E_1, \ldots, E_d$ the primitive idempotents of the adjacency algebra $\mathfrak{A}$. Then the following three conditions are equivalent:

(i) $\mathfrak{X}$ is imprimitive.

(ii) By a suitable rearrangement of indices $1, 2, \ldots, d$, there exists an index s $(0 < s < d)$ such that $A_i A_j$ is a linear combination of $A_0, A_1, \ldots, A_s$ for all i, j $(0 \le i \le s,\ 0 \le j \le s)$.

(iii) By a suitable rearrangement of indices $1, 2, \ldots, d$, there exists an index t $(0 < t < d)$ such that the Hadamard product $E_i \circ E_j$ is a linear combination of $E_0, E_1, \ldots, E_t$ for all i, j $(0 \le i \le t,\ 0 \le j \le t)$.

Proof. Suppose $\mathfrak{X}$ is imprimitive. Then some Γ_i is not connected. By Proposition 4.5, $\bigcup_{\nu=0}^{s} R_\nu$ is an equivalence relation $(0 < s < d)$, if we rearrange the indices $1, 2, \ldots, d$ suitably. So, by a suitable rearrangement of elements of X,

$$\sum_{\nu=0}^{s} A_\nu = I_q \otimes J_p, \tag{4.6}$$

where $p = \sum_{\nu=0}^{s} k_\nu$ (the cardinality of each equivalence class), $q = \frac{|X|}{p}$ (the number of equivalence classes), I_q is the identity matrix of degree q and J_p is the matrix of degree p whose entries are all 1. Set

$$M = \sum_{\nu=0}^{s} A_\nu .$$

Then, by (4.6), we get

$$\begin{aligned} &M^2 = pM, \\ &M \circ M = M, \qquad\qquad (4.7) \\ &M \neq I \text{ and } M \neq J. \end{aligned}$$

If we calculate M^2 directly from $M = \sum_{\nu=0}^{s} A_\nu$ by using $A_iA_j = \sum_k p_{ij}^k A_k$, we get

$$M^2 = \sum_{k=0}^{d} \left(\sum_{i,j=0}^{s} p_{ij}^k \right) A_k .$$

By (4.7), the coefficient $\sum_{i,j=0}^{s} p_{ij}^k$ of A_k is zero for $k > s$. Since p_{ij}^k is non-negative, we get $p_{ij}^k = 0$ for $0 \leq i \leq s$, $0 \leq j \leq s$, $s < k$. This implies that A_iA_j is a linear combination of $A_0, A_1, \ldots, A_s$ for $0 \leq i \leq s$, $0 \leq j \leq s$.

By (4.7), $\frac{1}{p}M$ is an idempotent of $\mathfrak{A}$. So $\frac{1}{p}M$ is a linear combination of the primitive idempotents $E_0, E_1, \ldots, E_d$ with coefficients either zero or one. By a suitable rearrangement of the indices 1, 2, ..., d, we may assume

$$\frac{1}{p}M = \sum_{\nu=0}^{t} E_\nu \qquad (0 < t < d).$$

The reason why E_0 appears in the above sum is that $\tau(M) \neq 0$ and $\tau(E_\nu) = 0$ for $\nu \neq 0$ by $M = \sum_{\nu=0}^{s} A_\nu$ and Proposition 3.3. If we calculate $(\frac{1}{p}M)\circ(\frac{1}{p}M)$ directly from $\frac{1}{p}M = \sum_{\nu=0}^{t} E_\nu$ by using $E_i \circ E_j = \frac{1}{|X|}\sum_k q_{ij}^k E_k$, we get

$$(\frac{1}{p}M)\circ(\frac{1}{p}M) = \frac{1}{|X|^2}\sum_{k=0}^{d}(\sum_{i,j=0}^{t} q_{ij}^k)E_k.$$

Since $(\frac{1}{p}M)\circ(\frac{1}{p}M) = \frac{1}{p^2}M$ by (4.7), the coefficient $\sum_{i,j=0}^{t} q_{ij}^k$ of E_k is zero for $k > t$. Since q_{ij}^k is non-negative by the Krein condition, we get $q_{ij}^k = 0$ for $0 \leq i \leq t$, $0 \leq j \leq t$, $t < k$. This implies that $E_i \circ E_j$ is a linear combination of $E_0, E_1, \ldots, E_t$ for $0 \leq i \leq t$, $0 \leq j \leq t$. Thus the condition (i) implies (ii) and (iii).

Suppose the condition (ii) holds. It is clear that Γ_i is disconnected for i $(0 \leq i \leq s)$.

Suppose (iii) holds. First we shall show

$$0 \leq \hat{i} \leq t \quad \text{for every } i \ (0 \leq i \leq t). \tag{4.8}$$

In the expression $E_i = \frac{1}{|X|}\sum_{\nu=0}^{d} q_i(\nu)A_\nu$, let $\lambda_0, \lambda_1, \ldots, \lambda_r$ be all the distinct numbers of $\{q_i(\nu) \mid \nu = 0, 1, \ldots, d\}$, and set

$$\Phi_\ell = \sum A_\nu \qquad (\ell = 0, 1, \ldots, r),$$

where the summation is over all ν such that $q_i(\nu) = \lambda_\ell$. Let E_i^h denote the product of h E_i's with respect to the Hadamard product. Then

$$E_i^h = \frac{1}{|X|^h} \sum_{\ell=0}^{r} \lambda_\ell^h \Phi_\ell \qquad (h = 0, 1, \ldots, r).$$

Since $\det(\lambda_\ell^h)_{0 \le \ell, h \le r} \neq 0$, Φ_ℓ is a linear combination of $E_0, E_i, E_i^2, \ldots, E_i^r$ $(0 \le \ell \le r)$. Since each E_i^h is a linear combination of $E_0, E_1, \ldots, E_t$ by the assumption (iii), so is Φ_ℓ $(0 \le \ell \le r)$. Now $E_{\hat{i}} = {}^tE_i = \overline{E}_i = \frac{1}{|X|} \sum_{\ell=0}^{r} \overline{\lambda}_\ell \Phi_\ell$ and hence $E_{\hat{i}}$ is a linear combination of $E_0, E_1, \ldots, E_t$. Since $E_{\hat{i}}$ is a primitive idempotent, $E_{\hat{i}} = E_j$ for some j $(0 \le j \le t)$, which proves (4.8).

The condition (iii) is equivalent to $q_{ij}^k = 0$ for $0 \le i \le t$, $0 \le j \le t$, $t < k$. Moreover it holds that $q_{ij}^k = 0$ for $0 \le i \le t$, $0 \le k \le t$, $t < j$ under the assumption (iii). This is because $q_{ij}^k = \frac{m_j}{m_k} q_{\hat{i}k}^j$ by Proposition 3.7(vi) and $0 \le \hat{i} \le t$ by (4.8). Thus

$$\begin{aligned}
\Big(\sum_{\nu=0}^{t} E_\nu\Big) \circ \Big(\sum_{\nu=0}^{t} E_\nu\Big) &= \frac{1}{|X|} \sum_{k=0}^{d} \Big(\sum_{i,j=0}^{t} q_{ij}^k\Big) E_k \\
&= \frac{1}{|X|} \sum_{k=0}^{t} \sum_{i=0}^{t} \Big(\sum_{j=0}^{d} q_{ij}^k\Big) E_k \\
&= \frac{1}{|X|} \Big(\sum_{i=0}^{t} m_i\Big) \sum_{k=0}^{t} E_k
\end{aligned}$$

(by Proposition 3.7(v)).

So $M = p \sum_{\nu=0}^{t} E_\nu$ satisfies (4.7) with $p = |X| / \sum_{i=0}^{t} m_i$.

Since $M \circ M = M$, M is a linear combination of $A_0, A_1, \dots, A_d$ with coefficients either zero or one. So, by a suitable rearrangement of indices $1, 2, \dots, d$, we may assume $M = \sum_{\nu=0}^{s} A_\nu$ $(0 < s < d)$. The reason why A_0 appears in the above sum is that $\mathrm{tr}\, M \neq 0$ and $\mathrm{tr} A_\nu = 0$ for $\nu \neq 0$. Since $M^2 = pM$ and p_{ij}^k's are all non-negative, $A_i A_j$ is a linear combination of $A_0, A_1, \dots, A_s$ for $0 \leq i \leq s$, $0 \leq j \leq s$. Thus the condition (iii) implies (ii). $\square$

Further properties of primitive commutative association schemes will be discussed in section 2.9. The following lemma is a key to the absolute bound condition.

LEMMA 4.7. Let A and B be matrices of the same size. Let m and ℓ be the rank of A and B respectively. Then

(i) the rank of $\underbrace{A \circ A \circ \dots \circ A}_{h} \leq \binom{m+h-1}{h}$ $(h = 1, 2, \dots)$,

(ii) the rank of $A \circ B \leq m\ell$.

Proof. Let A^h denote the product of h A's under the Hadamard product. Let the row space of A be spanned by row vectors $\mathbf{a}_1, \mathbf{a}_2, \dots, \mathbf{a}_m$. Since

$$\left(\sum_i c_i \mathbf{a}_i\right) \circ \left(\sum_j d_j \mathbf{a}_j\right) = \sum_{i,j} c_i d_j \mathbf{a}_i \circ \mathbf{a}_j$$

under the entry-wise product $\circ$, the row space of A^h is spanned by the vectors $\mathbf{a}_{i_1} \circ \mathbf{a}_{i_2} \circ \dots \circ \mathbf{a}_{i_h}$ $(1 \leq i_1 \leq m, 1 \leq i_2 \leq m, \dots, 1 \leq i_h \leq m)$. So the rank of A^h is no more than $\binom{m+h-1}{h}$, the number of ways of choosing h elements, allowing repetition from m elements.

The assertion (ii) can be proved similarly. □

THEOREM 4.8. Let $\mathcal{X} = (X, \{R_i\}_{0 \le i \le d})$ be a commutative association scheme of class d. Let m_i $(i = 0, 1, \dots, d)$ be the multiplicities of $\mathcal{X}$ and let q_{ij}^k be the Krein parameters of $\mathcal{X}$. Then for fixed i and j,

$$\Sigma\, m_k \le \begin{cases} \frac{1}{2} m_i(m_i + 1) & \text{if } i = j, \\ m_i m_j & \text{if } i \ne j, \end{cases}$$

where the summation is over all the k's such that q_{ij}^k is positive.

Proof. Since $E_i \circ E_j = \frac{1}{|X|} \sum_k q_{ij}^k E_k$, we get

$$\operatorname{rank}(E_i \circ E_j) = \sum_{k\,:\,q_{ij}^k > 0} \operatorname{rank} E_k = \sum_{k\,:\,q_{ij}^k > 0} m_k .$$

Apply Lemma 4.7 to $E_i \circ E_j$ to get the desired result. □

THEOREM 4.9. (The Absolute Bound Condition) Let $\mathcal{X} = (X, \{R_i\}_{0 \le i \le d})$ be a symmetric association scheme of class d. Let m_i $(i = 0, 1, \dots, d)$ be the multiplicities of $\mathcal{X}$. Suppose $\mathcal{X}$ is primitive. Then for each m_i $(i = 1, \dots, d)$,

$$|X| \le \binom{m_i + d - 1}{d} + \binom{m_i + d - 2}{d - 1} .$$

Proof. Let $E_0, E_1, \dots, E_d$ be the primitive idenpotents of the adjacency algebra and E_i^h denote the product of h

E_i's under the Hadamard product. We shall say E_j is a <u>component</u> of E_i^h if E_j appears in the linear combination expression of E_i^h with respect to the basis $E_0, E_1, \dots, E_d$.

Since $\mathfrak{X}$ is primitive, every E_j is a component of E_i^h for some h $(0 \le h \le d)$ by Theorem 4.6(iii) and the Krein condition. Let N_h be the set of indices j such that E_j is a component of E_i^h but not of E_i^ℓ $(0 \le \ell \le h-1)$. Then $N_0 = \{0\}$, $N_1 = \{i\}$ and

$$\{0,1,\dots,d\} = \bigcup_{h=0}^{d} N_h \quad \text{(a disjoint union).}$$

Since $\mathfrak{X}$ is symmetric, E_i^2 contains $E_0 = \frac{1}{|X|}J$ as a component. So each component of E_i^{h-2} is also a component of E_i^h. Thus

$$\sum_{j \in N_h} m_j = \sum_{j \in N_h} \operatorname{rank} E_j$$

$$\le \operatorname{rank} E_i^h - \operatorname{rank} E_i^{h-2} \qquad (2 \le h \le d)$$

and so

$$|X| = \sum_{h=0}^{d} \sum_{j \in N_h} m_j$$

$$\le \operatorname{rank} E_0 + \operatorname{rank} E_i + \sum_{h=2}^{d} (\operatorname{rank} E_i^h - \operatorname{rank} E_i^{h-2})$$

$$= \operatorname{rank} E_i^d + \operatorname{rank} E_i^{d-1}$$

$$\le \binom{m_i + d - 1}{d} + \binom{m_i + d - 2}{d-1} \quad \text{(by Lemma 4.7(i)).} \qquad \square$$

REMARKS. (4) Let $\mathfrak{X} = (X, \{R_i\}_{0 \le i \le d})$ be a symmetric association scheme. If a primitive idempotent E_i spans the adjacency algebra with respect to the Hadamard product in the sense of algebra, then $\mathfrak{X}$ can be realized in such a way that X is a subset of the unit sphere in the m_i-dimensional Euclidean space (m_i = rank E_i) and R_i $(0 \le i \le d)$ are the distance relations. In this realization, the equality of the absolute bound in Theorem 4.9 is attained if and only if X is a tight spherical 2d-design (Part II). Tight spherical 2d-designs are classified for $d \ge 3$ (Bannai and Damerell (1979, 1980)) and they exist very rarely. We shall discuss the case of $d = 2$ in section 2.8.

(5) If $\mathfrak{X}$ is non-symmetric, then we can get a better bound of $|X|$ in terms of an arbitrarily fixed m_i $(i \ne 0)$ than that of the symmetric case.

QUESTION. What is the reasonable absolute bound condition for non-symmetric commutative association schemes?

2.5 THE KAWADA-DELSARTE DUALITY IN C-ALGEBRAS

In section 2.3, we saw a duality between $\mathfrak{A}$ and $\hat{\mathfrak{A}}$, i.e., the Krein parameters q_{ij}^k behave as if there were an association scheme which has the intersection numbers q_{ij}^k (the Delsarte duality, Delsarte (1973b)). Before Delsarte, Y. Kawada (1942) was led to the concept of C-algebras, which is algebraically the same thing as what we call commutative association schemes nowadays, by formulating a finiteness version of the Tannaka duality. We shall review section 2.3, following Kawada (1942).

Let $\mathfrak{A}$ be an algebra over $\mathbb{C}$ with a basis $x_0, x_1, \ldots, x_d$ in the linear space sense. $\mathfrak{A}$ together with $x_0, x_1, \ldots, x_d$ is called a <u>C-algebra</u> (<u>Character-algebra</u>) if the following conditions (I)-(VI) hold:

(I) $\mathfrak{A}$ is a commutative algebra, i.e.,

$$x_i x_j = \sum_{k=0}^{d} p_{ij}^k x_k \quad \text{with} \quad p_{ij}^k = p_{ji}^k .$$

(II) $\mathfrak{A}$ has the identity element $e = x_0$, i.e., $p_{0j}^k = \delta_{jk}$.

(III) Every p_{ij}^k is a real number.

(IV) There exists a permutation $i \to i'$ $(i = 0, 1, \ldots, d)$ such that $(i')' = i$ and $p_{ij}^k = p_{i'j'}^{k'}$, i.e., the mapping $x_i \to x_{i'}$ $(i = 0, 1, \ldots, d)$ can be extended to an algebra isomorphism of $\mathfrak{A}$.

(V) $p_{ij}^0 = \delta_{ij'} k_i$ with $k_i > 0$ for all i, j.

(VI) The mapping $x_i \to k_i$ $(i = 0, 1, \ldots, d)$ is a linear representation of $\mathfrak{A}$.

To distinguish the basis $x_0, x_1, \ldots, x_d$, we use the notation $\mathfrak{A} = \langle x_i \mid 0 \le i \le d \rangle$ for a C-algebra $\mathfrak{A}$.

For example, the adjacency algebra $\mathfrak{A}$ together with the adjacency matrices A_i of a commutative association scheme is a C-algebra, and the dual $\hat{\mathfrak{A}}$ together with the nE_i's is a C-algebra. This is the reason why $\hat{\mathfrak{A}}$ behaves like $\mathfrak{A}$ algebraically, though $\hat{\mathfrak{A}}$ has no structure of association schemes in general, i.e., $\hat{\mathfrak{A}}$ is no longer realized by $\{0,1\}$-matrices.

In what follows, we assume $\mathfrak{A} = \langle x_i \mid 0 \le i \le d \rangle$ is a

C-algebra. We first remark that

(a) the mapping $i \to i'$ is uniquely determined by (V),

(b) $0' = 0$ by (II) and (V),

(c) $k_0 = 1$ by (II) and (V),

(d) $k_i = k_{i'}$ by (IV) and (V).

PROPOSITION 5.1.

$$k_\gamma p^\gamma_{\alpha\beta} = k_\beta p^\beta_{\alpha'\gamma} .$$

Proof. Express $(x_\alpha x_\beta)x_{\gamma'} = (x_\alpha x_{\gamma'})x_\beta$ as a linear combination of $x_0, x_1, \ldots, x_d$ and compare the coefficients of x_0. Then

$$(x_\alpha x_\beta)x_{\gamma'} = (\sum_\nu p^\nu_{\alpha\beta} x_\nu)x_{\gamma'} = p^\gamma_{\alpha\beta} k_\gamma x_0 + \cdots$$

and

$$(x_\alpha x_{\gamma'})x_\beta = (\sum_\nu p^\nu_{\alpha\gamma'} x_\nu)x_\beta = p^{\beta'}_{\alpha\gamma'} k_{\beta'} x_0 + \cdots$$

$$= p^\beta_{\alpha'\gamma} k_\beta x_0 + \cdots,$$

and we get the desired result. □

Rewrite the relation of Proposition 5.1 as

$$\sqrt{\frac{k_\gamma}{k_\beta}}\, p^\gamma_{\alpha\beta} = \sqrt{\frac{k_\beta}{k_\gamma}}\, p^\beta_{\alpha'\gamma} . \tag{5.1}$$

Let B_i be the matrix of degree $d + 1$ whose (j,k) entry is p^k_{ij} and $K = \mathrm{diag}(\sqrt{1}, \sqrt{k_1}, \sqrt{k_2}, \ldots, \sqrt{k_d})$. Then (5.1) means

$$\overline{{}^{t}K^{-1}B_iK} = K^{-1}B_{i'}K\ . \tag{5.2}$$

Let $\mathfrak{B}$ be the subalgebra of $M_{d+1}(\mathbb{C})$ spanned by $B_0, B_1, \ldots, B_d$. The left regular representation of $\mathfrak{A}$ with respect to the basis $\{x_0, x_1, \ldots, x_d\}$ gives an algebra isomorphism between $\mathfrak{A}$ and ${}^{t}\mathfrak{B}$ with the correspondence of x_i to ${}^{t}B_i$. Since $\mathfrak{B}$ is commutative, the anti-isomorphism $B_i \to {}^{t}B_i$ is an isomorphism. Thus we have:

THEOREM 5.2. Let $\mathfrak{A} = \langle x_i \mid 0 \le i \le d\rangle$ be a C-algebra. Then $\mathfrak{A}$ is, by the correspondence of x_i to B_i, isomorphic to the subalgebra $\mathfrak{B}$ of $M_{d+1}(\mathbb{C})$ spanned by $B_i = (p_{ij}^k)$ $(i = 0, 1, \ldots, d)$.

In particular, $\mathfrak{B}$ is commutative and so $K^{-1}B_iK$ $(i = 0, 1, \ldots, d)$ are normal matrices commuting with each other by (5.2). Therefore, there exists a unitary matrix U such that

$$U^{-1}K^{-1}B_iKU = \operatorname{diag}(p_i(0), p_i(1), \ldots, p_i(d)) \tag{5.3}$$
$$(i = 0, 1, \ldots, d).$$

Since x_0 is the identity, B_0 is the identity matrix and so

$$p_0(i) = 1 \qquad (i = 0, 1, \ldots, d). \tag{5.4}$$

By (5.2) and (5.3), we get

$$\overline{p_i(j)} = p_{i'}(j)\ . \tag{5.5}$$

Let P be the matrix whose (j,i) entry is $p_i(j)$. We call P the eigenmatrix of $\mathfrak{A}$. Then P is non-singular, because $U^{-1}K^{-1}B_iKU$ $(i = 0, 1, \ldots, d)$ are linearly independent. Let v_i be the ith row vector of P, i.e., $v_i = (p_0(i), p_1(i), \ldots, p_d(i))$. Obviously, $v_0, v_1, \ldots, v_d$ are linearly independent.

PROPOSITION 5.3. The column vector ${}^t v_i$ $(i = 0, 1, \ldots, d)$ is a common right eigenvector of $B_0, B_1, \ldots, B_d$, and any common right eigenvector of $B_0, B_1, \ldots, B_d$ is a scalar multiple of some ${}^t v_i$.

Proof. By (5.3), the mapping $\Delta_\alpha : B_i \to p_i(\alpha)$ $(i = 0, 1, \ldots, d)$ gives a linear representation of $\mathfrak{B}$. For every linear representation Δ of $\mathfrak{B}$, we want to show that

$$\Delta(B_i)\,{}^t v = B_i\,{}^t v \qquad (i = 0, 1, \ldots, d), \tag{5.6}$$

where $v = (\Delta(B_0), \Delta(B_1), \ldots, \Delta(B_d))$ and $\Delta(B_0) = 1$.

Apply Δ to $B_i B_j = \sum_k p_{ij}^k B_k$. Then $\Delta(B_i)\Delta(B_j) = \sum_k p_{ij}^k \Delta(B_k)$ and this means $\Delta(B_i)\,{}^t v = B_i\,{}^t v$. Since B_0 is the identity matrix, $\Delta(B_0) = 1$. This proves (5.6).

If we set $\Delta = \Delta_\alpha$, then $v = v_\alpha$, and ${}^t v_\alpha$ is a common right eigenvector of $B_0, B_1, \ldots, B_d$ by (5.6), i.e.,

$$p_i(\alpha)\,{}^t v_\alpha = B_i\,{}^t v_\alpha \tag{5.6)'$$

with $v_\alpha = (p_0(\alpha), p_1(\alpha), \ldots, p_d(\alpha))$, $p_0(\alpha) = 1$.

Let F_i be the projection matrix of degree $d + 1$ defined by $F_i\,{}^t v_\alpha = \delta_{i\alpha}\,{}^t v_\alpha$ $(\alpha = 0, 1, \ldots, d)$. Then

$$F_i F_j = \delta_{ij} F_i \tag{5.7}$$

$$F_0 + F_1 + \cdots + F_d = I, \text{ the identity matrix,}$$

and by (5.6)',

$$B_i = \sum_{j=0}^{d} p_i(j)F_j \,. \tag{5.8}$$

Since $P = (p_i(j))$ is non-singular, each F_j is a linear combination of $B_0, B_1, \ldots, B_d$:

$$F_i = \frac{1}{n} \sum_{j=0}^{d} q_i(j)B_j \tag{5.9}$$

with $n = \sum_{i=0}^{d} k_i$.

Obviously, ${}^t\mathbf{v}_0, {}^t\mathbf{v}_1, \ldots, {}^t\mathbf{v}_d$ are all the common right eigenvectors of $F_0, F_1, \ldots, F_d$ up to scalar multiple and hence they are all the common right eigenvectors of $B_0, B_1, \ldots, B_d$ up to scalar multiple. □

Let Δ be a linear representation of $\mathfrak{B}$ and let $\mathbf{v} = (\Delta(B_0), \ldots, \Delta(B_d))$. Then by (5.6), ${}^t\mathbf{v}$ is a common eigenvector of $B_0, B_1, \ldots, B_d$ and by Proposition 5.3, $\mathbf{v}$ coincides with some $\mathbf{v}_\alpha$. This means $\Delta_\alpha : B_i \to p_i(\alpha)$ $(\alpha = 0, 1, \ldots, d)$ are all the linear representations of $\mathfrak{B}$. Since $\mathfrak{A} \cong \mathfrak{B}$, we get the following proposition, allowing abuse of the notation Δ_α.

PROPOSITION 5.4. There are exactly $d + 1$ linear representations of $\mathfrak{A}$ and they are described by $\Delta_\alpha : x_i \to p_i(\alpha)$ $(\alpha = 0, 1, \ldots, d)$.

By (VI), we may assume $\Delta_0 : x_i \to k_i$, i.e.,

$$p_i(0) = k_i \qquad (i = 0, 1, \ldots, d). \tag{5.10}$$

Let $e_0, e_1, \ldots, e_d$ be the primitive idempotents of $\mathfrak{A}$ corresponding to $F_0, F_1, \ldots, F_d$ in $\mathfrak{B}$. Then

$$e_i e_j = \delta_{ij} e_i \quad \text{and} \quad e_0 + e_1 + \dots + e_d = x_0, \tag{5.7)'$$

$$x_i = \sum_{j=0}^{d} p_i(j) e_j, \tag{5.8)'$$

$$e_i = \frac{1}{n} \sum_{j=0}^{d} q_i(j) x_j \quad \text{with} \quad n = \sum_{i=0}^{d} k_i . \tag{5.9)'$$

By (5.8)' and (5.9)'

$$PQ = QP = nI \quad \text{with} \quad P = (p_i(j)) \quad \text{and} \tag{5.11}$$
$$Q = (q_i(j)).$$

REMARKS. (1) By (5.9)' and (5.11), we get $\Delta_\alpha : e_\beta \to \delta_{\alpha\beta}$.

(2) Proposition 5.4 is equivalent to saying that $\mathfrak{A}$ is semi-simple (see Curtis and Reiner (1962)). The following is a direct way to show $\mathfrak{A}$ is semi-simple. Let $\mathcal{J}$ be a non-zero ideal of $\mathfrak{A}$. Take a non-zero element $x = \sum_{i=0}^{d} a_i x_i \in \mathcal{J}$. Let $\bar{x} = \sum_{i=0}^{d} \overline{a_i} x_{i'}$. Then

$$\bar{x}x = \left(\sum_{i=0}^{d} \overline{a_i} a_i k_i\right) x_0 + \text{the terms in } x_1, \dots, x_d$$

and hence $\bar{x}x$ is a non-zero element of $\mathcal{J}$. Since $\overline{(yz)} = \bar{y}\,\bar{z}$ for all $y, z \in \mathfrak{A}$, $\overline{\bar{x}x} = x\bar{x} = \bar{x}x$. So from the beginning, we may assume $\bar{x} = x$, replacing x by $\bar{x}x$ if necessary. Then $x^2 = \bar{x}x \neq 0$, $x^4 = x^2 x^2 \neq 0$, ... and so the ideal $\mathcal{J}$ is not nilpotent. Thus $\mathfrak{A}$ is semi-simple, and Proposition 5.4, (5.7)', (5.8)', (5.9)' hold from general theory of semi-simple algebras.

Orthogonality relations hold for the matrices P and Q.

THEOREM 5.5. Let $\mathfrak{A}$ be a C-algebra with the basis $x_0, x_1, \ldots, x_d$ and the primitive idempotents $e_0, e_1, \ldots, e_d$. Let $x_i = \sum_{j=0}^{d} p_i(j) e_j$ and $e_i = \frac{1}{n} \sum_{j=0}^{d} q_i(j) x_j$ with $n = \sum_{i=0}^{d} k_i$. Then

(i) $q_j(i)/m_j = \overline{p_i(j)}/k_i$ with $k_i = p_i(0)$ and $m_j = q_j(0)$ $(m_j \neq 0$ for $j = 0, 1, \ldots, d)$.

(ii) (The First Orthogonality Relation)

$$\sum_{\nu=0}^{d} \frac{1}{k_\nu} p_\nu(i)\overline{p_\nu(j)} = \frac{n}{m_i} \delta_{ij} .$$

(iii) (The Second Orthogonality Relation)

$$\sum_{\nu=0}^{d} m_\nu p_i(\nu)\overline{p_j(\nu)} = nk_i \delta_{ij} .$$

Proof. Express $x_i x_{j'}$ as a linear combination of $x_0, x_1, \ldots, x_d$ and look at the coefficient of x_0:

$$\begin{aligned} x_i x_{j'} &= (\sum_\nu p_i(\nu) e_\nu)(\sum_\nu p_{j'}(\nu) e_\nu) \\ &= \sum_\nu p_i(\nu) p_{j'}(\nu) e_\nu \\ &= \frac{1}{n} \sum_{\alpha,\nu} p_i(\nu) p_{j'}(\nu) q_\nu(\alpha) x_\alpha \\ &= \frac{1}{n} (\sum_\nu p_i(\nu)\overline{p_j(\nu)} m_\nu) x_0 + \text{the terms in} \\ &\qquad x_1, \ldots, x_d, \end{aligned}$$

and on the other hand,

$$x_i x_{j'} = \sum_k p_{ij'}^k x_k$$

$$= \delta_{ij} k_i x_0 + \text{the terms in } x_1, \ldots, x_d .$$

So we get the second orthogonality relation.

Rewrite the second orthogonality relation in the matrix form:

$$^t\overline{P}\,\mathrm{diag}(m_0, m_1, \ldots, m_d) P = n \cdot \mathrm{diag}(k_0, k_1, \ldots, k_d). \quad (5.12)$$

By (5.11) and (5.12),

$$Q = \mathrm{diag}\left(\frac{1}{k_0}, \frac{1}{k_1}, \ldots, \frac{1}{k_d}\right) {}^t\overline{P}\,\mathrm{diag}(m_0, m_1, \ldots, m_d). \quad (5.13)$$

Comparing the (i,j) entries of (5.13), we get the relation (i). Since $\det Q \neq 0$ by (5.11), each $m_i \neq 0$ by (5.13). Put (5.13) in $PQ = nI$ to get

$$P\,\mathrm{diag}\left(\frac{1}{k_0}, \frac{1}{k_1}, \ldots, \frac{1}{k_\alpha}\right) {}^t\overline{P} = n\,\mathrm{diag}\left(\frac{1}{m_0}, \frac{1}{m_1}, \ldots, \frac{1}{m_d}\right). \quad (5.14)$$

Comparing the (i,j) entries of (5.14), we get the first orthogonality relation. □

Set $i = j$ in the first crthogonality relation to get the following:

COROLLARY 5.6.

$$m_i = n \Big/ \sum_{\nu=0}^{d} \frac{1}{k_\nu} |p_\nu(i)|^2 > 0 .$$

THEOREM 5.7. With the same notation as in Theorem 5.5, we have

$$p_{ij}^{k} = \frac{k_i k_j}{n} \sum_{\nu=0}^{d} \frac{1}{m_\nu^2} q_\nu(i) q_\nu(j) \overline{q_\nu(k)} .$$

Proof. Express $x_i x_j$ as a linear combination of $x_0, x_1, \ldots, x_d$ and compare the coefficients:

$$\begin{aligned} x_i x_j &= (\sum_\nu p_i(\nu) e_\nu)(\sum_\nu p_j(\nu) e_\nu) \\ &= \sum_\nu p_i(\nu) p_j(\nu) e_\nu \\ &= \frac{1}{n} \sum_k \sum_\nu p_i(\nu) p_j(\nu) q_\nu(k) x_k \\ &= \frac{k_i k_j}{n} \sum_k \sum_\nu \frac{1}{m_\nu^2} \overline{q_\nu(i)}\ \overline{q_\nu(j)}\ q_\nu(k)\ x_k \end{aligned}$$

by Theorem 5.5(i), and on the other hand,

$$x_i x_j = \sum_k p_{ij}^{k} x_k .$$

So we get

$$p_{ij}^{k} = \frac{k_i k_j}{n} \sum_\nu \frac{1}{m_\nu^2} \overline{q_\nu(i)}\ \overline{q_\nu(j)}\ q_\nu(k) . \tag{5.15}$$

Since p_{ij}^{k} and m_ν are real, we get the desired result by taking the complex conjugate of (5.15). □

PROPOSITION 5.8. The numbers p_{ij}^{k} satisfy the relations (i)-(vii) in Proposition 2.2.

Proof. The relations (i)-(iv) are trivial by the definition of C-algebras. The relation (vi) has been proved in Proposition 5.1. By expressing $x_k(x_i x_j) = (x_k x_i) x_j$

as a linear combination of $x_0, x_1, \ldots, x_d$ and comparing the coefficients, we get (vii). So we only need to show

(v) $\sum_{j=0}^{d} p_{ij}^k = k_i .$

Put $i = 0$ in Corollary 5.6 and use (5.10) to get

$$m_0 = 1. \tag{5.16}$$

Put $j = 0$ in Theorem 5.5(i) and use (5.10), (5.16) to get

$$q_0(i) = 1 \qquad (i = 0, 1, \ldots, d). \tag{5.17}$$

Since $e_0 = \frac{1}{n} \sum_i q_0(i) x_i$, we get

$$e_0 = \frac{1}{n}(x_0 + x_1 + \ldots + x_d). \tag{5.18}$$

Express $x_i e_0$ as a linear combination of $x_0, x_1, \ldots, x_d$:

$$\begin{aligned} x_i e_0 &= (\sum_j p_i(j) e_j) e_0 \\ &= p_i(0) e_0 \\ &= \frac{k_i}{n}(x_0 + x_1 + \ldots + x_d) \end{aligned}$$

by (5.10), (5.18). On the other hand,

$$\begin{aligned} x_i e_0 &= \frac{1}{n} x_i \sum_{j=0}^{d} x_j \\ &= \frac{1}{n} \sum_{k=0}^{d} \sum_{j=0}^{d} p_{ij}^k x_k . \end{aligned}$$

Thus, we get (v). □

Let $\hat{\mathfrak{A}}$ be the set of all linear mappings of $\mathfrak{A}$ into $\mathbb{C}$. Since linear mappings of $\mathfrak{A}$ are determined by the values on the basis $x_0, x_1, \ldots, x_d$, $\hat{\mathfrak{A}}$ is naturally identified with the set of all mappings from $\{x_0, x_1, \ldots, x_d\}$ to $\mathbb{C}$ and by this identification, $\hat{\mathfrak{A}}$ becomes an algebra over $\mathbb{C}$, i.e., $(fg)(x_i) = f(x_i)g(x_i)$ for $f, g \in \mathfrak{A}$. With each linear representation $\Delta_\alpha: x_i \to p_i(\alpha)$, we associate the linear mapping Δ_α^* defined by

$$\Delta_\alpha^*: x_i \to q_\alpha(i) = m_\alpha \overline{\Delta_\alpha\left(\frac{x_i}{k_i}\right)} \,. \tag{5.19}$$

Since $Q = (q_\alpha(i))$ is non-singular, $\Delta_0^*, \Delta_1^*, \ldots, \Delta_d^*$ are linearly independent and so form a basis of $\hat{\mathfrak{A}}$.

THEOREM 5.9. Let $\mathfrak{A} = \langle x_i \mid 0 \leq i \leq d \rangle$ be a C-algebra and let $P = (p_i(j))$ be the eigenmatrix of $\mathfrak{A}$. Let $Q = (q_i(j))$ be the matrix defined by $PQ = QP = nI$, where I is the identity matrix. Then $\hat{\mathfrak{A}}$ with the basis $\Delta_\alpha^*: x_i \to q_\alpha(i)$ $(\alpha = 0, 1, \ldots, d)$ is a C-algebra whose eigenmatrix is Q.

$\hat{\mathfrak{A}}$ with the basis Δ_α^* is called the <u>dual</u> <u>C-algebra</u> (or conjugate C-algebra).

<u>Proof</u>. Let f_α be the element of $\hat{\mathfrak{A}}$ defined by

$$f_\alpha: x_i \to \delta_{\alpha i} \qquad (i = 0, 1, \ldots, d). \tag{5.20}$$

Then $f_0, f_1, \ldots, f_d$ are the primitive idempotents of $\hat{\mathfrak{A}}$, i.e.,

$$f_\alpha f_\beta = \delta_{\alpha\beta} f_\alpha \quad \text{and} \quad f_0 + f_1 + \ldots + f_d = \mathrm{id}_{\hat{\mathfrak{A}}}, \tag{5.21}$$

where $\mathrm{id}_{\hat{\mathfrak{A}}}$ is the identity $x_i \to 1$ $(i = 0, 1, \ldots, d)$ of

$\hat{\mathfrak{A}}$. By (5.19),

$$\Delta_\alpha^* = \sum_{\beta=0}^{d} q_\alpha(\beta) f_\beta . \tag{5.22}$$

Since $PQ = nI$,

$$f_\alpha = \frac{1}{n} \sum_{\beta=0}^{d} p_\alpha(\beta) \Delta_\beta^* , \tag{5.23}$$

where $n = \sum_{i=0}^{d} m_i$, the relation $n = \sum_i m_i$ coming from the $(0,0)$ entry of $QP = nI$ together with the relations $q_i(0) = m_i$ and $p_0(i) = 1$.

(I) $\hat{\mathfrak{A}}$ is obviously a commutative algebra. Let

$$\Delta_\alpha^* \Delta_\beta^* = \sum_{\gamma=0}^{d} q_{\alpha\beta}^{\gamma} \Delta_\gamma^* , \tag{5.24}$$

i.e., $q_\alpha(i) q_\beta(i) = \sum_{\gamma=0}^{d} q_{\alpha\beta}^{\gamma} q_\gamma(i)$ for all i.

(II) Δ_0^* is the identity element of $\hat{\mathfrak{A}}$, since $q_0(i) = 1$ by (5.17).

(III) Express $\Delta_\alpha^* \Delta_\beta^*$ as a linear combination of $\Delta_0^*, \Delta_1^*, \ldots, \Delta_d^*$ by using (5.22), (5.23) to get a similar relation to (5.15), i.e.,

$$q_{\alpha\beta}^{\gamma} = \frac{m_\alpha m_\beta}{n} \sum_{\nu=0}^{d} \frac{1}{k_\nu^2} \overline{p_\nu(\alpha)}\, \overline{p_\nu(\beta)}\, p_\nu(\gamma) . \tag{5.25}$$

Since $\overline{p_\nu(\alpha)}\, \overline{p_\nu(\beta)}\, p_\nu(\gamma) = p_{\nu'}(\alpha) p_{\nu'}(\beta) \overline{p_{\nu'}(\gamma)}$ by (5.5), we get

$$q_{\alpha\beta}^{\gamma} = \frac{m_\alpha m_\beta}{n} \sum_{\nu=0}^{d} \frac{1}{k_\nu^2} p_\nu(\alpha) p_\nu(\beta) \overline{p_\nu(\gamma)} . \tag{5.26}$$

By (5.25) and (5.26), $q_{\alpha\beta}^{\gamma}$ is real.

(IV) For each linear representation $\Delta_\alpha: x_i \to p_i(\alpha)$, the mapping $x_i \to \overline{p_i(\alpha)}$ can be extended to a linear representation of $\mathfrak{A}$, because p_{ij}^{k}'s are real, and hence it coincides with one of $\Delta_0, \Delta_1, \ldots, \Delta_d$, say $\Delta_{\hat{\alpha}}$, by Proposition 5.4, i.e.,

$$\Delta_{\hat{\alpha}}: x_i \to \overline{p_i(\alpha)}, \quad \text{i.e.,} \quad p_i(\hat{\alpha}) = \overline{p_i(\alpha)}\,. \tag{5.27}$$

The correspondence $\wedge$ gives a permutation of $\{0, 1, \ldots, d\}$ and satisfies $\hat{\hat{\alpha}} = \alpha$ for all α.

By Corollary 5.6 and (5.27),

$$m_{\hat{\alpha}} = m_\alpha\,. \tag{5.28}$$

Now

$$\begin{aligned}\Delta_{\hat{\alpha}}^{*}(x_i) &= \overline{m_{\hat{\alpha}} \Delta_{\hat{\alpha}}\left(\frac{x_i}{k_i}\right)} && \text{(by (5.19))} \\ &= m_{\hat{\alpha}} \cdot \frac{1}{k_i}\, p_i(\alpha) && \text{(by (5.27))} \\ &= \overline{q_\alpha(i)} && \text{(by Theorem 5.5(i) and (5.28)).}\end{aligned}$$

So

$$\Delta_{\hat{\alpha}}^{*}: x_i \to \overline{q_\alpha(i)}, \quad \text{i.e.,} \quad q_{\hat{\alpha}}(i) = \overline{q_\alpha(i)}\,. \tag{5.29}$$

By (5.25), (5.26), (5.27) and (5.28),

$$q^{\hat{\gamma}}_{\hat{\alpha}\hat{\beta}} = \frac{m_{\hat{\alpha}}m_{\hat{\beta}}}{n}\sum_{\nu}\frac{1}{k_\nu^2}\,p_\nu(\hat{\alpha})p_\nu(\hat{\beta})\overline{p_\nu(\hat{\gamma})}$$

$$= \frac{m_\alpha m_\beta}{n}\sum_{\nu}\frac{1}{k_\nu^2}\,\overline{p_\nu(\alpha)}\;\overline{p_\nu(\beta)}\;p_\nu(\gamma)$$

$$= q^{\gamma}_{\alpha\beta}\ .$$

So we get the condition (IV).

(V) $$q^{0}_{\alpha\beta} = \frac{m_\alpha m_\beta}{m}\sum_{\nu}\frac{1}{k_\nu^2}\,p_\nu(\alpha)p_\nu(\beta)p_\nu(0) \qquad \text{(by (5.26))}$$

$$= \frac{m_\alpha m_\beta}{n}\sum_{\nu}\frac{1}{k_\nu}\,p_\nu(\alpha)\overline{p_\nu(\hat{\beta})} \qquad \text{(by (5.10), (5.27))}$$

$$= \frac{m_\alpha m_\beta}{n}\,\frac{n}{m_\alpha}\,\delta_{\alpha\hat{\beta}} \qquad \text{(by Theorem 5.5(ii))}$$

$$= \delta_{\alpha\hat{\beta}}\,m_\alpha$$

and $m_\alpha > 0$ by Corollary 5.6.

(VI) Since $\Delta^*_\alpha = \sum_{\beta=0}^{d} q_\alpha(\beta)f_\beta$, with $f_0, f_1, \ldots, f_d$ the primitive idempotents of $\hat{\mathfrak{A}}$,

$$\varphi_i\colon \Delta^*_\alpha \to q_\alpha(i) \text{ is a linear representation of } \hat{\mathfrak{A}} \quad (i = 0, 1, \ldots, d). \tag{5.30}$$

In particular, $\varphi_0\colon \Delta^*_\alpha \to q_\alpha(0) = m_\alpha$ is a linear representation. This completes the proof of the theorem. □

By (5.30), the dual C-algebra of $\hat{\mathfrak{A}}$ has the basis φ^*_i defined by

$$\varphi_i^*: \Delta_\alpha^* \to p_i(\alpha) = \Delta_\alpha(x_i) . \tag{5.31}$$

Calculate $\varphi_i^*\varphi_j^*$ as follows:

$$\begin{aligned}
\varphi_i^*\varphi_j^*(\Delta_\alpha^*) &= \varphi_i^*(\Delta_\alpha^*)\varphi_j^*(\Delta_\alpha^*) && \text{(by the definition of the product } \varphi_i^*\varphi_j^*) \\
&= \Delta_\alpha(x_i)\Delta_\alpha(x_j) && \text{(by (5.31))} \\
&= \Delta_\alpha(x_i x_j) && \text{(by Proposition 5.4)} \\
&= \Delta_\alpha(\sum_k p_{ij}^k x_k) \\
&= \sum_k p_{ij}^k \Delta_\alpha(x_k) \\
&= \sum_k p_{ij}^k \varphi_k^*(\Delta_\alpha^*) .
\end{aligned}$$

So we get

$$\varphi_i^*\varphi_j^* = \sum_{k=0}^{d} p_{ij}^k \varphi_k^* , \tag{5.32}$$

and hence the following theorem:

THEOREM 5.10. (Duality Theorem). Let $\mathfrak{A}$ be a C-algebra with the basis $x_0, x_1, \ldots, x_d$, and let $\hat{\mathfrak{A}}$ with $\Delta_0^*, \ldots, \Delta_d^*$ be the dual C-algebra of $\mathfrak{A}$. Let $\hat{\hat{\mathfrak{A}}}$ with $\varphi_0^*, \ldots, \varphi_d^*$ be the dual C-algebra of $\hat{\mathfrak{A}}$. Then $\mathfrak{A}$ is isomorphic to $\hat{\hat{\mathfrak{A}}}$ as C-algebras by the correspondence of x_i to φ_i^*.

Theorems 5.9 and 5.10 are called the Kawada-Delsarte duality.

Let $\mathfrak{A}$ be the adjacency algebra of an association scheme $\mathfrak{X} = (X, \{R_i\}_{0 \le i \le d})$ and let $E_0, E_1, \ldots, E_d$ be the primitive idempotents. $\mathfrak{A}$ with the basis A_i is a C-algebra, and the algebra $\hat{\mathfrak{A}}$ with the basis nE_i defined in section 2.3 is isomorphic to the dual C-algebra of $\mathfrak{A}$, since (5.21) - (5.24) hold with respect to the Hadamard product, replacing f_α and Δ_α^* by A_α and nE_α.

In section 2.3, the mappings tr, τ played a key role in the proof of orthogonality relations. They are interpreted well in terms of C-algebras as follows. Let $\mathfrak{A}$ be a C-algebra with the basis $x_0, x_1, \ldots, x_d$ and let $e_0, e_1, \ldots, e_d$ be the primitive idempotents of $\mathfrak{A}$. Define the Schur product $\circ$ in $\mathfrak{A}$ by $x_i \circ x_j = \delta_{ij} x_i$ and let $\hat{\mathfrak{A}}$ denote the algebra equipped with the Schur product on the underlying vector space $\mathfrak{A}$. Then since $(ne_i) \circ (ne_j) = \sum_k q_{ij}^k (ne_k)$ by (5.9)' and (5.24), $\hat{\mathfrak{A}}$ with the basis $ne_0, ne_1, \ldots, ne_d$ is isomorphic to the dual C-algebra of $\mathfrak{A}$. Now let Δ_0 (respectively φ_0) be the linear representation of $\mathfrak{A}$ (respectively $\hat{\mathfrak{A}}$): $x_i \to k_i$, i.e., $e_i \to \delta_{i0}$ (respectively $ne_i \to m_i$, i.e., $x_i \to \delta_{i0}$). Define the mappings tr and τ by

$$\mathrm{tr}(x) = n\varphi_0(x) \qquad (x \in \hat{\mathfrak{A}}),$$

$$\tau(x) = n\Delta_0(x) \qquad (x \in \mathfrak{A}).$$

Then obviously Proposition 3.3 holds. Moreover Lemma 3.2 holds:

$$\mathrm{tr}(x \cdot y') = \mathrm{tr}(x' \cdot y) = \tau(x \circ y),$$

where $x' = \sum_i c_i x_{i'}$ for $x = \sum_i c_i x_i$.

If a C-algebra $\mathfrak{A}$ comes from an association scheme, $\mathfrak{A}$ has the following properties:

(i) (the Integrality Condition) p_{ij}^k, m_i are integers and

(ii) (Krein Condition) $p_{ij}^k \geq 0$.

QUESTION. Let $\mathfrak{A}$ be a C-algebra with p_{ij}^k non-negative. When are the parameters q_{ij}^k all non-negative?

In section 2.9, we treat C-algebras with p_{ij}^k and q_{ij}^k non-negative and discuss C-subalgebras and quotient C-algebras.

C-algebras with $B_i = (p_{ij}^k)$ tridiagonal $(i \neq 0)$, i.e., $p_{ij}^k = 0$ for $|j - k| \geq 2$, is of particular interest and called of P-polynomial type (see section 3.6). If a C-algebra of P-polynomial type has all p_{ij}^k non-negative, what can we say about $B_i = (p_{ij}^k)$?

2.6 S-RINGS AND PERMUTATION GROUPS WITH REGULAR SUBGROUPS

In section 2.3 we observed that the Krein parameters q_{ij}^k behave like the intersection numbers p_{ij}^k of an association scheme $\mathfrak{X}$, and in the previous section we gave a reason for this observation. In this section, we shall show that if $\mathfrak{X}$ comes from a permutation group with a regular abelian subgroup, then there actually exists an association scheme whose intersection numbers are q_{ij}^k.

Let X be a finite group. Let $X_0, X_1, \ldots, X_d$ be non-empty subsets of X with the following properties:

(0) $X = X_0 \cup X_1 \cup \ldots \cup X_d$, $X_i \cap X_j = \emptyset$ $(i \neq j)$.

(i) $X_0 = \{1\}$.

(ii) Let $X_i^{-1} = \{a \in X \mid a^{-1} \in X_i\}$. Then $X_i^{-1} = X_{i'}$, for some i'.

(iii) $\mathbb{X}_i\mathbb{X}_j = \sum_{k=0}^{d} c_{ij}^k \mathbb{X}_k$, where $\mathbb{X}_\alpha = \sum_{a \in X_\alpha} a$.

The property (iii) implies c_{ij}^k's are integers. Let $\mathfrak{S}$ be the subalgebra of $\mathbb{C}[X]$ spanned by $\mathbb{X}_0, \mathbb{X}_1, \ldots, \mathbb{X}_d$, where $\mathbb{C}[X]$ is the group ring of X over the complex number field. Such an algebra $\mathfrak{S}$ is called an <u>S-ring</u> (Schur ring) over X. By (iii), $\dim \mathfrak{S} = d + 1$. Let R_i be the relation on X defined by

$$(x,y) \in R_i \text{ if and only if } yx^{-1} \in X_i. \tag{6.1}$$

The right action of X preserves R_i. The properties (0) - (iii) imply that $\mathfrak{X} = \mathfrak{X}(\mathfrak{S}) = (X, \{R_i\}_{0 \le i \le d})$ is a (non-commutative) association scheme. The correspondence of $\mathbb{X}_i$ to A_i gives an anti-isomorphism from $\mathfrak{S}$ onto the adjacency algebra of $\mathfrak{X}$, where A_i is the adjacency matrix with respect to R_i. If X is abelian, then an S-ring $\mathfrak{S}$ over X is a C-algebra and $\mathfrak{X}(\mathfrak{S})$ is a commutative association scheme.

Let G be a transitive permutation group on Ω ($|\Omega| = n$) and H the stabilizer of $\alpha \in \Omega$ in G. Let $\Lambda_0, \Lambda_1, \ldots, \Lambda_d$ be the orbits of G on $\Omega \times \Omega$, where Λ_0 is the diagonal. Let $\Lambda_{i'} = {}^t\Lambda_i$ and $\Lambda_i(\alpha) = \{\beta \in \Omega \mid (\alpha,\beta) \in \Lambda_i\}$. $\mathfrak{X} = (\Omega, \{\Lambda_i\}_{0 \le i \le d})$ is a (non-commutative) association scheme and the intersection numbers are given by

$$p_{ij}^k = \#\{\gamma \in \Lambda_i(\alpha) \text{ such that } \beta \in \Lambda_j(\gamma)\} \tag{6.2}$$

for $\beta \in \Lambda_k(\alpha)$.

Suppose G has a subgroup X with the property that

X is transitive on Ω and each non-identity element of X has no fixed points on Ω. Such a permutation group X is said to be <u>regular</u>. Fix $\alpha \in \Omega$. Then for each $\beta \in \Omega$, there is a unique element $b \in X$ such that $\alpha^b = \beta$, i.e., b has the cycle decomposition

$$b = (\alpha\beta\ldots)\ldots \quad . \tag{6.3}$$

So there is a one-to-one correspondence between Ω and X. Identifying X with Ω by the correspondence (6.3), G acts transitively on X:

$$b^g = (\alpha\beta^g\ldots)\ldots \in X \tag{6.4}$$

for $b = (\alpha\beta\ldots)\ldots \in X$ and $g \in G$. So

$$b^g = c \in X \text{ if and only if } \alpha^{bg} = \alpha^c \tag{6.5}$$

for $b \in X$, $g \in G$. In particular,

$$b^a = ba \text{ for } b,a \in X . \tag{6.6}$$

The orbits of $H = G_\alpha$ on Ω are $\Lambda_0(\alpha) = \{\alpha\}$, $\Lambda_1(\alpha),\ldots,\Lambda_d(\alpha)$. Let $X_0, X_1,\ldots,X_d$ be the corresponding orbits of H on X. Then

$$X_i = \{b \in X \mid \alpha^b \in \Lambda_i(\alpha)\}. \tag{6.7}$$

We want to show $X_0, X_1,\ldots,X_d$ have the properties (0) - (iii). The properties (0) and (i) trivially hold. Since

$$\alpha^b \in \Lambda_i(\alpha) \Leftrightarrow \alpha \in \Lambda_{i'}(\alpha^b) = \Lambda_{i'}(\alpha)^b$$
$$\Leftrightarrow \alpha^{b^{-1}} \in \Lambda_{i'}(\alpha),$$

the property (ii) holds. For $b \in X_k$, $c_{ij}^k =$

$\#\{(x,y) \in X_i \times X_j \mid xy = b\} = |X_i^{-1}b \cap X_j|$. Since $X_i = \{x \in X \mid (\alpha^b, \alpha^{x^{-1}b}) \in \wedge_{i'}\}$, $X_i^{-1}b = \{z \in X \mid (\alpha^b, \alpha^z) \in \wedge_{i'}\}$. So we have

$$c_{ij}^k = p_{ji}^k \,. \tag{6.8}$$

This implies the property (iii). Thus we get the following theorem.

THEOREM 6.1. Let G be a transitive permutation group on a finite set Ω with a regular subgroup X. Let H be the stabilizer of $\alpha \in \Omega$ and $\wedge_0(\alpha), \wedge_1(\alpha), \ldots, \wedge_d(\alpha)$ the orbits of H on Ω. Identifying X with Ω, let X_i be the orbit of H on X corresponding to $\wedge_i(\alpha)$. Then $\mathfrak{S}$ with $\mathbb{X}_0, \mathbb{X}_1, \ldots, \mathbb{X}_d$ is an S-ring over X, where $\mathbb{X}_i = \sum_{a \in X_i} a$. The (non-commutative) association scheme $(X, \{R_i\}_{0 \le i \le d})$ with R_i defined by (6.1) is isomorphic to $(\Omega, \{\wedge_i\}_{0 \le i \le d})$ by the correspondence of R_i to $\wedge_i$. $\mathfrak{S}$ is isomorphic to the adjacency algebra of $(\Omega, \{\wedge_i\})$ by the correspondence of $\mathbb{X}_i$ to $A_{i'}$, the adjacency matrix with respect to $\wedge_{i'}$.

Compare Theorem 6.1 with the following Proposition 6.2 in which the assumption of the existence of a regular subgroup is dropped.

PROPOSITION 6.2. Let G be a finite group and H a subgroup of G. Let $H = Ha_0, Ha_1, \ldots, Ha_{n-1}$ be the left cosets of G by H and $H = Hb_0H, Hb_1H, \ldots, Hb_dH$ the double cosets of G by H. Consider the action of G on $\Omega = \{Ha_0, Ha_1, \ldots, Ha_{n-1}\}$. Let $\mathfrak{A}$ be the adjacency algebra of the (non-commutative) association scheme $\mathfrak{X}$ defined by this action, and let A_i be the adjacency matrix of $\mathfrak{X}$ corresponding to Hb_iH, an orbit of H on Ω. Then $\mathfrak{A}$

is isomorphic to eRe as algebras by the correspondence of A_i, to $\mathfrak{X}_i$, where R is the group ring of G over $\mathbb{C}$,

$$e = \frac{1}{|H|} \sum_{a \in H} a, \quad \text{and} \quad \mathfrak{X}_i = \frac{1}{|H|} \sum_{a \in Hb_iH} a .$$

Proof. The element e is an idempotent, i.e., $e^2 = e$, and eR affords the permutation representation of G on Ω because $\{e, ea_1, \ldots, ea_{n-1}\}$ is a basis of eR. Let $\mathrm{Hom}_G(eR, eR)$ be the set of linear mappings from eR into eR which commute with the action of G. Such linear mappings are determined by the images of e, and the images of e belong to eRe by $e^2 = e$. Hence $\mathrm{Hom}_G(eR, eR) = \{\varphi_a \mid a \in eRe\}$, where $\varphi_a : x \to ax$ for $x \in eR$, and the correspondence of a to φ_a is an algebra anti-isomorphism from eRe onto $\mathrm{Hom}_G(eR, eR)$, because the action of $\mathrm{Hom}_G(eR, eR)$ is from the right. eRe has a basis $\mathfrak{X}_0 = e, \mathfrak{X}_1, \ldots, \mathfrak{X}_d$.

Express the right actions of G and $\mathrm{Hom}_G(eR, eR)$ as matrices with respect to the basis $\{e, ea_1, \ldots, ea_{n-1}\}$ of eR. Then by Theorem 1.3, $\mathrm{Hom}_G(eR, eR)$ is isomorphic to the adjacency algebra $\mathfrak{A}$. Since $\varphi_{\mathfrak{X}_i} : e \to \mathfrak{X}_i = \sum ea_\alpha$ with the summation over the left coset representatives $a_\alpha \in Hb_iH$, $\varphi_{\mathfrak{X}_i}$ and A_i have the same 0th row. Since the elements of $\mathfrak{A}$ are determined by the 0th row, $\varphi_{\mathfrak{X}_i} = A_i$. □

REMARK. The subalgebra $eRe \cong \mathrm{Hom}_G(eR, eR)$ gives another definition of the adjacency algebra (or the Hecke algebra) of a (non-commutative) association scheme which comes from a permutation group.

Let X be a finite abelian group. By Theorem 3.3 in Chapter I, every irreducible representation of X is of degree 1. Let Y be the set of irreducible representations of X. For $\Delta_\alpha, \Delta_\beta \in Y$, define the product $\Delta_\alpha\Delta_\beta$ by $(\Delta_\alpha\Delta_\beta)(a) = \Delta_\alpha(a)\Delta_\beta(a)$ for $a \in X$. Then $\Delta_\alpha\Delta_\beta$ is also an irreducible representation of X and hence Y becomes an abelian group. (In fact, it is well-known that Y is isomorphic to X.) Y is called the character group of X. Given an S-ring $\mathfrak{S}$ over X, we want to construct an S-ring $\mathfrak{S}^*$ over Y with the property that the intersection numbers p_{ij}^k of the association scheme $\mathfrak{X}(\mathfrak{S}^*)$ coincide with the Krein parameters q_{ij}^k of $\mathfrak{X}(\mathfrak{S})$. Namely, we will prove the following:

THEOREM 6.3. (The Duality for S-rings over Abelian Groups) Let $\mathfrak{S}$ be an S-ring over a finite abelian group X and let Y be the character group of X. Let $\sim$ be the equivalence relation on Y defined by $\Delta_\alpha \sim \Delta_\beta$ if and only if the restrictions of Δ_α and Δ_β to $\mathfrak{S}$ coincide. Let $Y_0, Y_1, \ldots, Y_d$ be the equivalence classes, and let $\mathbb{Y}_i = \sum_{\Delta_\alpha \in Y_i} \Delta_\alpha$. Then the subalgebra $\mathfrak{S}^*$ (of $\mathbb{C}[Y]$), spanned by $\mathbb{Y}_0, \mathbb{Y}_1, \ldots, \mathbb{Y}_d$, becomes an S-ring with the property that $\dim \mathfrak{S} = \dim \mathfrak{S}^*$ and the intersection numbers of $\mathfrak{X}(\mathfrak{S}^*)$ are the Krein parameters of $\mathfrak{X}(\mathfrak{S})$. ($\mathfrak{S}^*$ is called the dual S-ring of $\mathfrak{S}$.)

Proof. We keep the notation in section 2.5.

We begin with a general remark. Let $\mathfrak{A}$ be a C-algebra with the basis $x_0, x_1, \ldots, x_d$ and the primitive idempotents $e_0, e_1, \ldots, e_d$. Define the Schur product $\circ$ on $\mathfrak{A}$ by

$$x_i \circ x_j = \delta_{ij} x_j \,. \tag{6.9}$$

Then as we saw at the end of section 2.5, it holds that

$$(ne_i) \circ (ne_j) = \sum_k q_{ij}^k (ne_k), \tag{6.10}$$

and $\mathfrak{A} = \langle ne_i \mid 0 \leq i \leq d\rangle$ with respect to the Schur product is isomorphic to the dual C-algebra $\hat{\mathfrak{A}} = \langle \Delta_i^* \mid 0 \leq i \leq d\rangle$ by the correspondence of ne_i to Δ_i^*.

Let $\mathfrak{S}$ be an S-ring over a finite abelian group X ($|X| = n$) and $\mathbf{X}_0, \mathbf{X}_1, \ldots, \mathbf{X}_d$ the basis of $\mathfrak{S}$. The group ring $\mathbb{C}[X]$ of X is a C-algebra with the basis $\{a \mid a \in X\}$. Let $\mathbb{C}[X]^*$ be the algebra equipped with the Schur product on the underlying vector space $\mathbb{C}[X]$. So $a \circ b = \delta_{ab} a$ for $a, b \in X$. $\mathfrak{S}$ is a subalgebra both of $\mathbb{C}[X]$ and $\mathbb{C}[X]^*$, i.e., closed both under the ordinary and Schur multiplications. Let $e_0, e_1, \ldots, e_{n-1}$ be the primitive idempotents of $\mathbb{C}[X]$. Then the primitive idempotents of $\mathfrak{S} \subseteq \mathbb{C}[X]$ can be written as $e_{S_0}, e_{S_1}, \ldots, e_{S_d}$, where

$$e_{S_i} = \sum_{\alpha \in S_i} e_\alpha \tag{6.11}$$

and

$$S_0 \cup S_1 \cup \ldots \cup S_d = \{0, 1, \ldots, n-1\},$$

$$S_i \cap S_j = \emptyset \qquad (i \neq j).$$

The primitive idempotents of $\mathfrak{S} \subseteq \mathbb{C}[X]^*$ are $\mathbf{X}_0, \mathbf{X}_1, \ldots, \mathbf{X}_d$. $\mathfrak{S} = \langle ne_{S_i} \mid 0 \leq i \leq d\rangle \subseteq \mathbb{C}[X]^*$ is the dual C-algebra of $\mathfrak{S} = \langle \mathbf{X}_i \mid 0 \leq i \leq d\rangle \subseteq \mathbb{C}[X]$. By (6.10),

$$(ne_{S_i}) \circ (ne_{S_j}) = \sum_{k=0}^{d} q_{ij}^k (ne_{S_k}), \tag{6.10'}$$

where q_{ij}^k's are the Krein parameters of the association scheme $\mathfrak{X}(\mathfrak{S})$ attached to $\mathfrak{S}$.

Let Y be the character group of X. Then the dual

C-algebra of $\mathbb{C}[X]$ is $\mathbb{C}[Y]$ with the basis $\{\Delta_\alpha^* \mid \Delta_\alpha \in Y\}$. The algebra $\mathbb{C}[X]^*$ with respect to the Schur product $\circ$ is isomorphic to $\mathbb{C}[Y]$ by the correspondence of ne_i to Δ_i^*. This isomorphism carries ne_{S_i} to $\mathbb{Y}_i$, where

$$\mathbb{Y}_i = \sum_{\alpha \in S_i} \Delta_\alpha^*, \qquad (6.11)'$$

and it holds that

$$\mathbb{Y}_i \mathbb{Y}_j = \sum_{k=0}^{d} q_{ij}^k \mathbb{Y}_k . \qquad (6.10)''$$

Since $\Delta_\alpha^*(a) = \overline{\Delta_\alpha(a)}$ by (5.19), Δ_i^* is a character of X and so $Y = \{\Delta_0^*, \Delta_1^*, \ldots, \Delta_{n-1}^*\}$. Since $ne_0 = \sum_{x \in X} x = \sum_{i=0}^{d} \mathfrak{X}_i$, e_0 is an element of $\mathfrak{S}$ and so we may assume $e_{S_0} = e_0$ in (6.11). Then $\mathbb{Y}_0 = \Delta_0$, which is the identity character of X.

Since $q_{ij}^0 = \delta_{i\hat{j}} m_i$, $\mathbb{Y}_i^{-1} = \mathbb{Y}_{\hat{i}}$ follows from (6.10)".

Therefore the subalgebra $\mathfrak{S}^*$ of $\mathbb{C}[Y]$ spanned by $\mathbb{Y}_0, \mathbb{Y}_1, \ldots, \mathbb{Y}_d$ is an S-ring. By (6.10)", the intersection numbers of $\mathfrak{X}(\mathfrak{S}^*)$ are the Krein parameters of $\mathfrak{X}(\mathfrak{S})$.

Let $\sim$ be the equivalence relation on Y defined by $\Delta_\alpha \sim \Delta_\beta$ if and only if the restrictions of Δ_α and Δ_β to $\mathfrak{S}$ coincide. Then the equivalence classes are $Y_0, Y_1, \ldots,$ where $Y_i = \{\Delta_\alpha \mid \alpha \in S_i\}$. This is because $\Delta_\alpha(e_\beta) = \delta_{\alpha\beta}$ and $\Delta_\alpha(e_{S_i}) = 1, 0,$ according to $\alpha \in S_i$, $\alpha \notin S_i$. This completes the proof of the theorem. □

As an immediate consequence of Theorem 6.3, we have the following:

THEOREM 6.4. Let G be a transitive permutation group on a finite set Ω with an abelian regular subgroup. Let q_{ij}^k be the Krein parameters of the association scheme defined by the action of G on Ω. Then there exists an association scheme whose intersection numbers are q_{ij}^k.

EXAMPLE 6.5. Let G be a permutation group on Ω with a regular normal elementary abelian subgroup $X = E_{p^\ell}$ of order p^ℓ. Then $G = X \cdot G_\alpha$, the semidirect product of X by G_α, and the stabilizer group G_α $(\alpha \in \Omega)$ is a subgroup of $GL(\ell,p)$. Let ${}^tG_\alpha = \{{}^ty \in GL(\ell,p) \mid y \in G_\alpha\}$ and consider the permutation group $\hat{G} = X \cdot {}^tG_\alpha$ on Ω, identifying X with Ω. Then the intersection numbers p_{ij}^k of the association scheme defined by the action of $\hat{G}$ on Ω coincide with the Krein parameters q_{ij}^k of the association scheme defined by the action of G on Ω.

<u>Proof</u>. Let $X_0 = \{0\}, X_1, \ldots, X_d$ be the orbits of G_α on X. Let $\hat{X}$ be the character group of X. ${}^tG_\alpha$ acts on $\hat{X}$ by $\Delta^{{}^ta}(x) = \Delta(x^a)$ $(\Delta \in \hat{X},\ a \in G_\alpha,\ x \in X)$. Let $\hat{X}_0 = \{\Delta_0\}, \hat{X}_1, \ldots, \hat{X}_d$ be the orbits of ${}^tG_\alpha$ on $\hat{X}$, where Δ_0 is the identity character of X. Let $\sim$ be the equivalence relation on $\hat{X}$ defined by $\Delta \sim \Delta'$ if and only if the restrictions of Δ and Δ' to $\mathfrak{S} = \langle \mathbb{X}_0, \mathbb{X}_1, \ldots, \mathbb{X}_d\rangle$ coincide. Then $\hat{X}_0, \hat{X}_1, \ldots, \hat{X}_d$ are the equivalence classes because $\Delta(\sum_{a \in G_\alpha} x^a) = (\sum_{a \in {}^tG_\alpha} \Delta^a)(x)$.

Associated with the inner product $\langle\ ,\ \rangle$ of X, there is an isomorphism $\varphi\colon x \to \varphi_x$ from X to $\hat{X}$, where $\varphi_x(y) = \langle x,y\rangle$ for $x,y \in X$. Since $\langle x^{{}^ta}, y\rangle =$

$\langle x, y^a \rangle$ for $a \in G_\alpha$, φ commutes with the action of ${}^tG_\alpha$, i.e., $\varphi_{x^b} = (\varphi_x)^b$ for $b = {}^ta \in {}^tG_\alpha$. Thus we can identify the action of ${}^tG_\alpha$ on X with that on $\hat{X}$. □

For applications of S-rings, see Wielandt (1964, Chapter IV).

2.7 AN INTERPRETATION OF THE CHARACTER TABLE OF A FINITE GROUP

Let G be a finite group and $C_0 = \{1\}, C_1, \ldots, C_d$ the conjugacy classes of G. Let $\mathbb{F}_\alpha$: $a \to F_\alpha(a)$ $(\alpha = 0, 1, \ldots, d)$ be the non-equivalent irreducible representations of G and χ_α the character of $\mathbb{F}_\alpha$. We may assume $\mathbb{F}_0$ is the identity representation $\mathbb{1}_G$.

Let $a_0 = 1, a_1, \ldots, a_d$ be representatives of the conjugacy classes and T the matrix of degree $d + 1$ whose (α, i) entry is $\chi_\alpha(a_i)$:

$$T = (\chi_\alpha(a_i)). \tag{7.1}$$

T is the character table of G.

As we saw in section 1.6 of Chapter I, the product $\mathbb{F}_\alpha \otimes \mathbb{F}_\beta$ is a representation of G and has character $\chi_\alpha \chi_\beta$. So

$$\mathbb{F}_\alpha \otimes \mathbb{F}_\beta \sim \sum_{\gamma=0}^{d} c_{\alpha\beta}^{\gamma} \mathbb{F}_\gamma, \tag{7.2}$$

$$\chi_\alpha \chi_\beta = \sum_{\gamma=0}^{d} c_{\alpha\beta}^{\gamma} \chi_\gamma . \tag{7.3}$$

The <u>representation graph</u> with respect to $\mathbb{F}_\alpha$ has, by definition, vertices $\{0,1,\ldots,d\}$ and $c_{\alpha\beta}^{\gamma}$ directed edges from β to γ for $\beta,\gamma \in \{0,1,\ldots,d\}$.

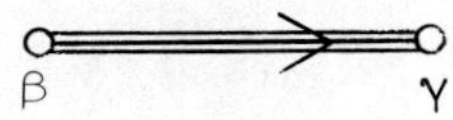

$c_{\alpha\beta}^{\gamma}$ directed edges

If $c_{\alpha\beta}^{\gamma}$ is large, we simply attach the number $c_{\alpha\beta}^{\gamma}$ to the arrow from β to γ.

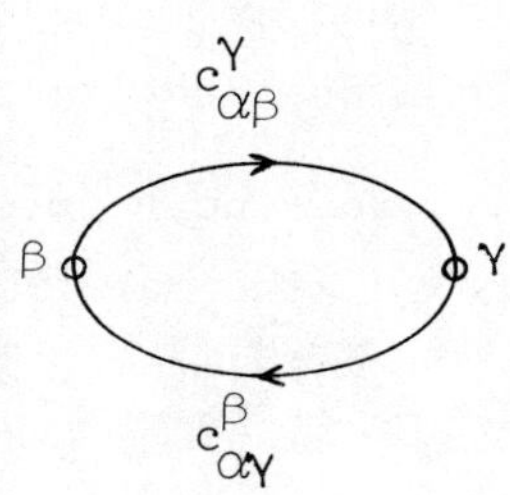

If $c_{\alpha\beta}^{\gamma} = c_{\alpha\gamma}^{\beta} = 1$, we use the following abbreviation.

Let W_α be the matrix of degree $d + 1$ whose (β,γ) entry is $c_{\alpha\beta}^{\gamma}$:

$$W_\alpha = (c^\gamma_{\alpha\beta}) . \tag{7.4}$$

W_α is called the incidence matrix for the representation graph with respect to $\mathbb{F}_\alpha$. We shall prove the following theorem.

THEOREM 7.1. Let G be a finite group. Let T be the character table of G and let $W_0, W_1, \ldots, W_d$ be the incidence matrices of the representation graphs of G. Then every column of T is a common right eigenvector of $W_0, W_1, \ldots, W_d$, and any common right eigenvector of $W_0, W_1, \ldots, W_d$ is a scalar multiple of some column of T. (By taking the standardized common eigenvectors, the character table T of G is uniquely determined, up to arrangement of rows and columns, by $W_0, W_1, \ldots, W_d$.)

Define the relation R_i by

$$(x,y) \in R_i \text{ if and only if } yx^{-1} \in C_i, \tag{7.5}$$

and the permutation $i \to i'$ by

$$C_{i'} = \{x \in G \mid x^{-1} \in C_i\}. \tag{7.6}$$

Then $\mathfrak{X} = (G, \{R_i\}_{0 \le i \le d})$ is a commutative association scheme (section 2.2, Example 2.1(2)). The valences are

$$k_i = |C_i| . \tag{7.7}$$

Let $\mathbb{C}_i$ be the formal sum of the elements of C_i:

$$\mathbb{C}_i = \sum_{a \in C_i} a . \tag{7.8}$$

Then as we saw in section 2.2, Example 2.1(2),

$$\mathbb{C}_i\mathbb{C}_j = \sum_{k=0}^{d} p^k_{ij}\mathbb{C}_k \tag{7.9}$$

with p_{ij}^k the intersection numbers of $\mathfrak{X}$. Let $\mathfrak{A}$ be the algebra generated by $\mathbb{C}_0, \mathbb{C}_1, \ldots, \mathbb{C}_d$ over the complex number field. ($\mathfrak{A}$ is a subalgebra of the group ring of G.) By (7.9), $\mathfrak{A}$ with the basis $\{\mathbb{C}_0, \mathbb{C}_1, \ldots, \mathbb{C}_d\}$ is isomorphic to the adjacency algebra of $\mathfrak{X}$ as C-algebras. By (4.12), (4.13) of Chapter I, the mapping

$$\mathbb{C}_i \to \omega_i = \frac{k_i \chi(a_i)}{f} \qquad (f = \chi(1)) \tag{7.10}$$

is a linear representation of $\mathfrak{A}$ for each irreducible character χ. Let $P = (p_i(\alpha))$ be the eigenmatrix of $\mathfrak{A}$. Then by Proposition 5.4, the linear representation (7.10) coincides with some Δ_α: $\mathbb{C}_i \to p_i(\alpha)$. By the first orthogonality relation of characters, different irreducible characters give different linear representations (7.10) of $\mathfrak{A}$. So with a suitable rearrangement of the indices 1, 2, ..., d, we may assume

$$\frac{k_i \chi_\alpha(a_i)}{f_\alpha} = p_i(\alpha) \qquad (f_\alpha = \chi_\alpha(1)), \tag{7.11}$$

i.e.,

$$\mathrm{diag}\left(\frac{1}{f_0}, \ldots, \frac{1}{f_d}\right) T \,\mathrm{diag}(k_0, \ldots, k_d) = P\,. \tag{7.12}$$

Put (5.13) and (7.12) into $PQ = nI$. Then

$$\mathrm{diag}\left(\frac{1}{f_0}, \ldots, \frac{1}{f_d}\right) T \,\mathrm{diag}(k_0, \ldots, k_d) \cdot \tag{7.13}$$

$${}^t\overline{T}\,\mathrm{diag}\left(\frac{1}{f_0}, \ldots, \frac{1}{f_d}\right) \mathrm{diag}(m_0, \ldots, m_d) = nI.$$

The first orthogonality relation of characters is

$$T \,\mathrm{diag}(k_0, k_1, \ldots, k_d)\, {}^t\overline{T} = nI. \tag{7.14}$$

Put (7.14) into (7.13) to get $(\mathrm{diag}(\frac{1}{f_0},\ldots,\frac{1}{f_d}))^2 \cdot \mathrm{diag}(m_0,\ldots,m_d) = I$, i.e.,

$$f_\alpha = \sqrt{m_\alpha}\,. \tag{7.15}$$

By (7.12), (7.15) and (5.13) we get the following theorem.

THEOREM 7.2. Let G be a finite group and T the character table of G. Let $\mathfrak{X}$ be the association scheme defined by (7.5), (7.6) and let $P = (p_i(\alpha))$, $Q = (q_i(\alpha))$, k_i, m_i be the first and second eigenmatrices, valencies, multiplicities of $\mathfrak{X}$ respectively. Then

$$T = \mathrm{diag}(\sqrt{m_0},\sqrt{m_1},\ldots,\sqrt{m_d})\,P\,\mathrm{diag}(\frac{1}{k_0},\frac{1}{k_1},\ldots,\frac{1}{k_d})$$

$$= \mathrm{diag}(\frac{1}{\sqrt{m_0}},\frac{1}{\sqrt{m_1}},\ldots,\frac{1}{\sqrt{m_d}})\,{}^t\overline{Q}\,.$$

The sizes of the conjugacy classes are $k_0,k_1,\ldots,k_d$ and the degrees of the irreducible characters are $\sqrt{m_0},\sqrt{m_1},\ldots,\sqrt{m_d}$.

Rewrite (7.3) by using the relation of Theorem 7.2

$$x_\alpha(a_i) = \frac{1}{\sqrt{m_\alpha}}\overline{q_\alpha(i)}, \tag{7.16}$$

to get

$$\frac{1}{\sqrt{m_\alpha}}\frac{1}{\sqrt{m_\beta}}\overline{q_\alpha(i)}\;\overline{q_\beta(i)} = \sum_\gamma \frac{1}{\sqrt{m_\gamma}}c^\gamma_{\alpha\beta}\,\overline{q_\gamma(i)}\,,$$

i.e.,

$$q_\alpha(i)q_\beta(i) = \sum_{\gamma=0}^{d}\sqrt{\frac{m_\alpha m_\beta}{m_\gamma}}\,c^\gamma_{\alpha\beta}q_\gamma(i)\,. \tag{7.17}$$

Comparing (7.17) with (5.24), we get

$$c_{\alpha\beta}^{\gamma} = \sqrt{\frac{m_\gamma}{m_\alpha m_\beta}}\, q_{\alpha\beta}^{\gamma} . \tag{7.18}$$

Let $\hat{B}_\alpha$ be the matrix whose (β,γ) entry is $q_{\alpha\beta}^{\gamma}$. Then by (7.18),

$$W_\alpha = \frac{1}{\sqrt{m_\alpha}} \operatorname{diag}\left(\frac{1}{\sqrt{m_0}},\ldots,\frac{1}{\sqrt{m_d}}\right)\hat{B}_\alpha \cdot \operatorname{diag}(\sqrt{m_0},\ldots,\sqrt{m_d}). \tag{7.19}$$

Applying Proposition 5.3 to the dual C-algebra $\hat{\mathfrak{A}}$, the columns of ${}^tQ$ are common right eigenvectors of $\hat{B}_0,\ldots,\hat{B}_d$ and they are all the common right eigenvectors of $\hat{B}_0,\ldots,\hat{B}_d$ up to scalar multiple. In the dual C-algebra $\hat{\mathfrak{A}}$, the relations (5.5) and (5.27) become

$$\overline{q_i(j)} = q_{\hat{i}}(j) \quad \text{and} \quad \overline{q_i(j)} = q_i(j'), \tag{7.20}$$

and this means the complex conjugation induces a permutation on the rows of $Q = (q_i(j))$ as well as the columns of Q. So the set of columns of ${}^tQ$ coincides with that of ${}^t\overline{Q}$. Therefore, the columns of T are common right eigenvectors of $W_0,\ldots,W_d$ and are the only common right eigenvectors up to scalar multiple by Theorem 7.2 and (7.19). This completes the proof of Theorem 7.1.

Let G be the group generated by x, y, z with the defining relations $x^\ell = y^m = z^n = xyz$. As is well known (cf. Coxeter and Moser (1957)) G is finite if and only if $\frac{1}{\ell} + \frac{1}{m} + \frac{1}{n} > 1$, i.e., $(\ell,m,n) = (2,2,n)$, $(2,3,3)$, $(2,3,4)$ or $(2,3,5)$. These finite groups G are called <u>binary polyhedral groups</u> (binary dihedral, tetradhedral, octahedral and icosahedral groups, respectively), and $G/Z \cong D_{2n}$, A_4, S_4, A_5, respectively, where

Z is the center of G. Each binary polyhedral group G has an irreducible representation $\mathbb{F}$ of degree 2. Recently, McKay (1980) observed that the representation graph of G with respect to $\mathbb{F}$ is a Euclidean type Dynkin diagram $\tilde{D}_n$, $\tilde{E}_6$, $\tilde{E}_7$, $\tilde{E}_8$, respectively.

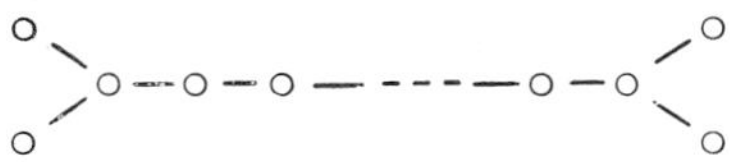
$\tilde{D}_n$

n + 1 vertices

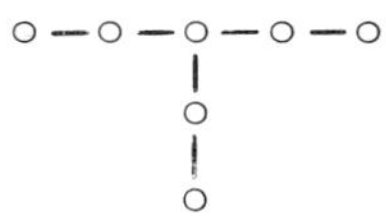
$\tilde{E}_6$

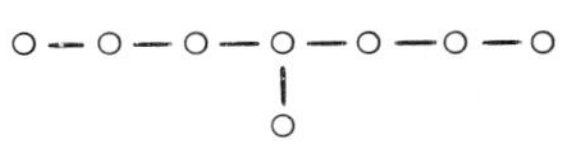
$\tilde{E}_7$

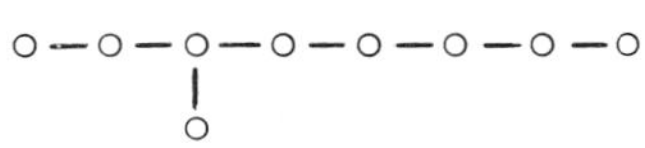
$\tilde{E}_8$

Also the diagram

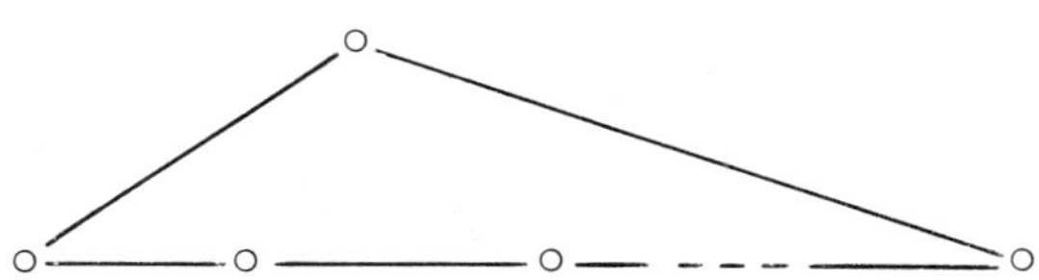
$\tilde{A}_n$

n + 1 vertices

appears as a representation graph (with respect to non-irreducible representation of degree 2) of some cyclic group. So all the single bond Euclidean type Dynkin diagrams appear here.

Happel, Preiser and Ringel (1980) generalized McKay's observation to finite groups with faithful selfdual representations of degree 2. Namely, the representation graph of any finite subgroup of $SU_2(\mathbb{C})$ with respect to the faithful natural representation of degree 2 is one of $\tilde{A}_n$, $\tilde{D}_n$, $\tilde{E}_6$, $\tilde{E}_7$, $\tilde{E}_8$. Also, some other Euclidean type Dynkin diagrams with multiple bond appear as representation graphs of certain 2-dimensional finite linear groups over non-splitting fields for the groups (cf. Happel, Preiser and Ringel (1980), Iwahori and Yokonuma (1982), etc.).

QUESTION. (1) Which graphs can be representation graphs? In particular, if G has a representation of small degree, what representation graphs exist?

Let us discuss more properties of character tables.

The relations (5.5) and (5.27) imply that the complex conjugation — induces a permutation both on the columns and the rows of P. Let a C-algebra $\mathfrak{A}$ come from a commutative association scheme $\mathfrak{X} = (X, \{R_i\}_{0 \leq i \leq d})$. Then p_{ij}^k are all integers. Let L be the field generated by $p_i(j)$ $(0 \leq i \leq d,\ 0 \leq j \leq d)$ over the rational field. Let H be a subgroup of the Galois group of L over the rational field. We assume each element of H commutes with the complex conjugation —. For a linear representation $\Delta_\alpha: x_i \to p_i(\alpha)$ of $\mathfrak{A}$ and an element $\sigma \in H$, the mapping $\Delta_\alpha^\sigma: x_i \to p_i(\alpha)^\sigma$ can be extended to a linear representation of $\mathfrak{A}$, because p_{ij}^k's are all rational. Hence, by Proposition 5.4, $\Delta_\alpha^\sigma = \Delta_\beta$ for some β, i.e., σ induces a permutation $\pi(\sigma)$ on $\{0,1,\ldots,d\}$ such that

$$p_i(\alpha)^\sigma = p_i(\alpha^{\pi(\sigma)}) \quad \text{for all } i, \alpha. \tag{7.21}$$

Assume the following:

q_{ij}^k's are all rational. (H)

Then by applying the same argument to the dual algebra $\hat{\mathfrak{A}}$, there exists a permutation $\rho(\sigma)$ on $\{0,1,\ldots,d\}$ such that

$$q_i(\alpha)^\sigma = q_i(\alpha^{\rho(\sigma)}) \quad \text{for all } i, \alpha. \tag{7.22}$$

By Corollary 5.6 and (7.21),

$$m_i = m_{i^{\pi(\sigma)}}, \tag{7.23}$$

and dually

$$k_i = k_{i^{\rho(\sigma)}}. \tag{7.24}$$

By Theorem 5.5(i), (7.22), (7.23), (7.24) and the assumption that σ commutes with the complex conjugation —,

$$p_i(\alpha)^\sigma = p_{i^{\rho(\sigma)}}(\alpha) \quad \text{for all } i, \alpha. \tag{7.25}$$

Let σ act on $P = (p_i(j))$ by $P^\sigma = (p_i(j)^\sigma)$. Then σ induces the permutation $\pi(\sigma)$ on the rows of P and the permutation $\rho(\sigma)$ on the columns of P:

$$P^\sigma = (p_i(j^{\pi(\sigma)})) = \varpi(\sigma)P$$

$$= (p_{i^{\rho(\sigma)}}(j)) = P\wp(\sigma),$$

where $\varpi(\sigma)$ and $\wp(\sigma)$ are the permutation matrices corresponding to $\pi(\sigma)$ and $\rho(\sigma)$. So

$$\varpi(\sigma) = P\wp(\sigma)P^{-1}.$$

Therefore,

$$\operatorname{tr}\varpi(\sigma) = \operatorname{tr}\wp(\sigma). \tag{7.26}$$

Therefore, the number of rows of P fixed by $\pi(\sigma)$ equals that of columns of P fixed by $\rho(\sigma)$. By (7.26),

$$\sum_{\sigma \in H} \operatorname{tr}\pi(\sigma) = \sum_{\sigma \in H} \operatorname{tr}\rho(\sigma),$$

so by Example 5.6 of Chapter I, H has the same number of orbits on the rows and on the columns of P. Thus, we get the following theorem.

THEOREM 7.3. Let $P = (p_i(j))$ be the eigenmatrix of an association scheme $\mathfrak{X}$. Let L be the field generated by all $p_i(j)$ over the rational field. Let H be a subgroup of the Galois group of L over the rational field such that each element of H commutes with the complex conjugation —. Suppose the Krein parameters are all rational. Then for $\sigma \in H$, the action $P^\sigma = (p_i(j)^\sigma)$ induces permutations on the columns and on the rows of P. These two permutation representations are equivalent. In particular,

(i) σ fixed the same number of rows and columns of P,

(ii) H has the same number of orbits on the rows and on the columns of P.

Let $\mathfrak{X}$ be a commutative association scheme coming from a group G by (7.5), (7.6). Then by (7.15), (7.18) and the fact that $c_{\alpha\beta}^{\gamma}$'s are integers, $q_{\alpha\beta}^{\gamma}$'s are rational. Hence, Theorem 7.3 holds for $\mathfrak{X}$. Such a commutative association scheme $\mathfrak{X}$ has a further property:

(iii) $p_i(j)$'s belong to a cyclotomic number field because of the expression (7.11) and the fact that the character values of a finite group are all in a cyclotomic filed (see (6.3) of Chapter I).

REMARK. The generalized 6-gon with 126 vertices at the end

of section 2.3 gives an example in which q_{ij}^k are not all rational numbers and the conclusion of Theorem 7.3 is not true.

QUESTION. (2) Are there any association schemes which are counterexamples for the property (iii), that is, $p_i(j)$ are not all in a cyclotomic number field?

2.8 NORTON ALGEBRAS AND PERMUTATION GROUPS

In this section, we assume that association schemes admit group actions, but most of the arguments in this section hold without the group actions (see Remark (4)).

Let G be a transitive permutation group on $\Omega = \{1,2,\ldots,n\}$, θ the permutation character of G on Ω and $\mathfrak{X}$ the (non-commutative) association scheme defined by the action of G on Ω. Assume $\mathfrak{X}$ is a commutative association scheme, i.e., θ is multiplicity-free by Theorem 1.4. We are interested in the relation between $\chi_i\chi_j$ and $E_i \circ E_j$, where χ_i's are the irreducible characters of G appearing in θ and E_i's are the primitive idempotents of the adjacency algebra of $\mathfrak{X}$. We shall prove the following theorem.

THEOREM 8.1. (Scott (1977)) Let G be a transitive permutation group on $\Omega = \{1,2,\ldots,n\}$ and θ the permutation character of G on Ω. Assume θ is multiplicity-free: $\theta = \chi_0 + \chi_1 + \cdots + \chi_d$ with $\chi_0,\chi_1,\ldots,\chi_d$ distinct irreducible characters of G. Let q_{ij}^k be the Krein parameters of the association scheme $\mathfrak{X} = (\Omega,\{\Lambda_i\}_{0\leq i\leq d})$, where Λ_i are the orbits of G on $\Omega \times \Omega$. Then

(i) χ_k appears in $\chi_i\chi_j$, i.e., $(\chi_k,\chi_i\chi_j) \neq 0$, whenever $q_{ij}^k \neq 0$, and

(ii) χ_k appears in $\mathrm{Sym}^2\chi_i$, whenever $q_{ii}^k \neq 0$, where $\mathrm{Sym}^2\chi_i$ is the symmetric part of χ_i^2.

Let $\mathfrak{A}$ be the adjacency algebra of $\mathfrak{X}$, A_i $(0 \leq i \leq d)$ the incidence matrices and E_i $(0 \leq i \leq d)$ the primitive idempotents of $\mathfrak{A}$. For $\alpha \in \Omega = \{1,\ldots,n\}$, x_α denotes the row vector $(0,\ldots,0,1,0,\ldots,0)$ with 1 in the αth place. Let $X = \{x_\alpha \mid \alpha \in \Omega\}$ and let $V = \mathbb{C}^n$ be the Hermitian space with X an orthonormal basis. Let V_i be the image of V by the projection E_i. So

$$V = V_0 + V_1 + \cdots + V_n, \quad V_i \perp V_j \qquad (8.1)$$

(perpendicular for $i \neq j$),

$$uE_i = \begin{cases} u & \text{if } u \in V_i, \\ 0 & \text{if } u \in V_j \quad (j \neq i). \end{cases}$$

Let $\mathbb{A}: a \to A(a)$ be the permutation representation of G:

$$x_\alpha A(a) = x_\beta \quad \text{if } \beta = \alpha^a \qquad (\alpha \in \Omega,\ a \in G). \qquad (8.2)$$

By Theorem 1.3, $\mathfrak{A}$ is the set of matrices commuting with all $A(a)$ $(a \in G)$. Let $\mathrm{Hom}(V,V)$ be the set of linear mappings from V to V. Regarding $A(a)$ and E_i as elements of $\mathrm{Hom}(V,V)$ by forgetting the basis X of V, we use the following notation:

$$uA(a) = ua \qquad (a \in G,\ u \in V), \qquad (8.3)$$

$$uE_i = u\pi_i \qquad (u \in V).$$

The subspace V_i is invariant by the action of G, because $ua = u\pi_i a = ua\pi_i \in V_i$ for $u \in V_i$, $a \in G$. Let $\mathbb{F}_i: a \to F_i(a)$ be the representation of G afforded by the action of G on V_i. Then

$$\mathbb{A} \sim \mathbb{F}_0 + \mathbb{F}_1 + \cdots + \mathbb{F}_d \tag{8.4}$$

and $\mathbb{F}_0, \mathbb{F}_1, \ldots, \mathbb{F}_d$ are non-equivalent irreducible representations of G because $\theta = \chi_0 + \chi_1 + \cdots + \chi_d$ with $\chi_i \neq \chi_j$ $(i \neq j)$. We may assume χ_i is the character of $\mathbb{F}_i$. Thus, χ_i corresponds to E_i. The complex conjugate character $\overline{\chi_i}$ corresponds to $\overline{E_i} = E_{\hat{i}}$. This is because if $\{u_1, u_2, \ldots, u_r\}$ is the basis of V_i diagonalizing $A(a)$ such that $u_i A(a) = \lambda_i u_i$ $(\lambda_i \in C)$, then $\{\overline{u}_1, \overline{u}_2, \ldots, \overline{u}_r\}$ is the basis of $V_{\hat{i}}$ such that $\overline{u}_i A(a) = \overline{\lambda}_i \overline{u}_i$, where $\overline{u}$ is the complex conjugate row vector of u.

Let $V \otimes V$ be the tensor product of V, i.e., the n^2-dimensional vector space with the relation $(\sum_\alpha \lambda_\alpha u_\alpha) \otimes (\sum_\beta \mu_\beta v_\beta) = \sum_{\alpha,\beta} \lambda_\alpha \mu_\beta (u_\alpha \otimes v_\beta)$ for $\lambda_\alpha, \mu_\beta \in C$, $u_\alpha, v_\beta \in V$. Define a Hermitian form $\langle\ ,\ \rangle$ on $V \otimes V$ in such a way that $\{x \otimes y \mid x, y \in x\}$ is an orthonormal basis. Then

$$\langle u_1 \otimes v_1, u_2 \otimes v_2 \rangle = \langle u_1, u_2 \rangle \langle v_1, v_2 \rangle \tag{8.5}$$

for $u_i, v_i \in V$ $(i = 1, 2)$. Similarly, $V \otimes V \otimes V$ becomes a Hermitian space with $\langle u_1 \otimes v_1 \otimes w_1, u_2 \otimes v_2 \otimes w_2 \rangle = \langle u_1, u_2 \rangle \langle v_1, v_2 \rangle \langle w_1, w_2 \rangle$ for $u_i, v_i, w_i \in V$ $(i = 1, 2)$.

G acts on $V \otimes V$ by

$$(u \otimes v)a = (ua) \otimes (va) \tag{8.6}$$

for $u,v \in V$, $a \in G$. The tensor product $V_i \otimes V_j$ (i.e., the subspace of $V \otimes V$ generated by $u \otimes v$ ($u \in V_i$, $v \in V_j$)) is invariant by the action of G and affords the representation $\mathbb{F}_i \otimes \mathbb{F}_j$ of G. (To see this conventionally, choose the bases of V_i and V_j to diagonalize the action of a on V_i and V_j and check that the trace of the action of a ($a \in G$) is $\chi_i(a)\chi_j(a)$.)

LEMMA 8.2. E_i is the Gram matrix of $\{x\pi_i \mid x \in X\}$, i.e.,

$$(E_i)_{xy} = \langle x\pi_i, y\pi_i \rangle \quad \text{for } x,y \in X,$$

where $(E_i)_{xy}$ is the (α,β) entry of E_i such that $x = x_\alpha$, $y = x_\beta$ $(\alpha,\beta \in \Omega)$.

Proof. Since $x\pi_i$ and $y\pi_i$ are the αth and β^{th} rows of E_i, $\langle x\pi_i, y\pi_i \rangle = (E_i {}^t\overline{E}_i)_{xy}$. But ${}^t\overline{E}_i = E_i$ and $E_i^2 = E_i$, so we get the desired result. $\square$

Let σ_{ij}^k be the linear mapping from $V_i \otimes V_j$ to V_k defined by

$$\sigma_{ij}^k : u \otimes v \to \sum_{x \in X} \langle u \otimes v, x\pi_i \otimes x\pi_j \rangle x\pi_k \tag{8.7}$$

$$(u \in V_i,\ v \in V_j).$$

PROPOSITION 8.3. Let q_{ij}^k be the Krein parameters of $\mathfrak{X}$. Then

(i) $\sigma_{ij}^k = 0$ if and only if $q_{ij}^k = 0$, and

(ii) $\sigma_{ij}^k(x\pi_i \otimes x\pi_j) = \frac{1}{n} q_{ij}^k x\pi_k$ for $x \in X$.

Proof. (i) For $u \in V_i$, $v \in V_j$, $w \in V_k$,

$$\langle (u \otimes v)\sigma_{ij}^k, w\rangle = \sum_{x \in X} \langle u, x\pi_i\rangle\langle v, x\pi_j\rangle\overline{\langle w, x\pi_k\rangle}.$$

Let $\overline{w}$ and $\overline{x\pi_k}$ be the complex conjugate row vectors of w and $x\pi_k$. Then $\overline{x\pi_k} = x\overline{E_k} = xE_{\hat{k}} = x\pi_{\hat{k}}$ and $\overline{\langle w, x\pi_k\rangle} = \langle \overline{w}, \overline{x\pi_k}\rangle = \langle \overline{w}, x\pi_{\hat{k}}\rangle$. So,

$$\begin{aligned}\langle (u \otimes v)\sigma_{ij}^k, w\rangle &= \sum_{x \in X} \langle u, x\pi_i\rangle\langle v, x\pi_j\rangle\langle \overline{w}, x\pi_{\hat{k}}\rangle \\ &= \sum_{x \in X} \langle u \otimes v \otimes \overline{w}, x\pi_i \otimes x\pi_j \otimes x\pi_{\hat{k}}\rangle \\ &= \langle u \otimes v \otimes \overline{w}, \sum_{x \in X} x\pi_i \otimes x\pi_j \otimes x\pi_{\hat{k}}\rangle.\end{aligned}$$

Hence,

$$\sigma_{ij}^k = 0 \text{ if and only if } \sum_{x \in X} x\pi_i \otimes x\pi_j \otimes x\pi_{\hat{k}} = 0. \quad (8.8)$$

But

$$\begin{aligned}&\langle \sum_{x \in X} x\pi_i \otimes x\pi_j \otimes x\pi_{\hat{k}}, \sum_{y \in X} y\pi_i \otimes y\pi_j \otimes y\pi_{\hat{k}}\rangle \\ &= \sum_{x,y \in X} \langle x\pi_i, y\pi_i\rangle\langle x\pi_j, y\pi_j\rangle\langle x\pi_{\hat{k}}, y\pi_{\hat{k}}\rangle \\ &= \sum_{x,y \in X} (E_i)_{xy}(E_j)_{xy}(E_{\hat{k}})_{xy} \qquad \text{(by Lemma 8.2)} \\ &= \sum_{x,y \in X} (E_i \circ E_j \circ E_{\hat{k}})_{xy} = \tau(E_i \circ E_j \circ E_{\hat{k}}) \\ &= \frac{m_k}{n} q_{ij}^k \qquad \text{(by Theorem 3.6(i))}.\end{aligned}$$

So,

$$\sum_{x \in X} x\pi_i \otimes x\pi_j \otimes x\pi_{\hat{k}} = 0 \text{ if and only if } \tag{8.9}$$

$$q_{ij}^k = 0 .$$

By (8.8) and (8.9), we get the assertion (i).

(ii) Let $x = x_\alpha$ $(\alpha \in \Omega)$ and consider the αth row of $\frac{1}{n} q_{ij}^k E_k = (E_i \circ E_j)E_k$. Then

$$\frac{1}{n} q_{ij}^k xE_k = x(E_i \circ E_j)E_k$$

$$= (\sum_{y \in X} (E_i \circ E_j)_{xy} y)E_k$$

$$= \sum_{y \in X} (E_i)_{xy}(E_j)_{xy} yE_k$$

$$= \sum_{y \in X} \langle x\pi_i, y\pi_i \rangle \langle x\pi_j, y\pi_j \rangle y\pi_k \quad \text{(by Lemma 8.2)}$$

and we get the desired result. □

For $u,v \in V_i$, define the product $u * v$ by

$$u * v = (u \otimes v)\sigma_{ii}^i . \tag{8.10}$$

By this multiplication, V_i becomes a commutative (but non-associative) algebra. This algebra is nontrivial if $q_{ii}^i \neq 0$ by Proposition 8.3 and called the <u>Norton algebra</u>.

Let W_{ij} be the subspace of $V_i \otimes V_j$ generated by $\{x\pi_i \otimes x\pi_j \mid x \in X\}$. For $x \in X$ and $a \in G$, $(x\pi_i \otimes x\pi_j)a = (x\pi_j a) \otimes (x\pi_j a) = (xa\pi_i) \otimes (xa\pi_j) = y\pi_i \otimes y\pi_j$, where $y = x_\beta$ with $x = x_\alpha$ and $\beta = \alpha^a$. So W_{ij} is invariant by the action of G. The mapping σ_{ij}^{k} from W_{ij} to V_k commutes with the action of G because for $x \in X$,

$$\begin{aligned}
(x\pi_i \otimes x\pi_j)\sigma_{ij}^{k}a &= \frac{1}{n}\, q_{ij}^{k} x\pi_k a \qquad \text{(by Proposition 8.3)} \\
&= \frac{1}{n}\, q_{ij}^{k} xa\pi_k \\
&= (xa\pi_i \otimes xa\pi_j)\sigma_{ij}^{k} \qquad \text{(by Proposition 8.3 and } xa \in X) \\
&= (x\pi_i a \otimes x\pi_j a)\sigma_{ij}^{k} \\
&= (x\pi_j \otimes x\pi_j)a\sigma_{ij}^{k}\,.
\end{aligned}$$

If $\sigma_{ij}^{k} \neq 0$, then the irreducible representation of G afforded by V_k appears in that afforded by W_{ij} because of Schur's lemma. Since $W_{ij} \subseteq V_i \otimes V_j$, χ_k appears in $\chi_i\chi_j$ whenever $\sigma_{ij}^{k} \neq 0$, i.e., $q_{ij}^{k} \neq 0$. Let $i = j$. Then $W_{ii} \subseteq \mathrm{Sym}^2(V_i)$, where $\mathrm{Sym}^2(V_i)$ is the subspace of $V_i \otimes V_i$ generated by $u \otimes v + v \otimes u$ $(u,v \in V_i)$. Hence, if $q_{ij}^{k} \neq 0$, χ_k appears in $\mathrm{Sym}^2\chi_i$ which is the character of the representation afforded by $\mathrm{Sym}^2(V_i)$. This completes the proof of Theorem 8.1. □

PROPOSITION 8.4. Let W_{ij} be the subspace of $V_i \otimes V_j$ spanned by $x\pi_i \otimes x\pi_j$ $(x \in X)$. W_{ij} is isometrically embedded in V by the G-mapping (i.e., the mapping which commutes with the action of G):

$$x\pi_i \otimes x\pi_j \to \frac{1}{\sqrt{n}} \sum_{k=0}^{d} \sqrt{q_{ij}^{k}}\; x\pi_k\,.$$

In particular, the image of this mapping is the sum of

those V_k for which $q_{ij}^k \neq 0$, and so $\dim W_{ij} = \operatorname{rank}(E_i \circ E_j)$.

Proof. This mapping is the sum of $\sqrt{\frac{n}{q_{ij}^k}}\,\sigma_{ij}^k$ over k with $q_{ij}^k \neq 0$ by Proposition 8.3 and commutes with the action of G. This mapping is isometry because

$$\langle x\pi_i \otimes x\pi_j, y\pi_i \otimes y\pi_j \rangle = \langle x\pi_i, y\pi_i \rangle \langle x\pi_j, y\pi_j \rangle$$

$$= (E_i)_{xy}(E_j)_{xy} \qquad \text{(by Lemma 8.2)}$$

$$= (E_i \circ E_j)_{xy}$$

$$= \frac{1}{n} \sum_{k=0}^{d} q_{ij}^k (E_k)_{xy}$$

$$= \frac{1}{n} \sum_{k=0}^{d} q_{ij}^k \langle x\pi_k, y\pi_k \rangle \qquad \text{(by Lemma 8.2)}$$

$$= \langle \frac{1}{\sqrt{n}} \sum_{k=0}^{d} \sqrt{q_{ij}^k}\; x\pi_k , \frac{1}{\sqrt{n}} \sum_{k=0}^{d} \sqrt{q_{ij}^k}\; y\pi_k \rangle$$

by $x\pi_k, y\pi_k \in V_k$ and $V_k \perp V_\ell$ for $k \neq \ell$. The dimension of W_{ij} is $\sum_{k\,:\,q_{ij}^k \neq 0} m_k = \operatorname{rank}(E_i \;\; E_j)$. □

QUESTIONS. (1) To what extent is the converse of Theorem 8.1 true? In other words, find a criterion for V_k to belong to $V_i \otimes V_j$ but not to W_{ij}. (Perhaps W_{ij} covers almost all irreducible components $V_i \otimes V_j$.)

(2) Given the Krein parameters, can we determine $(\chi_k, \chi_i\chi_j)$? There are some results treating lower bounds

for $(\chi_k, \chi_i\chi_j)$, but it seems that good bounds have not been found yet (cf. Scott (1977)).

REMARK. (1) The converse of Theorem 8.1 is not always true. A counterexample is the action of S_{2d} on $J(2d,d)$ (Example 2.1(4)) with $\chi_i = \chi^{(2d-1,1)}$ and $\chi_j = \chi_k = \chi^{(2d-2,2)}$ where $\chi^{(\ell_1, \ell_2, \ldots, \ell_r)}$ denotes the irreducible character of S_ℓ $(\ell = \ell_1 + \ldots + \ell_r)$ corresponding to the Young diagram of type $[\ell_1, \ell_2, \ldots, \ell_r]$.

COROLLARY 8.5 TO THEOREM 8.1. With the notation of Theorem 8.1,

$$(\chi_i, \chi_j)_H \geq \#\{k \mid q^k_{\hat{i}j} \neq 0\},$$

where H is the stabilizer of α $(\alpha \in \Omega)$ in G.

<u>Proof</u>. $(\chi_i, \chi_j)_H = (1_H, \bar{\chi}_i\chi_j)_H = (1_H, \chi_{\hat{i}}\chi_j)_H$

$$= (1_H^G, \chi_{\hat{i}}\chi_j)_G$$

by the Frobenius reciprocity. Since $1_H^G = \theta = \chi_0 + \chi_1 + \ldots + \chi_d$, we get the desired result by Theorem 8.1. $\square$

QUESTION. (3) (Cameron, Goethals and Seidel) What is the combinatorial version of the Frobenius reciprocity? They discussed an analogue of the Frobenius reciprocity in their paper "Strongly regular graphs having strongly regular constituents" (Cameron, Goethals and Seidel (1978a)).

REMARKS. (2) The (non-commutative) association scheme $\mathfrak{X}$ defined by the action of G on Ω is symmetric if and only if the permutation character θ is multiplicity free and each irreducible constitutent χ of θ takes the values of G in the real number field, because $\overline{\chi}_i = \chi_{\hat{i}}$ with $\overline{E}_i = E_{\hat{i}}$. As seen in Example 2.1(i), $\mathfrak{X}$ is symmetric if and only if G is generously transitive, i.e., for every distinct pair $\alpha, \beta \in \Omega$, there exists an element a of G such that $\alpha^a = \beta$, $\beta^a = \alpha$.

(3) Let G be a generously transitive permutation group on $\Omega = \{1,2,\ldots,n\}$. Suppose G is primitive and of rank 3, i.e., the association scheme is primitive and of class 2. Let $\theta = 1 + \chi_1 + \chi_2$ be the permutation character with χ_1, χ_2 distinct irreducible characters of G. Since G is primitive, $q_{11}^2 \neq 0$ by Theorem 4.6(iii) and obviously $q_{11}^0 \neq 0$. So by Theorem 8.1, $1 + \chi_2$ is a component (constituent) of $\mathrm{Sym}^2\chi_1$. Thus

$$\begin{aligned} n &= 1 + \chi_1(1) + \chi_2(1) \\ &\leq \chi_1(1) + (\mathrm{Sym}^2\chi_1)(1) \\ &= \chi_1(1) + \frac{1}{2}\chi_1(1)(\chi_1(1) + 1) \\ &= \frac{1}{2}\chi_1(1)(\chi_1(1) + 3). \end{aligned}$$

Since $\chi_1(1) = \dim V_1 = m_1$, we get

$$n \leq \frac{1}{2} m_1(m_1 + 3) . \tag{8.11}$$

This is nothing but the absolute bound condition ($d = 2$ in Theorem 4.9). In addition if $q_{11}^1 \neq 0$, then χ_1 is a component of $\mathrm{Sym}^2\chi_1$ and so we get

$$n \leq \frac{1}{2} m_1(m_1 + 1) . \tag{8.12}$$

This is the new bound of Neumaier and can be directly proved by $n = \mathrm{rank}(E_1 \circ E_1)$ and Lemma 4.7. Since

$W_{11} \subseteq \mathrm{Sym}^2(V_1)$ and $\dim W_{11} = \mathrm{rank}\,(E_1 \circ E_1)$ by Proposition 8.4, the equality holds in (8.12) if and only if $\mathrm{rank}(E_1 \circ E_1) = \frac{1}{2} m_1(m_1+1)$, i.e., $W_{11} = \mathrm{Sym}^2(V_1)$.

(4) At the beginning of this section, we assumed that $\mathfrak{X}$ comes from a permutation group G on Ω, but without this assumption, Lemma 8.2, Proposition 8.3 and Proposition 8.4 hold for every commutative association scheme.

PROBLEMS. (1) As mentioned in Remark (4) of section 2.4, it is well known that the equality holds in (8.12) if and only if there exists a tight 4-spherical design (see Part II). Tight 2s-spherical designs with $s \geq 3$ have been classified by Bannai and Damerell (1979, 1980). However, the classification of tight 4-spherical designs is not yet completed. So, classify them.

(2) Classify primitive symmetric association schemes of class 2 (i.e., primitive strongly regular graphs) which have $q_{11}^1 > 0$ and satisfy the equality in (8.12). The only example we know is $J(v,2)$ (Example 2.1(4)):

$$B_1 = \begin{pmatrix} 0 & 1 & 0 \\ 2(v-2) & v-2 & 4 \\ 0 & v-3 & 2(v-4) \end{pmatrix},$$

eigenvalues of B_1	$k = 2(v-2)$	-2	$n-4$
multiplicities	$m_0 = 1$	$m_1 = \frac{v(v-3)}{2}$	$m_2 = v-1$

$$q_{22}^2 = 1 \neq 0 \quad \text{and} \quad n = \frac{v(v-1)}{2} = \frac{1}{2} m_2(m_2+1).$$

COROLLARY 8.6 TO THEOREM 8.1. The notation is the same as in Theorem 8.1. Let $E_0, E_1, \ldots, E_d$ be the primitive idempotents of the incidence algebra of $\mathfrak{X}$ and let E_i^h

denote the hth power of E_i under the Hadamard product. E_j is said to be a component of E_i^h if E_j appears in the expression of E_i^h as a linear combination of $E_o, E_1, \ldots, E_d$.

(i) If E_j is a component of E_i^h, then χ_j is a component of χ_i^h.

(ii) Let χ be primitive. Then for any j, i $(i \neq 0)$, χ_j is a component of χ_i^h for some h $(0 \leq h \leq d)$.

<u>Proof</u>. (i) E_j is a component of $E_i \circ E_k$ for some component E_k of E_i^{h-1}. So $q_{ik}^j \neq 0$. By Theorem 8.1, χ_j is a component of $\chi_i \chi_k$. Since χ_k is a component of χ_i^{h-1} by induction, χ_j is a component of χ_i^h.

(ii) By Theorem 4.6(iii), E_j is a component of E_i^h for some h $(0 \leq h \leq d)$. By the assertion (i), χ_j is a component of χ_i^h. □

2.9 PRIMITIVITY AND IMPRIMITIVITY OF ASSOCIATION SCHEMES

Let G be a permutation group on Ω $(|\Omega| = n)$. In this section, G is always assumed to be transitive on Ω. G is said to be <u>imprimitive</u> if there is a subset S of Ω with the property that

$$S = S^a \quad \text{or} \quad S \cap S^a = \emptyset \quad \text{for all} \quad a \in G \tag{9.1}$$

and

$$1 \lneqq |S| \lneqq n,$$

where $S^a = \{\alpha^a \mid \alpha \in S\}$. Such an S is called a non-trivial block. $\{S^a \mid a \in G\}$ gives a partition of Ω and is called the system of imprimitivity. If there is no non-trivial block, G is said to be primitive.

PROPOSITION 9.1. Let G be a transitive permutation group on Ω ($|\Omega| = n$) and let H be the stabilizer of α ($\alpha \in \Omega$) in G. Then G is primitive if and only if H is a maximal subgroup of G.

Proof. Suppose G is imprimitive with the system of imprimitivity Σ. Let S ($S \in \Sigma$) be the block containing α and set $K = \{a \in G \mid S^a = S\}$. Then K is a subgroup of G and $G \gneqq K \gneqq H$.

Suppose H is not maximal. Let K be a subgroup of G such that $G \gneqq K \gneqq H$. Set $S = \{\alpha^a \mid a \in K\}$. Since $1 \lneqq |K:H| \lneqq |G:H|$, $1 \lneqq |S| \lneqq n$. Suppose $S \cap S^g \ni \beta$ ($g \in G$). Then $\beta = \alpha^a = \alpha^{bg}$ for some $a,b \in K$. So $bga^{-1} \in H$, $g \in b^{-1}Ha \subseteq K$ and hence $S^g = S$. □

THEOREM 9.2. (Sims (1967)) Let G be a transitive permutation group on Ω ($|\Omega| = n$) and $\Lambda_0, \Lambda_1, \ldots, \Lambda_d$ the orbits of G on $\Omega \times \Omega$ with Λ_0 the diagonal. Let Γ_i be the directed graph with the vertix set Ω and the edge set Λ_i. Then G is primitive if and only if each Γ_i ($1 \leq i \leq d$) is connected, i.e., for any $\alpha, \beta \in \Omega$ ($\alpha \neq \beta$), there exists a sequence of vertices $\alpha_0 = \alpha, \alpha_1, \ldots, \alpha_{r-1}, \alpha_r = \beta$ such that $(\alpha_\nu, \alpha_{\nu+1}) \in \Lambda_i$ ($0 \leq \nu \leq r-1$). (Such a sequence is called a path of length r from α to β in Γ_i.)

Proof. First we remark that when considering the connectedness of Γ_i, we can neglect the directions of edges of Γ_i. This follows from Proposition 4.5, but in our case, we have a direct easy proof. By the finiteness of Ω, there is a loop $\alpha_0, \alpha_1, \ldots, \alpha_r$ with $\alpha_0 = \alpha_r$, $(\alpha_\nu, \alpha_{\nu+1}) \in \Lambda_i$, but since G is transitive on Λ_i, we may assume $\alpha_0 = \alpha$, $\alpha_1 = \beta$. Then $\alpha_1, \alpha_2, \ldots, \alpha_r$ is a path from β to α.

Suppose G is imprimitive and let S be a non-trivial block. Choose arbitrary distinct vertices α, β of S and let $(\alpha, \beta) \in \Lambda_i$. For $a \in H = G_\alpha$, β^a belongs to S, because $S \cap S^a \ni \alpha$ and so $S = S^a \ni \beta^a$. Since H is transitive on $\Lambda_i(\alpha) = \{\gamma \mid (\alpha, \gamma) \in \Lambda_i\}$, $\Lambda_i(\alpha)$ is contained in S. By repeating this argument, any path in Γ_i starting from α ends in S and hence Γ_i is disconnected.

Suppose some Γ_i is disconnected. Let S be a connected component of Γ_i. Then $1 \lneqq |S| \lneqq n$. If there is a path from α to β in Γ_i, there is also a path from α^a to β^a in Γ_i for $a \in G$. So S^a is also a connected component of Γ_i and hence $S \cap S^a = \emptyset$ or S^a. This completes the proof of the theorem. □

In short, G is primitive if and only if $\mathfrak{X} = (\Omega, \{\Lambda_i\}_{0 \le i \le d})$ is primitive.

REMARK. (1) (Glauberman (1969)) Let G be a transitive permutation group on Ω and let $\Lambda_0, \Lambda_1, \ldots, \Lambda_d$ be the orbits of G on $\Omega \times \Omega$. Then the graph $\Gamma_i = (\Omega, \Lambda_i)$ is connected if and only if G is generated by a stabilizer group G_α and an element $a \in G$ such that $(\alpha, \alpha^a) \in \Lambda_i$.

For square matrices M_1, M_2, we write $M_1 \sim M_2$ if M_1 can be transformed to M_2 by rearranging the rows and columns simultaneously, i.e., if $P^{-1} M_1 P = M_2$ for some permutation matrix P.

THEOREM 9.3. Let $\mathfrak{X} = (X, \{R_i\}_{0 \le i \le d})$ be a commutative association scheme. Let A_i be the adjacency matrix with respect to R_i and let E_i be the primitive idempotent of the adjacency algebra $\mathfrak{A}$ $(0 \le i \le d)$. Let p_{ij}^k and q_{ij}^k be the intersection numbers and the Krein parameters, respectively. Then the following (i)-(v) are equivalent to each other:

(i) $A_i \sim \begin{pmatrix} * & * \\ 0 & * \end{pmatrix}$ for some i $(i \ne 0)$.

(ii) $\sum_{i=0}^{s} A_i = p \sum_{i=0}^{t} E_i \sim I_q \otimes J_p$ for some s, t $(1 \le s \le d-1,\ 1 \le t \le d-1)$, where I_q is the identity matrix of degree $q = \sum_{i=0}^{t} m_i$ and J_p is the matrix of degree $p = \sum_{i=0}^{s} k_i$ with the entries all one.

(iii) $p_{ij}^k = 0$ for $0 \le i \le s,\ 0 \le j \le s,\ s+1 \le k \le d$ $(1 \le s \le d-1)$.

(iv) $q_{ij}^k = 0$ for $0 \le i \le t,\ 0 \le j \le t,\ t+1 \le k \le d$ $(1 \le t \le d-1)$.

(v) There exists E_i $(i \ne 0)$ which has the same x^{th} and y^{th} rows for some x, y $(x \ne y)$.

The numbers s in (ii) and (iii) are the same and the numbers t in (ii) and (iv) are the same. The conditions (ii), (iii), (iv) are meant to hold by a suitable rearrangement of the indices $1, 2, \ldots, d$ for A_i and another suitable rearrangement of the indices $1, 2, \ldots, d$ for E_i.

Proof. First we notice that by applying τ (see section

2.3) and tr to (ii), we get $p = \sum_{i=0}^{s} k_i$ and $q = \sum_{i=0}^{t} m_i$. The implications (ii) ⇒ (i), (iii), (iv) are trivial. By Proposition 4.5, (i) implies (ii).

(iv) ⇒ (ii): Suppose (iv) holds. Set $M = p \sum_{i=0}^{t} E_i$ with $p = n/\sum_{i=0}^{t} m_i$. Then by the proof of Theorem 4.6, $M = \sum_{i=0}^{s} A_i$ and

$$\left.\begin{array}{l} M^2 = pM, \quad M \circ M = M \\ {}^tM = M, \quad \text{any diagonal entry of } M = 1. \end{array}\right\} \qquad (9.2)$$

It is easy to see that for any matrix M satisfying (9.2), $M \sim I_q \otimes J_p$.

(iii) ⇒ (ii): Suppose (iii) holds. Set $M = \sum_{i=0}^{s} A_i$. Then by a similar argument to the proof of Theorem 4.6, M satisfies (9.2) and hence $M \sim I_q \otimes I_p$.

(ii) ⇒ (v): Suppose (ii) holds. Then $\sum_{i=0}^{t} E_i$ has the same x^{th} and y^{th} rows for some distinct x and y. So there exist $e_x = (0,\ldots,0,1,0,\ldots,0)$ with 1 in the xth place and $e_y = (0,\ldots,0,1,0,\ldots,0)$ with 1 in the yth place such that $e_x \sum_{i=0}^{t} E_i = e_y \sum_{i=0}^{t} E_i$. Let V_i be the image of E_i. Then $V_i \cap V_j = \{0\}$ $(i \neq j)$. So $e_xE_i = e_yE_i$ $(0 \leq i \leq t)$ and E_i have the same x^{th} and y^{th} rows.

(v) ⇒ (ii): Suppose (v) holds. Let $\sim$ be an equivalence relation on X defined by

$x \sim y$ if and only if E_i has the same x^{th} and y^{th} rows. (9.3)

We want to show $\sim$ is a union of association relations R_i $(0 \le i \le s)$. Let $E_i = \frac{1}{n} \sum_{\alpha=0}^{d} q_i(\alpha) A_\alpha$. Then the (x,z) and (y,z) entries of E_i are $\frac{1}{n} q_i(\alpha)$ and $\frac{1}{n} q_i(\beta)$, where $(x,z) \in R_\alpha$ and $(y,z) \in R_\beta$. Hence E_i has the same x^{th} and y^{th} rows with $(x,y) \in R_\gamma$ if and only if $q_i(\alpha) = q_i(\beta)$ whenever there exists z with $(x,z) \in R_\alpha$, $(y,z) \in R_\beta$, i.e.,

$$q_i(\alpha) = q_i(\beta) \quad \text{whenever} \quad p^{\gamma}_{\alpha\beta'} \ne 0. \tag{9.4}$$

The condition (9.4) does not depend on the choice of $(x,y) \in R_\gamma$. So $\sim$ is a union of association relations R_i $(0 \le i \le s)$ and hence $\sum_{i=0}^{s} A_i \sim I_q \otimes J_p$. If $s = d$, E_i has the same rows and by ${}^t\overline{E}_i = E_{i'}$, E_i has the same columns, which is a contradiction to $i \ne 0$. This completes the proof of the theorem. □

A commutative association scheme $\mathfrak{X} = (X, \{R_i\}_{0 \le i \le d})$ is said to be <u>imprimitive</u> if one of (i)-(v) of Theorem 9.3 holds. It is easy to see that the directed graph Γ_i with respect to R_i is disconnected if and only if the incidence matrix A_i satisfies (i) of Theorem 9.3.

Let $\mathfrak{X} = (X, \{R_i\}_{0 \le i \le d})$ be imprimitive and $\sum_{i=0}^{s} A_i \sim I_q \otimes J_p$. Then $\bigcup_{i=0}^{s} R_i$ is an equivalence relation. An equivalence class S of $\bigcup_{i=0}^{s} R_i$ is called a block. Obviously, $|S| = p = \sum_{i=0}^{s} k_i$. The set Σ of equivalence classes of $\bigcup_{i=0}^{s} R_i$ is called the system of imprimitivity. Obviously, $|\Sigma| = n/p = q = \sum_{i=0}^{t} m_i$. For a block S, the association scheme $\mathfrak{X}_0 = (S, \{R_i\}_{0 \le i \le s})$ on S is called an association subscheme of $\mathfrak{X}$.

We shall define a quotient association scheme $\mathfrak{X}/\mathfrak{X}_0$ on the system of imprimitivity.

THEOREM 9.4. We keep the notation of Theorem 9.3. Let $\mathfrak{X} = (X, \{R_i\}_{0 \le i \le d})$ be an imprimitive commutative association scheme with Σ the system of imprimitivity. So $\sum_{i=0}^{s} A_i = p \sum_{i=0}^{t} E_i = I_q \otimes J_p$, arranging the rows and columns according to the blocks of Σ. Let $\sim$ be the relation on $\{0,1,2,\dots,d\}$ defined by $i \sim j$ if and only if $p_{i\alpha}^{j} \neq 0$ for some $0 \le \alpha \le s$. Then $\sim$ is an equivalence relation. Let $T_0, T_1, \dots, T_r$ be the equivalence classes and $D_i = \sum_{\alpha \in T_i} A_\alpha$. Then $r = t$ and there exists a matrix $\imath(D_i)$ of degree q such that $D_i = \frac{1}{p}\imath(D_i) \otimes J_p$, and $\frac{1}{p}\imath(D_i)$ $(0 \le i \le t)$ are the incidence matrices of an association scheme on Σ. (This is called the quotient association scheme of $\mathfrak{X}$.) Moreover, $E_0, E_1, \dots, E_t$ are the primitive idempotents of the algebra spanned by $D_0, D_1, \dots, D_t$.

Proof. We first notice that $0 \le i' \le s$ for $0 \le i \le s$, because ${}^tM = M$ and ${}^tM = \sum_{i=0}^{s} A_{i'}$. Since $p_{i0}^{i} = 1$ and $k_j p_{i\alpha}^{j} = k_i p_{j\alpha'}^{i}$, by Proposition 2.2, the relation $\sim$ is reflective and symmetric. Let $p_{i\alpha}^{j} \neq 0$ and $p_{j\beta}^{k} \neq 0$

$(0 \leq \alpha \leq s,\ 0 \leq \beta \leq s)$, i.e., there exists w, x, y, z satisfying the following diagram:

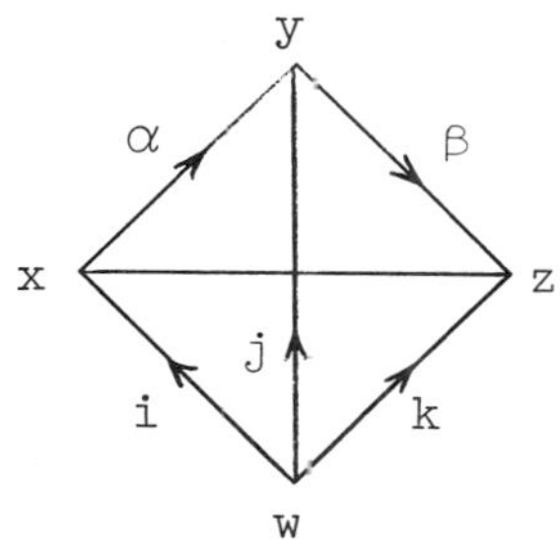

Then $(x,z) \in R_\gamma$ for $0 \leq \gamma \leq s$ and so $p_{i\gamma}^{k} \neq 0$. So $\sim$ is transitive.

The equivalence class containing 0 is $\{0,1,\ldots,s\}$. We may assume

$$T_0 = \{0,1,\ldots,s\} . \tag{9.5}$$

Since $p_{i\alpha}^{j} = p_{i'\alpha'}^{j'}$, it holds that $i' \sim j'$ whenever $i \sim j$ and so $T_i' \equiv \{\beta' \mid \beta \in T_i\}$ is an equivalence class. So

$$T_i' = T_{i'} \quad \text{for some } i' . \tag{9.6}$$

We want to show that for any blocks S_1 and S_2, the following holds:

If $(x_0,y_0) \in R_\alpha$ for some $(x_0,y_0) \in S_1 \times S_2$, then for any $x \in S_1$, there exists some $y \in S_2$ such that $(x,y) \in R_\alpha$. (9.7)

Let $(x,x_0) \in R_\beta$ $(0 \leq \beta \leq s)$ and $(x,y_0) \in R_\gamma$. Then $p_{\beta\alpha}^{\gamma} \neq 0$ and so $p_{\alpha\beta}^{\gamma} \neq 0$, i.e., there exists $y \in X$ such that $(x,y) \in R_\alpha$ and $(y,y_0) \in R_\beta$. Since $0 \leq \beta \leq s$, $y \in S_2$.

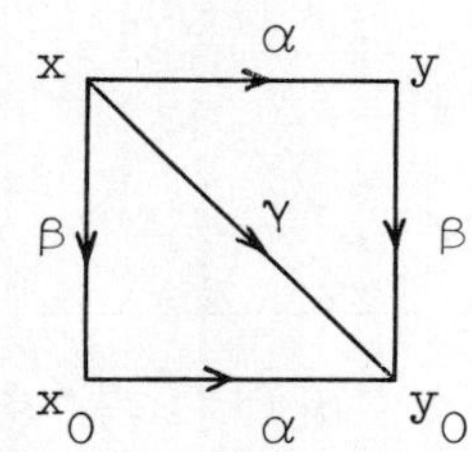

By (9.7), we get that for any blocks S_1, S_2,

$$\{\alpha \mid R_\alpha \cap (S_1 \times S_2) \neq \emptyset\} = T_i \quad \text{for some } i . \tag{9.8}$$

This means $D_i = \sum_{\alpha \in T_i} A_\alpha$ has the same (x_1, x_2) and (y_1, y_2) entries for x_j, y_j in the same block ($j = 1, 2$). So we have

$$D_i = \frac{1}{p} \iota(D_i) \otimes J_p \tag{9.9}$$

for some $\frac{1}{p} \iota(D_i)$ whose entries are 0 or 1.

If a Kronecker product $C \otimes J_p$ is a linear combination of A_α $(0 \le \alpha \le d)$, where C is a matrix of degree q, then the coefficients of A_α and A_β are the same for $\alpha, \beta \in T_k$ $(0 \le k \le r)$ by (9.8) and hence $C \otimes J_p$ is a linear combination of D_k $(0 \le k \le r)$. In particular, $D_i D_j = \frac{1}{p} \iota(D_i)\iota(D_j) \otimes J_p$ is a linear combination of D_k $(0 \le k \le r)$. By (9.5), (9.6), $D_0 = I_q \otimes J_p$ and ${}^tD_i = D_{i'}$. Thus $\frac{1}{p} \iota(D_i)$ $(0 \le i \le r)$ satisfy the condition (2.1') and so are the adjacency matrices of an association scheme on Σ.

E_i $(0 \le i \le t)$ has the same x^{th} and y^{th} rows for x, y in the same block by the proof of Theorem 9.3. Since ${}^tM = M$ and ${}^tM = p \sum_{i=0}^{t} E_{\hat{i}}$, we have $0 \leq \hat{i} \leq t$ for $0 \leq i \leq t$ and so $E_i = {}^tE_{\hat{i}}$ $(0 \leq i \leq t)$ has the same x^{th} and y^{th} columns for x, y in the same block. Hence E_i has the same (x_1, x_2) and (y_1, y_2) entries for x_j, y_j in the same block $(j = 1, 2)$. This implies that there exists a matrix $\iota(E_i)$ of degree q such that $E_i = \frac{1}{p}\iota(E_i) \otimes J_p$. Thus $E_i = \frac{1}{n}\sum_\alpha q_i(\alpha)A_\alpha$ is a linear combination of D_j $(0 \leq j \leq r)$ for $0 \leq i \leq t$.

Let $\mathfrak{A}_0$ be the algebra spanned by D_i $(0 \leq i \leq r)$. Since $D_i = \frac{1}{p}\iota(D_i) \otimes J_p$, we can define a mapping ι from $\mathfrak{A}_0$ to $M_q(\mathbb{C})$ (the full matrix algebra of order q) by $D = \frac{1}{p}\iota(D) \otimes J_p$ for $D \in \mathfrak{A}_0$. Then it is easy to see ι is an algebra isomorphism. Since $\frac{1}{p}\iota(D_i)$ $(0 \leq i \leq r)$ are the incidence matrices of an association scheme, $\mathfrak{A}_0$ is a C-algebra with the basis $\frac{1}{p}D_i$ $(0 \leq i \leq r)$. By (5.4), $\frac{1}{p}D_0$ is the sum of the primitive idempotents of $\mathfrak{A}_0$. But $\frac{1}{p}D_0 = \frac{1}{p}I_q \otimes J_p = \sum_{i=0}^{t} E_i$. Hence E_i $(0 \leq i \leq t)$ are all of the primitive idempotents of $\mathfrak{A}_0$ and $r = t$. This completes the proof of the theorem. □

EXAMPLE 9.5. Let G be an imprimitive permutation group on a finite set X (G is assumed to be transitive on X) and let Σ be the system of imprimitivity. Let Λ_i $(i = 0, 1, \ldots, d)$ be the orbits of G on $X \times X$. Suppose the association scheme $\mathfrak{X} = (X, \{\Lambda_i\}_{0 \leq i \leq d})$ is commutative. It is clear that $\mathfrak{X}$ is imprimitive with Σ the system of imprimitivity. The quotient association scheme of $\mathfrak{X}$ coincides with the one which is obtained by the action of G on Σ.

Proof. Let T_i $(0 \leq i \leq t)$ be the equivalence classes on $\{0,1,\ldots,d\}$ defined in Theorem 9.4 and let $K = \{a \in G \mid S_0^a = S_0\}$ for $S_0 \in \Sigma$. Then $\bigcup_{\alpha \in S_0} \bigcup_{j \in T_i} \Lambda_j(\alpha) = \bigcup_{a \in K} S^a$ for some $S \in \Sigma$. □

REMARK. (2) d is not always bounded by a function of s and t (cf. Cameron-Goethals-Seidel (1978b)).

QUESTION. Let G be a transitive permutation group on Ω and let $\theta = 1 + \chi_1 + \cdots + \chi_d$ be the permutation character with χ_i distinct irreducible characters of G. By using the information only of characters χ_i $(0 \leq i \leq d)$, can we determine whether G is primitive or not? For example, if we know the part of the character table T of G which corresponds to the irreducible components $\chi_0,\chi_1,\ldots,\chi_d$ of the permutation character of G on Ω, then can we determine whether G is primitive or not?

We shall review Theorem 9.4 in terms of C-algebras. Let $\mathfrak{A}$ be a C-algebra with the basis $x_0,x_1,\ldots,x_d$ and let $e_0,e_1,\ldots,e_d$ be the primitive idempotents of $\mathfrak{A}$. We identify the dual C-algebra $\hat{\mathfrak{A}}$ of $\mathfrak{A}$ with the C-algebra $\mathfrak{A} = \langle ne_i \mid 0 \leq i \leq d\rangle$ with respect to the Schur product $\circ$: $x_i \circ x_j = \delta_{ij}x_i$. We use the usual notation: $x_ix_j = \sum_k p_{ij}^k x_k$, $(ne_i)\circ(ne_j) = \sum_k q_{ij}^k(ne_k)$, $x_i = \sum_j p_i(j)e_j$, $ne_i = \sum_j q_i(j)x_j$, $k_i = p_{ii'}^0 = p_i(0)$, $m_i = q_{i\hat{i}}^0 = q_i(0)$ and $n = \sum_i k_i = \sum_i m_i$. In what follows, we always assume

$$p_{ij}^k \geq 0 \text{ and } q_{ij}^k \geq 0 .$$

For a subset $T \subseteq \{0,1,\ldots,d\}$, $\langle x_\alpha \mid \alpha \in T\rangle$ denotes the vector space spanned by x_α $(\alpha \in T)$. If T satisfies

(I) $p_{ij}^{k} = 0$ for $i,j \in T$, $k \notin T$, i.e., $x_i x_j \in \langle x_\alpha \mid \alpha \in T\rangle$ for $i,j \in T$,

(II) $0 \in T$,

(III) $i' \in T$ for $i \in T$,

then $\langle x_\alpha \mid \alpha \in T\rangle$ is a C-algebra with the basis $\{x_\alpha \mid \alpha \in T\}$. Such $\langle x_\alpha \mid \alpha \in T\rangle$ is called a <u>C-subalgebra</u> of $\mathfrak{A}$. A C-subalgebra $\langle x_\alpha \mid \alpha \in T\rangle$ is said to be <u>proper</u> if $\{0\} \subsetneq T \subsetneq \{0,1,\ldots,d\}$.

The conditions (I), (III) imply (II) because $p_{ii'}^{0} = k_i \neq 0$.

LEMMA 9.6. The condition (I) implies (III).

<u>Proof</u>. The proof is similar to that of (4.8) because $\langle x_\alpha \mid \alpha \in T\rangle$ is closed both under the multiplications of $\mathfrak{A}$ and $\hat{\mathfrak{A}}$. □

LEMMA 9.7. (i) Let $\langle x_\alpha \mid \alpha \in T\rangle$ be a C-subalgebra. Then the element $x = \sum_{\alpha \in T} x_\alpha$ satisfies $x \cdot x = px$ and $x \circ x = x$ with $p = \sum_{\alpha \in T} k_\alpha$.

(ii) If $x \in \mathfrak{A}$ $(x \neq 0)$ satisfies $x \cdot x = ax$ and $x \circ x = bx$ for some $a,b \in \mathbb{C}$, then $x = a \sum_{\alpha \in U} e_\alpha = b \sum_{\alpha \in T} x_\alpha$ for some subsets U, T of $\{0,1,\ldots,d\}$, and $\langle x_\alpha \mid \alpha \in T\rangle$, $\langle ne_\alpha \mid \alpha \in U\rangle$ are C-subalgebras of $\mathfrak{A}$, $\hat{\mathfrak{A}}$ respectively.

Moreover $n = (\sum_{\alpha \in T} k_\alpha)(\sum_{\alpha \in U} m_\alpha)$ and $a = b \sum_{\alpha \in T} k_\alpha$.

<u>Proof</u>. (i) The identity $x \circ x = x$ is trivial. The identity $x \cdot x = px$ follows from (5.18) because

$\langle x_\alpha \mid \alpha \in T\rangle$ is a C-algebra.

(ii) The identity $x \cdot x = ax$ means that $\frac{1}{a}x$ is an idempotent of $\mathfrak{A}$ and so $\frac{1}{a}x = \sum_{\alpha \in U} e_\alpha$ for some U, while the identity $x \circ x = bx$ means that $\frac{1}{b}x$ is an idempotent of $\hat{\mathfrak{A}}$ and so $\frac{1}{b}x = \sum_{\alpha \in T} x_\alpha$ for some T. Since p_{ij}^k's are assumed nonnegative, the identity $x \cdot x = ax$ implies $\langle x_\alpha \mid \alpha \in T\rangle$ is closed under the multiplication of $\mathfrak{A}$ and hence a C-subalgebra by Lemma 9.6. Similarly, $\langle ne_\alpha \mid \alpha \in U\rangle$ is a C-subalgebra of $\hat{\mathfrak{A}}$.

Apply the linear representation of $\mathfrak{A}$: $x_i \to k_i$ to $x \cdot x = ax$. Then $a = b \sum_{\alpha \in T} k_\alpha$. Apply the linear representation of $\hat{\mathfrak{A}}$: $ne_i \to m_i$ to $x \circ x = bx$. Then $b = \frac{a}{n} \sum_{\alpha \in U} m_\alpha$. $\square$

A C-subalgebra $\langle x_\alpha \mid \alpha \in T\rangle$ of $\mathfrak{A}$ is proper if and only if the element $\sum_{\alpha \in T} x_\alpha$ is neither x_0 nor ne_0 because $\sum_{i=0}^{d} x_i = ne_0$. Dually, a C-subalgebra $\langle ne_\alpha \mid \alpha \in U\rangle$ of $\mathfrak{A}$ is proper if and only if the element $\sum_{\alpha \in U} ne_\alpha$ is neither ne_0 nor nx_0. Thus as a corollary to Lemma 9.7, we have the following:

THEOREM 9.8. Let $\mathfrak{A}$ be a C-algebra with the property that $p_{ij}^k \geq 0$, $q_{ij}^k \geq 0$. Then $\mathfrak{A}$ has a proper C-subalgebra if and only if the dual $\hat{\mathfrak{A}}$ has a proper C-subalgebra.

Compare the above with Theorem 9.3. The C-algebra version of Theorem 9.4 is as follows.

THEOREM 9.9. Let $\mathfrak{A}$ be a C-algebra with the property that $p_{ij}^k \geq 0$, $q_{ij}^k \geq 0$, and let $\langle x_\alpha | \alpha \in T\rangle$ be a C-subalgebra of $\mathfrak{A}$. Let $\sim$ be the relation on $\{0,1,\ldots,d\}$ defined by $i \sim j$ if and only if $p_{i\alpha}^j \neq 0$ for some $\alpha \in T$. Then $\sim$ is an equivalence relation. Let $T_0 = T, T_1, \ldots, T_t$ be the equivalence classes and let $x_{T_i} = \sum_{\alpha \in T_i} x_\alpha$. Then the vector space spanned by x_{T_i} $(i = 0, 1, \ldots, t)$ over $\mathbb{C}$ is (i) a C-algebra with the basis $\frac{1}{p} x_{T_i}$ and the multiplication induced from $\mathfrak{A}$ and (ii) a C-subalgebra of $\hat{\mathfrak{A}}$ with the basis $\{ne_\alpha | \alpha \in U\}$, where $\sum_{\alpha \in T} x_\alpha = p \sum_{\alpha \in U} e_\alpha$ and $p = \sum_{\alpha \in T} k_\alpha$. The C-algebra (ii) is the dual of (i).

Furthermore, for $\alpha \in T_i$, $x_\alpha x_T = \lambda x_{T_i}$ for some $\lambda \in \mathbb{C}$. Thus $\{x \cdot x_T | x \in \mathfrak{A}\}$ is the vector space spanned by x_{T_i} over $\mathbb{C}$. (The C-algebra (i) is called the <u>quotient</u> <u>C-algebra</u> and denoted by $\mathfrak{A}/\langle x_\alpha | \alpha \in T\rangle$.)

<u>Proof</u>. The conditions (I), (II), (III) for T imply that $\sim$ is transitive, reflective and symmetric, respectively.

Let $x_T = \sum_{\alpha \in T} x_\alpha$. Then $x_T = p \sum_{\alpha \in U} e_\alpha$ and $\langle ne_\alpha | \alpha \in U\rangle$ is a C-subalgebra of $\hat{\mathfrak{A}}$ by Lemma 9.7. Idempotents of $\hat{\mathfrak{A}}$ have the form $x_V = \sum_{\alpha \in V} x_\alpha$ for some $V \subseteq \{0,1,\ldots,d\}$. Let $x_{V_0}, x_{V_1}, \ldots, x_{V_s}$ be the primitive idempotents of $\langle ne_\alpha | \alpha \in U\rangle$ with respect to the multiplication of $\hat{\mathfrak{A}}$. The identity of $\langle ne_\alpha | \alpha \in U\rangle$ with respect to the multiplication of $\mathfrak{A}$ is $\frac{1}{p} x_T = \sum_{\alpha \in U} e_\alpha$. So $\frac{1}{p} x_{V_i} x_T = x_{V_i}$.

Since $e_\beta x_T = 0$ for $\beta \notin U$ and $x_{V_i} x_T = p x_{V_i}$, the set $\{x \cdot x_T \mid x \in \mathfrak{A}\}$ is a linear span of $x_{V_0}, x_{V_1}, \ldots, x_{V_s}$. This implies that $x_\alpha x_T = \lambda x_{V_i}$ for $\alpha \in V_i$, because $x_{V_i} x_T = p x_{V_i}$ and $p_{ij}^k \geq 0$. Hence, V_i is a single equivalence class of $\sim$. Since $\langle ne_\alpha \mid \alpha \in U\rangle$ is a C-subalgebra of $\hat{\mathfrak{A}}$, the element $ne_0 = \sum_{i=0}^{d} x_i$ is contained in $\langle ne_\alpha \mid \alpha \in U\rangle$. So $V_0 \cup V_1 \cup \ldots \cup V_s = \{0,1,\ldots,d\}$. Thus $s = t$ and $V_i = T_i$.

If $i \sim j$, then $i' \sim j'$. So the set $\{\alpha' \mid \alpha \in T_i\}$ is an equivalence class of $\sim$. We denote this equivalence class by $T_{i'}$. Then

$$(\frac{1}{p} x_{T_i})(\frac{1}{p} x_{T_j}) = (\frac{1}{p} \sum_{\alpha \in T_i} k_\alpha)\delta_{i'j} \frac{1}{p} x_T + \cdots \quad .$$

The mapping $\frac{1}{p} x_{T_i} \to \frac{1}{p} \sum_{\alpha \in T_i} k_\alpha$ is the restriction of the representation $x_\alpha \to k_\alpha$ of $\mathfrak{A}$ to $\langle \frac{1}{p} x_{T_i} \mid i = 0, 1, \ldots, t\rangle$. Thus we get the C-algebra (i).

The parameter n for the C-algebra (i) is $\frac{1}{p} \sum_{i=0}^{t} \sum_{\alpha \in T_i} k_\alpha = \sum_{\alpha \in U} m_\alpha$, the equality coming from Lemma 9.7. Since $\frac{1}{p} x_{T_i} = \sum_{\alpha \in T_i} \sum_{j \in U} \frac{1}{p} p_\alpha(j) e_j$, the first eigenmatrix of the C-algebra (i) has (i,j) entry

$$\frac{1}{p} \sum_{\alpha \in T_i} p_\alpha(j) \; . \tag{9.10}$$

Since e_i $(i \in U)$ is a linear combination of x_{T_j},

$q_i(\alpha) = q_i(\beta)$ for $\alpha, \beta \in T_j$. So the second eigenmatrix has (i,j) entry

$$q_i(\alpha) \qquad (\alpha \in T_j). \tag{9.11}$$

Thus the second eigenmatrix of (i) coincides with the first of (ii) and hence the C-algebra (ii) is the dual of (i). □

REMARK. (3) The first and second eigenmatrices of the quotient C-algebra $\langle \frac{1}{p} x_{T_i} \mid 0 \le i \le t \rangle$ of $\mathfrak{A}$ are given by (9.10) and (9.11) in terms of $p_i(j)$, $q_i(j)$ of $\mathfrak{A}$. In particular, the degrees and the multiplicities are k_{T_i}/k_{T_0} and m_i, where $k_{T_i} = \sum_{\alpha \in T_i} k_\alpha$.

Consider the linear mapping $x_\alpha \to \frac{1}{k_{T_0}} x_\alpha x_{T_0}$ from a C-algebra $\mathfrak{A} = \langle x_\alpha \mid 0 \le \alpha \le d \rangle$ to a quotient C-algebra $\langle \frac{1}{k_{T_0}} x_{T_i} \mid 0 \le i \le t \rangle$ of $\mathfrak{A}$. This is an algebra homomorphism. By Theorem 9.9, $x_\alpha x_{T_0} = \lambda_\alpha x_{T_i}$ for some $\lambda_\alpha \in \mathbb{C}$, where $\alpha \in T_i$. Applying the linear representation $x_\alpha \to k_\alpha$, we have $\lambda_\alpha = k_\alpha k_{T_0}/k_{T_i}$. So the homomorphism carries x_α to $\frac{k_\alpha}{k_{T_i}} x_{T_i}$, where $\alpha \in T_i$.

Let $\mathfrak{A} = \langle x_\alpha \mid 0 \le \alpha \le d \rangle$ and $\mathfrak{A}' = \langle x'_i \mid 0 \le i \le d' \rangle$ be C-algebras. Let φ be an algebra homomorphism from $\mathfrak{A}$ to $\mathfrak{A}'$ with the property that $\varphi(x_0) = x'_0$ and $\varphi(x_\alpha) = \mu_\alpha x'_i$ for some positive $\mu_\alpha \in \mathbb{R}$ and some i $(0 \le \alpha \le d)$. Such φ is called a <u>C-algebra homomorphism</u>. Let $T_i = \{\alpha \mid \varphi(x_\alpha) = \mu_\alpha x'_i\}$. We may assume T_i $(0 \le i \le t)$ are all the non-empty sets. $\langle \alpha \mid \alpha \in T_0 \rangle$ is a C-subalgebra of $\mathfrak{A}$ and $\langle \frac{1}{k_{T_0}} x_{T_i} \mid 0 \leq i \leq t \rangle$ is the quotient C-algebra $\mathfrak{A}/\langle \alpha \mid \alpha \in T_0 \rangle$.

We want to show $\varphi(\frac{1}{k_{T_0}} x_{T_i}) = x_i'$ $(0 \leq i \leq t)$. Write $\varphi(\frac{1}{k_{T_0}} x_{T_i}) = \gamma_i x_i'$ $(\gamma_i > 0)$. Since $\langle \gamma_i x_i' \mid 0 \leq i \leq t \rangle$ is the image of $\langle \frac{1}{k_{T_0}} x_{T_i} \rangle$, it is a C-algebra. By (VI) of the definition of C-algebras, the mapping $\gamma_i x_i \to \gamma_i \gamma_{i'} k_i'$ is a linear representation of $\langle \gamma_i x_i' \mid 0 \leq i \leq t \rangle$ because $(\gamma_i x_i')(\gamma_{i'} x_{i'}') = \gamma_i \gamma_{i'} k_i' x_0' + \cdots$. The mapping $\gamma_i x_i \to \gamma_i k_i'$ is also a linear representation. By Proposition 5.4, the vectors $(\gamma_0 \gamma_{0'} k_0', \ldots, \gamma_t \gamma_{t'} k_t')$ and $(\gamma_0 k_0', \ldots, \gamma_t k_t')$ are rows of the first eigenmatrix of $\langle \gamma_i x_i' \mid 0 \leq i \leq t \rangle$. By the orthogonality relation of eigenmatrices, these vectors have to coincide and hence $\gamma_i = 1$ for all i. So we have the following:

THEOREM 9.10. Let $\mathfrak{A} = \langle x_\alpha \mid 0 \leq \alpha \leq d \rangle$ and $\mathfrak{A}' = \langle x_i' \mid 0 \leq i \leq d' \rangle$ be C-algebras whose structure constants and dual structure constants are non-negative.

(1) Let φ be a C-algebra homomorphism from $\mathfrak{A}$ to $\mathfrak{A}'$. Then $\ker \varphi$ and $\operatorname{Im} \varphi$ are C-subalgebras of $\mathfrak{A}$ and $\mathfrak{A}'$ respectively, and $\mathfrak{A}/\ker \varphi$ and $\operatorname{Im} \varphi$ are isomorphic as C-algebras, where $\ker \varphi = \langle x_\alpha \mid \varphi(x_\alpha) = \mu_\alpha x_0' \rangle$ and $\operatorname{Im} \varphi = \langle x_i' \mid \varphi(x_\alpha) = \mu_\alpha x_i'$ for some $\alpha \rangle$. More precisely, let $T_i = \{\alpha \mid \varphi(x_\alpha) = \mu_\alpha x_i'\}$ and $k_{T_i} = \sum_{\alpha \in T_i} k_\alpha$, where k_α are the valencies of $\mathfrak{A}$. Then $\mathfrak{A}/\ker \varphi = \langle \frac{1}{k_{T_0}} x_{T_i} \mid T_i \neq \emptyset \rangle$, and the correspondence of $\frac{1}{k_{T_0}} x_{T_i}$ to x_i' gives the isomorphism between $\mathfrak{A}/\ker \varphi$ and $\operatorname{Im} \varphi$.

(2) Let $\mathfrak{A}_1 = \langle x_\alpha \mid \alpha \in S \rangle$ and $\mathfrak{A}_2 = \langle x_\alpha \mid \alpha \in T \rangle$ be C-subalgebras of $\mathfrak{A}$. Then $\mathfrak{A}_1\mathfrak{A}_2/\mathfrak{A}_2$ and $\mathfrak{A}_1/(\mathfrak{A}_1 \cap \mathfrak{A}_2)$ are isomorphic as C-algebras, where $\mathfrak{A}_1 \cap \mathfrak{A}_2 = \langle x_\alpha \mid \alpha \in S \cap T \rangle$ and $\mathfrak{A}_1\mathfrak{A}_2 = \langle x_\gamma \mid p^\gamma_{\alpha\beta} \neq 0$ for some $\alpha \in S$, $\beta \in T \rangle$.

Let $\mathfrak{A} = \langle x_\alpha \mid 0 \leq \alpha \leq d \rangle$ be a C-algebra and let $\mathfrak{A} = \mathfrak{A}_0 \supset \mathfrak{A}_1 \supset \ldots \supset \mathfrak{A}_r = \{x_0\}$ be a sequence of C-subalgebras of $\mathfrak{A}$. If there are no C-subalgebras $\mathfrak{B}$ of $\mathfrak{A}$ such that $\mathfrak{A}_i \supsetneq \mathfrak{B} \supsetneq \mathfrak{A}_{i+1}$ $(i = 0, 1, \ldots, r-1)$, the sequence is said to be a <u>composition series</u> of $\mathfrak{A}$ and the quotient C-algebras $\mathfrak{A}_i/\mathfrak{A}_{i+1}$ $(i = 0, 1, \ldots, r-1)$ are called <u>composition factors</u>. Composition factors have no proper C-subalgebras. By the finiteness of d, the existence of a composition series of $\mathfrak{A}$ is trivial. The set of composition factors is unique up to isomorphism by the following theorem:

THEOREM 9.11. (Rao, Ray-Chaudhuri and Singhi (1982)) Let $\mathfrak{A}$ be a C-algebra whose structure constants and dual structure constants are non-negative. Let $\mathfrak{A} = \mathfrak{A}_0 \supset \mathfrak{A}_1 \supset \ldots \supset \mathfrak{A}_r$ and $\mathfrak{A} = \mathfrak{B}_0 \supset \mathfrak{B}_1 \supset \ldots \supset \mathfrak{B}_s$ be composition series of $\mathfrak{A}$. Then $r = s$, and $\mathfrak{A}_i/\mathfrak{A}_{i+1}$ and $\mathfrak{B}_{i^\sigma}/\mathfrak{B}_{i^\sigma+1}$ $(i = 0, 1, \ldots, r-1)$ are isomorphic as C-algebras by a suitable permutation σ of $\{0, 1, \ldots, s\}$.

<u>Proof</u>. We use the induction on $\dim \mathfrak{A}$. We may assume $\mathfrak{A}_1 \neq \mathfrak{B}_1$. Then $\mathfrak{A} = \mathfrak{A}_1\mathfrak{B}_1$ and $\mathfrak{A}/\mathfrak{A}_1 \cong \mathfrak{B}_1/(\mathfrak{B}_1 \cap \mathfrak{A}_1)$, $\mathfrak{A}/\mathfrak{B}_1 \cong \mathfrak{A}_1/(\mathfrak{A}_1 \cap \mathfrak{B}_1)$.

For a proper C-subalgebra $\mathfrak{B}$ of $\mathfrak{A}$, let $\mathrm{Comp}(\mathfrak{B})$ be the set of composition factors, allowing repetition. $\mathrm{Comp}(\mathfrak{B})$ is independent of the choice of composition series by the induction hypothesis. Then, allowing repetition,

$$\begin{aligned}
&\{\mathfrak{B}_i/\mathfrak{B}_{i+1} \mid 0 \le i \le s-1\} \\
&= \{\mathfrak{A}/\mathfrak{B}_1\} \cup \mathrm{Comp}(\mathfrak{B}_1) \\
&= \{\mathfrak{A}/\mathfrak{B}_1,\ \mathfrak{B}_1/(\mathfrak{B}_1 \cap \mathfrak{A}_1)\} \cup \mathrm{Comp}(\mathfrak{A}_1 \cap \mathfrak{B}_1) \\
&= \{\mathfrak{A}/\mathfrak{B}_1,\ \mathfrak{A}/\mathfrak{A}_1\} \cup \mathrm{Comp}(\mathfrak{A}_1 \cap \mathfrak{B}_1) \\
&= \{\mathfrak{A}/\mathfrak{A}_1,\ \mathfrak{A}_1/(\mathfrak{A}_1 \cap \mathfrak{B}_1)\} \cup \mathrm{Comp}(\mathfrak{A}_1 \cap \mathfrak{B}_1) \\
&= \{\mathfrak{A}/\mathfrak{A}_1\} \cup \mathrm{Comp}(\mathfrak{A}_1) \\
&= \{\mathfrak{A}_i/\mathfrak{A}_{i+1} \mid 0 \le i \le r-1\}. \qquad \square
\end{aligned}$$

REMARK. (4) Let $\mathfrak{X} = (X, \{R_i\}_{0 \le i \le d})$ be a commutative association scheme and $\mathfrak{A}$ the adjacency algebra of $\mathfrak{X}$. Let $\mathfrak{A} = \mathfrak{A}_0 \supset \mathfrak{A}_1 \supset \ldots \supset \mathfrak{A}_r$ be a composition series of $\mathfrak{A}$. Then there exists a sequence $\mathfrak{X} = \mathfrak{X}_0 \supset \mathfrak{X}_1 \supset \ldots \supset \mathfrak{X}_r = (\{x\}, R_0)$ of association subschemes of $\mathfrak{X}$ such that the quotients $\mathfrak{X}_i/\mathfrak{X}_{i+1}$ $(i = 0, 1, \ldots, r-1)$ have adjacency algebras $\mathfrak{A}_i/\mathfrak{A}_{i+1}$. In particular, $\mathfrak{X}_i/\mathfrak{X}_{i+1}$ are primitive. We may call the above sequence a composition series of $\mathfrak{X}$ and the quotients $\mathfrak{X}_i/\mathfrak{X}_{i+1}$ composition factors of $\mathfrak{X}$.

It would be very nice if the composition factors $\mathfrak{X}_i/\mathfrak{X}_{i+1}$ were unique up to isomorphism as association schemes, but this fails in general. For example, the generalized 6-gon with 126 vertices (see section 3.3, Question) is bipartite, but the bipartite halves are primitive and not isomorphic to each other.

Naturally, we can ask the following question.

QUESTION. (2) Let Σ be a system of imprimitivity of $\mathfrak{X}$. For $S_1, S_2 \in \Sigma$, when are the association subschemes

$(S_1, \{R_i\}_{0 \leq i \leq s})$ and $(S_2, \{R_i\}_{0 \leq i \leq s})$ isomorphic?

2.10 TANNAKA-KREIN DUALITY FOR NON-COMMUTATIVE ASSOCIATION SCHEMES

Let us start with the original Tannaka duality (Tannaka (1938)) for compact groups. Let G be a compact group. Let G' be the set of all finite dimensional representations of G:

$$G' = \{\mathbb{D} \mid \mathbb{D}: G \to GL(n,\mathbb{C}) \text{ a homomorphism as topological groups}\}.$$

We distinguish between equivalent but different representations. G' admits the following operations:

(i) $\mathbb{D}_1 \otimes \mathbb{D}_2$ (the tensor product),

(ii) $\mathbb{D}_1 + \mathbb{D}_2 = \begin{pmatrix} \mathbb{D}_1 & 0 \\ 0 & \mathbb{D}_2 \end{pmatrix}$ (the direct sum),

(iii) $P^{-1}\mathbb{D}P$ for every non-singular matrix P (an equivalent representation),

(iv) $\overline{\mathbb{D}}$ (the complex conjugate).

A representation of G' is a mapping $\mathbb{A}: \mathbb{D} \to A(\mathbb{D})$ from G' to matrices over $\mathbb{C}$ such that $\mathbb{D}$ and $A(\mathbb{D})$ are of the same degree and

(i) $A(\mathbb{D}_1 \otimes \mathbb{D}_2) = A(\mathbb{D}_1) \otimes A(\mathbb{D}_2)$,

(ii) $A(\mathbb{D}_1 + \mathbb{D}_2) = A(\mathbb{D}_1) + A(\mathbb{D}_2) := \begin{pmatrix} A(\mathbb{D}_1) & 0 \\ 0 & A(\mathbb{D}_2) \end{pmatrix}$,

(iii) $\mathbb{A}(P^{-1}\mathbb{D}P) = P^{-1}\mathbb{A}(\mathbb{D})P$,

(iv) $\mathbb{A}(\overline{\mathbb{D}}) = \overline{\mathbb{A}(\mathbb{D})}$.

Let G" be the set of all representations of G'. Define the product of $\mathbb{A}_1, \mathbb{A}_2 \in G''$ by $\mathbb{A}_1\mathbb{A}_2(\mathbb{D}) = \mathbb{A}_1(\mathbb{D})\mathbb{A}_2(\mathbb{D})$ and give G" the weakest topology under which the function $\hat{\mathbb{D}}: \mathbb{A} \to \mathbb{A}(\mathbb{D})$ from G" to the full matrix algebra $M_d(\mathbb{C})$ with d the degree of $\mathbb{D}$ is continuous for every $\mathbb{D} \in G'$. Then the product $\mathbb{A}_1\mathbb{A}_2$ is again a representation of G' and G" becomes a compact group.

THE TANNAKA DUALITY THEOREM. Let G be a compact group. Then G is isomorphic to G" as topological groups (by the correspondence of $x \in G$ to $\mathbb{A} \in G''$, where $\mathbb{A}(\mathbb{D}) = \mathbb{D}(x)$ for all $\mathbb{D} \in G'$).

Krein (1949) formulated the Tannaka duality theorem as follows. We shall first define a <u>Krein algebra</u> $\mathfrak{A}$ (or <u>a square block algebra</u>) (see Hewitt and Ross (1970), section 30). Let $\mathfrak{A}$ be a commutative algebra over $\mathbb{C}$ with multiplicative unit e and let there be a mapping $x \to \overline{x}$ from $\mathfrak{A}$ onto $\mathfrak{A}$ such that for all $x, y \in \mathfrak{A}$ and $\lambda \in \mathbb{C}$, the following identities hold: $\overline{x + y} = \overline{x} + \overline{y}$, $\overline{\lambda x} = \overline{\lambda}\,\overline{x}$, $\overline{xy} = \overline{x}\,\overline{y}$, $\overline{\overline{x}} = x$, where $\overline{\lambda}$ is the complex conjugate of λ. Let $\mathfrak{A}$ contain a basis (in the linear space sense) which is the union of pairwise disjoint sets $\{u_{jk}^{(\nu)} \mid 1 \le j \le n_\nu, 1 \le k \le n_\nu\}$, where ν runs through a non-empty index set N and each n_ν is a positive integer ($|N|$ may be finite or infinite). With each set $\{u_{jk}^{(\nu)} \mid 1 \le j \le n_\nu, 1 \le k \le n_\nu\}$, we associate the square matrix $U_\nu = (u_{jk}^{(\nu)})_{1 \le j \le n_\nu, 1 \le k \le n_\nu}$ whose entries are elements of $\mathfrak{A}$. U_ν is called a <u>block</u> and n_ν is called the <u>order</u> of the block U_ν. Set $\mathcal{U} = \{U_\nu \mid \nu \in N\}$. Let $\mathcal{U}$ satisfy the following properties:

(i) In $\mathcal{U}$, there is the 1-dimensional matrix

(e), where e is the identity element of $\mathfrak{A}$.

(ii) For each block U_ν of order n_ν, there exists a unitary matrix S of order n_ν such that $S^{-1}\overline{U}_\nu S \in \mathcal{U}$, where $\overline{U}_\nu = (\overline{u_{jk}^{(\nu)}})$ for $U_\nu = (u_{jk}^{(\nu)})$.

(iii) For each two blocks $U = U_{\nu_1}$, $V = U_{\nu_2} \in \mathcal{U}$ of order m and n, there exists a unitary matrix $S^{(U,V)}$ of degree mn such that

$$U \otimes V = (S^{(U,V)})^{-1}(W_1 \oplus \ldots \oplus W_r)S^{(U,V)}$$

for some $W_i \in \mathcal{U}$ $(i = 1, \ldots, r)$.

(iv) In the decomposition of $U \otimes V$ in (iii), the block (e) appears at most once, and it appears if and only if $V = S^{-1}\overline{U}S$ for some unitary matrix S.

(v) For each block $U_\nu = (u_{jk}^{(\nu)}) \in \mathcal{U}$,

$$U_\nu U_\nu^* = (\delta_{jk}e)_{1 \leq j \leq n_\nu,\, 1 \leq k \leq n_\nu} := \begin{pmatrix} e & & \\ & \ddots & \\ & & e \end{pmatrix},$$

where $U_\nu^* = {}^t\overline{U}_\nu = (\overline{u_{kj}^{(\nu)}})$.

Then $\mathfrak{A}$ is called a Krein algebra or a square block algebra.

REMARK. (1) In (iv), it is unnecessary to assume the latter half of (iv): the statement that the block (e) appears in the decomposition of $U \otimes V$ in (iii) if and only if $V = S^{-1}\overline{U}S$ for some unitary matrix S can be derived from the rest of the assumptions in (i),...,(v).

Let G be a compact group and let N be the set of all finite dimensional irreducible unitary representations of G up to equivalence. Note that every irreducible representation of a compact group is equivalent to a unitary representation of finite dimension. For each $\nu \in N$, we fix a unitary matrix representation $x \to (u_{jk}^{(\nu)}(x))_{1 \leq j \leq n_\nu, 1 \leq k \leq n_\nu}$ of G. Let U_ν denote the matrix $(u_{jk}^{(\nu)})_{1 \leq j \leq n_\nu, 1 \leq k \leq n_\nu}$. Let $\mathfrak{A} = \mathfrak{A}(G)$ be the set of all finite complex linear combinations of $u_{jk}^{(\nu)}$ ($\nu \in N$, $1 \leq j \leq n_\nu$, $1 \leq k \leq n_\nu$). For $f, g \in \mathfrak{A}$, define the product fg by pointwise multiplication and let $\overline{f}$ be the complex conjugate of f. Then $\mathfrak{A}$ is an algebra and $\{u_{jk}^{(\nu)} \mid \nu \in N, 1 \leq j \leq n_\nu, 1 \leq k \leq n_\nu\}$ is a basis of $\mathfrak{A}$. $\mathfrak{A}$ with blocks U_ν ($\nu \in N$) becomes a Krein algebra and is called <u>the Krein algebra of G</u>.

Given an arbitrary Krein algebra $\mathfrak{A}$, let $H_\mathfrak{A}$ be the set of all multiplicative linear functionals $\varphi: \mathfrak{A} \to \mathbb{C}$, i.e., $\varphi(xy) = \varphi(x)\varphi(y)$ and $\varphi \neq 0$. For $\varphi, \psi \in H_\mathfrak{A}$, define the product $\varphi\psi$ by

$$\varphi\psi(u_{jk}^{(\nu)}) = \sum_{r=1}^{n_\nu} \varphi(u_{jr}^{(\nu)})\psi(u_{rk}^{(\nu)}).$$

Then $\varphi\psi$ is an element of $H_\mathfrak{A}$. Let us give $H_\mathfrak{A}$ the weakest topology under which the function $\hat{x}: \varphi \to \varphi(x)$ from $H_\mathfrak{A}$ to $\mathbb{C}$ is continuous for every $x \in \mathfrak{A}$. Then $H_\mathfrak{A}$ becomes a topological group. Let $G_\mathfrak{A}$ be the set of multiplicative linear functional φ on $\mathfrak{A}$ such that $\varphi(\overline{x}) = \overline{\varphi(x)}$. Then $G_\mathfrak{A}$ becomes a compact subgroup of $H_\mathfrak{A}$.

THE KREIN DUALITY THEOREM. Let $\mathfrak{A}$ be a Krein algebra. Then $\mathfrak{A}$ is isomorphic to the Krein algebra $\mathfrak{A}(G_\mathfrak{A})$ of $G_\mathfrak{A}$ as Krein algebras.

THE TANNAKA-KREIN DULAITY THEOREM. Let G be a compact group and $\mathfrak{A} = \mathfrak{A}(G)$, the Krein algebra of G. Then G is isomorphic to $G_\mathfrak{A}$ as topological groups by the correspondence $x \in G$ to $\hat{x} \in G_{\mathfrak{A}}$, where $\hat{x}(f) = f(x)$ for all $f \in \mathfrak{A}$.

As an immediate consequence of the Tannaka-Krein Duality Theorem, compact groups are determined by their irreducible unitary representations.

We shall formulate the Tannaka-Krein duality for (non-commutative) association schemes. Let $\mathfrak{X} = (X, \{R_i\}_{0 \le i \le d})$ be an association scheme which is not necessarily commutative. Let $A_0, A_1, \ldots, A_d$ be the adjacency matrices of $\mathfrak{X}$ and let $\mathfrak{A}$ be the algebra spanned by $A_0, A_1, \ldots, A_d$ over $\mathbb{C}$ (the adjacency algebra or the Bose-Mesner algebra). Then $\dim_{\mathbb{C}} \mathfrak{A} = d + 1$.

We want to show $\mathfrak{A}$ is semi-simple. Let $\mathcal{J}$ be an arbitrary non-trivial left ideal of $\mathfrak{A}$. For $0 \neq A \in \mathcal{J}$, ${}^t\overline{A}A$ is an element of $\mathcal{J}$. In general, if $A \neq C$, ${}^t\overline{A}A$ is non-zero and a non-negative definite Hermitian matrix. So $({}^t\overline{A}A)^k$ is a non-zero element of $\mathcal{J}$ for every integer k, which contradicts the nilpotency of $\mathcal{J}$. Therefore $\mathfrak{A}$ has no non-trivial nilpotent left ideals, i.e., $\mathfrak{A}$ is semi-simple.

By Wedderburn's theorem, $\mathfrak{A}$ is a direct sum of subalgebras C_ν isomorphic to the full matrix algebra $M_{n_\nu}(\mathbb{C})$ over $\mathbb{C}$ $(0 \le \nu \le r)$:

$$\mathfrak{A} = C_0 \oplus C_1 \oplus \cdots \oplus C_r, \qquad C_\nu \cong M_{n_\nu}(C)$$

with $\dim_{\mathbb{C}} \mathfrak{A} = d + 1 = \sum_{\nu=0}^{r} n_\nu^2$. The algebra $\mathfrak{A}$ is completely reducible, i.e., there is a non-singular (unitary) matrix $S \in M_n(C)$ $(n = |X|)$ such that

$$S^{-1}AS = \mathrm{diag}(\underbrace{\Delta_0(A)}_{e_0}, \underbrace{\Delta_1(A), \ldots, \Delta_1(A)}_{e_1}, \ldots, \underbrace{\Delta_r(A), \ldots, \Delta_r(A)}_{e_r}) \tag{10.1}$$

for all $A \in \mathfrak{A}$, where $\Delta_0, \Delta_1, \ldots, \Delta_r$ are inequivalent irreducible representations of $\mathfrak{A}$, Δ_ν being afforded by C_ν and of degree n_ν. So we have $n = \sum_{\nu=0}^{r} e_\nu n_\nu$. Since any faithful representation of $\mathfrak{A}$ contains each of $\Delta_0, \Delta_1, \ldots, \Delta_r$ as a component, each e_ν is non-zero. The 1-dimensional space spanned by $(1,1,\ldots,1)$ is a common eigenspace of the basis $A_0, A_1, \ldots, A_d$ of $\mathfrak{A}$ and hence affords a representation of $\mathfrak{A}$ of degree 1. We may assume that Δ_0 is the above-defined representation of degree 1. Then $e_0 = 1$, because the matrix J whose entries are all one belongs to $\mathfrak{A}$ and $\Delta_0(J) = n$, while the eigenvalues of J are n and 0 with multiplicities 1 and $n - 1$ respectively.

We write

$$\Delta_\nu(A) = (u_{jk}^{(\nu)}(A))_{1 \leq j \leq n_\nu,\, 1 \leq k \leq n_\nu} \in M_{n_\nu}(\mathbb{C}) \qquad (10.2)$$

for $A \in \mathfrak{A}$. Let $Y = \{A_0, A_1, \ldots, A_d\}$. Since Y is a basis of $\mathfrak{A}$, the function $u_{jk}^{(\nu)}$ on $\mathfrak{A}$ is determined by the values $u_{jk}^{(\nu)}(A)$. Let $\mathfrak{A}^{\#}$ be the set of all complex linear combinations of $u_{jk}^{(\nu)}$ $(0 \leq \nu \leq r,\ 1 \leq j \leq n_\nu,$ $1 \leq k \leq n_\nu)$. Then by a theorem of Frobenius and Schur (see Curtis and Reiner (1962, section 27)), or Theorem 10.2 below, the $u_{jk}^{(\nu)}$ are linearly independent as functions on Y, and so by $d + 1 = \sum_{\nu=0}^{r} n_\nu^2$, $\mathfrak{A}^{\#}$ is the set of all functions on Y. $\mathfrak{A}^{\#}$ becomes a commutative algebra with respect to pointwise multiplication. The algebra $\mathfrak{A}^{\#}$ together with the basis $\{u_{ij}^{(\nu)}\}$ plays the same role as the Krein algebra in the Tannaka-Krein dulaity.

For an arbitrary algebra $\mathcal{A}$, a non-zero linear function φ from $\mathcal{A}$ to $\mathbb{C}$ is said to be a <u>linear</u>

<u>multiplicative</u> <u>functional</u> if

$$\varphi(xy) = \varphi(x)\varphi(y)$$

for all $x,y \in \mathcal{A}$. Let $H_{\mathcal{A}}$ denote the set of all linear multiplicative functionals of $\mathcal{A}$.

Now our duality theorem is formulated as follows.

THEOREM 10.1. Let $\mathfrak{A} = \langle A_0,A_1,\ldots,A_d\rangle$ be the adjacency algebra of a (non-commutative) association scheme of class d. Let $\mathfrak{A}^{\#}$ with the specified basis u_{jk}^{ν} $(0 \leq \nu \leq r,\ 1 \leq j \leq n_{\nu},\ 1 \leq k \leq n_k)$ be the above-defined algebra for $\mathfrak{A}$. Let $H_{\mathfrak{A}^{\#}}$ be the set of all linear multiplicative functionals of $\mathfrak{A}^{\#}$. Then $H_{\mathfrak{A}^{\#}}$ is identified with the set $Y = \{A_0,A_1,\ldots,A_d\}$ by the correspondence of $\varphi_i \in H_{\mathfrak{A}^{\#}}$ to $A_i \in Y$, where $\varphi_i(u) = u(A_i)$ for all $u \in \mathfrak{A}^{\#}$. For φ and ψ in $H_{\mathfrak{A}^{\#}}$, let us define the product $\varphi\psi$ by

$$\varphi\psi(u_{jk}^{(\nu)}) = \sum_{h=1}^{n_\nu} \varphi(u_{jh}^{(\nu)})\psi(u_{hk}^{(\nu)}).$$

Then $\varphi\psi$ is a linear combination of elements in $H_{\mathfrak{A}^{\#}}$, and this algebra structure of $H_{\mathfrak{A}^{\#}}$ is isomorphic to that of Y in the original adjacency algebra $\mathfrak{A}$. Precisely speaking, if A_i, A_j, A_k are elements of Y corresponding to φ_i, φ_j, φ_k respectively, then

$$A_iA_j = \sum_{k=0}^{d} p_{ij}^{k}A_k$$

if and only if

$$\varphi_i\varphi_j = \sum_{k=0}^{d} p_{ij}^k\varphi_k.$$

Proof. (1) For any $A_i \in Y$, the point evaluation $u \to u(A_i)$ $(\forall u \in \mathfrak{A}^\#)$ is a linear multiplicative functional of $\mathfrak{A}^\#$. We want to show the converse, namely for any linear multiplicative functional φ of $\mathfrak{A}^\#$, there exists a unique $A_i \in Y$ such that $\varphi(u) = u(A_i)$ $(\forall u \in \mathfrak{A}^\#)$.

Let u_i be the function from Y to $\mathbb{C}$ defined by $u_i(A_j) = \delta_{ij}$ $(\forall A_j \in Y)$. Then $u_0, u_1, \ldots, u_d$ are the primitive idempotents of $\mathfrak{A}^\#$. For any ideal $\mathcal{J}$ of $\mathfrak{A}^\#$, there exists a subset L of $\{0,1,\ldots,d\}$ such that $\mathcal{J}$ is generated by u_i $(i \in L)$. So $\mathcal{J} = \langle u_i \mid i \in L\rangle = \{u \in \mathfrak{A}^\# \mid u(A_j) = 0 \ (\forall j \notin L)\}$, i.e., there is a one-to-one correspondence between the ideals $\mathcal{J}$ of $\mathfrak{A}^\#$ and the subsets L of $\{0,1,\ldots,d\}$ in such a way that $\mathcal{J}$ is the set of functions $u \in \mathfrak{A}^\#$ which vanish outside L. For $i \in \{0,1,\ldots,d\}$, let $\mathfrak{m}_i = \{u \in \mathfrak{A}^\# \mid u(A_i) = 0\}$. Then $\mathfrak{m}_0, \mathfrak{m}_1, \ldots, \mathfrak{m}_d$ are all the maximal ideals of $\mathfrak{A}^\#$.

Let φ be a linear multiplicative functional of $\mathfrak{A}^\#$. Then the kernel of φ is a maximal ideal of $\mathfrak{A}^\#$ and so coincides with some $\mathfrak{m}_i$. This implies φ is the point evaluation at A_i: $\varphi(u) = u(A_i)$ $(\forall u \in \mathfrak{A}^\#)$. (In the original Tannaka-Krein duality theorem, owing to the complexity of topology, we need a lengthy and elaborate argument to show that φ with $\varphi(\overline{u}) = \overline{\varphi(u)}$ is a point evaluation. The reader is referred to Appendix C in (Hewitt and Ross (1963)).

(2) Let $\varphi_i \in H_{\mathfrak{A}^\#}$ be the point evaluation at $A_i \in Y$: $\varphi_i(u) = u(A_i)$ $(\forall u \in \mathfrak{A}^\#)$. We want to show that

$$\varphi_i\varphi_j = \sum_{k=0}^{d} p_{ij}^k\varphi_k,$$ where p_{ij}^k's are the intersection numbers of $\mathfrak{A}$.

Apply the νth irreducible representation $\Delta_\nu = (u_{\alpha\beta}^{(\nu)})$ to $A_iA_j = \sum_{k=0}^{d} p_{ij}^k A_k$ and consider the (α,β) entry. The right hand side is

$$\sum_{k=0}^{d} p_{ij}^k u_{\alpha\beta}^{(\nu)}(A_k) = \sum_{k=0}^{d} p_{ij}^k \varphi_k(u_{\alpha\beta}^{(\nu)})$$

by the definition of φ_k, while the left hand side is

$$(\Delta_\nu(A_i)\Delta_\nu(A_j))_{\alpha\beta} = \sum_{\gamma=1}^{n_\nu} u_{\alpha\gamma}^{(\nu)}(A_i)u_{\gamma\beta}^{(\nu)}(A_j)$$

$$= \sum_{\gamma=1}^{n_\nu} \varphi_i(u_{\alpha\gamma}^{(\nu)})\varphi_j(u_{\gamma\beta}^{(\nu)})$$

$$= \varphi_i\varphi_j(u_{\alpha\beta}^{(\nu)})$$

by the definition of the product $\varphi_i\varphi_j$. Thus, we get $\sum_{k=0}^{d} p_{ij}^k\varphi_k = \varphi_i\varphi_j$. This completes the proof of Theorem 10.1. □

In the following we shall formulate the orthogonal relations for (non-commutative) association schemes. We introduce the following notation. Let

$$Z = \text{diag}\{e_0 I_{n_0^2}, e_1 I_{n_1^2}, \ldots, e_r I_{n_r^2}\}$$

$$(\text{size } \sum_{\nu=0}^{r} n_\nu^2 = d + 1),$$

where I_ℓ is the identity matrix of size ℓ, and

$$T = \mathrm{diag}\{k_0, k_1, \ldots, k_d\},$$

where k_ℓ is the ℓth valency of the (non-commutative) association scheme $\mathfrak{X} = (X, \{R_i\}_{0 \le i \le d})$, i.e., $k_\ell = p^0_{\ell\ell'}$. Also, let

$$F = (u^{(\nu)}_{\alpha\beta}(A_i))_{(\nu,\alpha,\beta),i} \qquad (\sum_{\nu=0}^{r} n_\nu^2 = d+1)$$

and

$$F^* = (u^{(\nu)}_{\beta\alpha}(A_{i'}))_{(\nu,\alpha,\beta),i},$$

where the triples (ν,α,β) are the row indices in their lexicographic order $(0 \le \nu \le r,\ 1 \le \alpha \le n_\nu,\ 1 \le \beta \le n_\nu)$ and the i's are the column indices $(0 \le i \le d)$. Then we have the following:

THEOREM 10.2. (The Schur Relations)

(i) ${}^tFZF^* = nT$ $(n = |X|)$, i.e.,

$$\sum_{(\nu,\alpha,\beta)} e_\nu u^{(\nu)}_{\alpha\beta}(A_i) u^{(\nu)}_{\beta\alpha}(A_{j'}) = nk_i\delta_{ij}.$$

(ii) $F^*T^{-1}\,{}^tF = nZ^{-1}$, i.e.,

$$\sum_{i=0}^{d} \frac{1}{k_i} u^{(\mu)}_{\beta q}(A_{i'}) u^{(\nu)}_{p\alpha}(A_i) = \frac{n}{e_\nu}\delta_{\nu\mu}\delta_{pq}\delta_{\alpha\beta}.$$

Proof. By (10.1), $S\,\mathrm{diag}(\underbrace{\Delta_0}_{e_0}, \underbrace{\Delta_1, \ldots, \Delta_1}_{e_1}, \ldots, \underbrace{\Delta_r, \ldots, \Delta_r}_{e_r})S^{-1}$

is the identity map from $\mathfrak{A}$ to $\mathfrak{A}$. Applying the above map to $A_iA_{j'}$ and taking the trace, we get

$$\mathrm{tr}(A_iA_{j'}) = \sum_{\nu=0}^{r} e_\nu \mathrm{tr}(\Delta_\nu(A_iA_{j'})).$$

Since $tr(A_iA_{j'}) = nk_i\delta_{ij}$ and

$$tr(\Delta_\nu(A_iA_{j'})) = \sum_{\alpha,\beta=1}^{n_\nu} u_{\alpha\beta}^{(\nu)}(A_i)u_{\beta\alpha}^{(\nu)}(A_{j'}),$$

we get the relation (i).

Since ${}^tFZF^* = nT$ implies $F^*T^{-1}\,{}^tF = nZ^{-1}$, we get the relation (ii). □

Compare Theorem 10.2 with Theorem 3.5.

COROLLARY 10.3. $u_{jk}^{(\nu)}$ $(0 \leq \nu \leq r,\ 1 \leq j \leq n_\nu,\ 1 \leq k \leq n_\nu)$ are linearly independent as functions on $Y = \{A_0, A_1, \ldots, A_d\}$.

Proof. By Theorem 10.2, the matrix F is non-singular. This implies the linear independence of $\{u_{jk}^{(\nu)}\}$. □

THEOREM 10.4.

$$p_{ij}^{\ell} = \frac{1}{nk_\ell}\sum_{\nu=0}^{r}\sum_{\alpha,\beta,\gamma=1}^{n_\nu} e_\nu u_{\alpha\beta}^{(\nu)}(A_i)u_{\beta\gamma}^{(\nu)}(A_j)u_{\gamma\alpha}^{(\nu)}(A_{\ell'}).$$

Proof. Apply $S\,\mathrm{diag}(\underbrace{\Delta_0}_{e_0},\underbrace{\Delta_1,\ldots,\Delta_1}_{e_1},\ldots,\underbrace{\Delta_r,\ldots,\Delta_r}_{e_r})S^{-1}$,

which is the identity map, to $A_iA_jA_{\ell'} = \sum_{m=0}^{d} p_{ij}^{m}A_mA_{\ell'}$ and take the trace. □

Compare Theorem 10.4 with Theorem 3.6.

REMARKS. (2) If the adjacency algebra $\mathfrak{A}$ is commutative, then $n_\nu = 1$ for all $\nu = 0, 1, \ldots, r$ and we get the Kawada-Delsarte duality. Exactly speaking, we must twist the duality mentioned here a bit to get the exact form of the Kawada-Delsarte duality. Namely, let $\mathfrak{A}^\# = \langle\tilde{u}_{\alpha\beta}^{(\nu)}\rangle$

with $\tilde{u}^{(\nu)}_{\alpha\beta}$ defined by $\tilde{u}^{(\nu)}_{\alpha\beta}(A_i) = \frac{e_\nu}{k_i}\overline{u^{(\nu)}_{\alpha\beta}(A_i)}$ (instead of $\mathfrak{A}^\# = \langle u^{(\nu)}_{\alpha\beta}\rangle$) and let $H_{\mathfrak{A}^\#} = \langle\hat{\varphi}_i\rangle$ with $\hat{\varphi}_i$ defined by $\hat{\varphi}_i(\tilde{u}^{(\nu)}_{\alpha\beta}) = \frac{k_i}{e_\nu}\overline{\tilde{u}^{(\nu)}_{\alpha\beta}(A_i)}$ (instead of $H_{\mathfrak{A}^\#} = \langle\varphi_i\rangle$ with φ_i defined by $\varphi_i(u^{(\nu)}_{\alpha\beta}) = u^{(\nu)}_{\alpha\beta}(A_i)$).

(3) Let G be a finite group and let H be a subgroup of G. Let $\mathfrak{X}$ be the (non-commutative) association scheme attached to the transitive permutation group G on $H\backslash G$. The adjacency algebra $\mathfrak{A}$ of $\mathfrak{X}$ is isomorphic to the Hecke algebra $C(H\backslash G/H)$ (cf. section 2.11). The original Tannaka-Krein duality is essentially our duality for $C(H\backslash G/H)$ when we specialize to $H = 1$.

(4) As Kawada (1942) proved his duality theorem for C-algebras, our duality theorem can also be formulated for algebras which do not necessarily come from a (non-commutative) association scheme.

QUESTION. (1) Formulate similar duality theorems for $H\backslash G/H$, where G is a non-finite compact topological group G and H is a closed subgroup of G.

REMARKS. (5) We want to discuss imprimitivity and quotients for a non-commutative association scheme $\mathfrak{X} = (X, \{R_i\}_{0\le i\le d})$. Let A_i $(i = 0, 1, \ldots, d)$ be the adjacency matrices of $\mathfrak{X}$ and let p^k_{ij} and k_i be the intersection numbers and valencies of $\mathfrak{X}$. As in Theorem 9.3, the following three conditions are equivalent.

(i) $A_i \sim \begin{pmatrix} * & * \\ 0 & * \end{pmatrix}$ for some i $(i \ne 0)$.

(ii) $\sum_{\alpha\in T} A_\alpha \sim I_q \otimes J_p$ for some $T \subset \{0,1,\ldots,d\}$ such that $0 \in T$ and $2 \le |T| \le d - 1$, where $p = \sum_{\alpha\in T} k_\alpha$ and $q = |X|/p$.

(iii) $p_{ij}^k = 0$ $(i,j \in T,\ k \notin T)$ for some $T \subset \{0,1,\dots,d\}$ such that $0 \in T$ and $2 \leq |T| \leq d-1$.

The subsets T in (ii) and (iii) coincide. If one of the above conditions holds, $\mathfrak{X}$ is called imprimitive. Then $\bigcup_{\alpha \in T} R_\alpha$ becomes an equivalence relation and the set Σ of equivalence classes is called the system of imprimitivity. For $S \in \Sigma$, $(S, \{R_\alpha\}_{\alpha \in T})$ becomes an association subscheme of $\mathfrak{X}$.

Let T be the set in the condition (ii) and let Σ be the system of imprimitivity with respect to T. The quotient association scheme on Σ can be defined only when Σ is normal. Σ is said to be normal, if the statement of (9.7) holds for any distinct $S_1, S_2 \in \Sigma$, or equivalently if

$$A_i\Big(\sum_{\alpha \in T} A_\alpha\Big) = \Big(\sum_{\alpha \in T} A_\alpha\Big)A_i \tag{10.3}$$

for all $i = 0, 1, \dots, d$. (Note that the (x,y) entry of (S_1,S_2) block in $A_i(\sum_{\alpha \in T} A_\alpha)$ (respectively $(\sum_{\alpha \in T} A_\alpha)A_i$) is the xth row sum (respectively y^{th} column sum) of the (S_1,S_2) block in A_i, which is either 0 or $\lambda = \sum_{\alpha \in T} p_{i\alpha}^i$ (respectively either 0 or $\lambda = \sum_{\alpha \in T} p_{i'\alpha}^{i'}$).) As in Theorem 9.4, let $\sim$ be the equivalence relation on $\{0,1,\dots,d\}$ defined by $i \sim j$ if and only if $p_{i\alpha}^j \neq 0$ for some $\alpha \in T$, and let $T_0 = T, T_1, \dots, T_t$ be the equivalence classes. If T is normal, then we have

$$A_i\Big(\sum_{\alpha \in T} A_\alpha\Big) = \lambda\Big(\sum_{\beta \in T_i} A_\beta\Big) \tag{10.4}$$

for some λ, where T_i is the equivalence class containing i. Let $D_i = \sum_{\alpha \in T_i} A_\alpha$. If T is normal, then by the same proof as in Theorem 9.4, we get that

$$D_i = \frac{1}{p} \iota(D_i) \otimes J_p, \qquad (10.5)$$

arranging the rows and columns according to the blocks of Σ, and $\frac{1}{p} \iota(D_i)$ $(0 \leq i \leq t)$ are the incidence matrices of an association scheme on Σ. This is called the <u>quotient association scheme</u> of $\mathfrak{X}$. If we assume an imprimitive group action on $\mathfrak{X}$ with Σ the system of imprimitivity, then Σ is always normal and the quotient association scheme is the one obtained by the group action on Σ.

(6) To generalize Remark (5), let us consider quotients of a non-commutative C-algebra $\mathfrak{A} = \langle x_i \mid 0 \leq i \leq d \rangle$, i.e., a C-algebra without the commutativity of multiplication. A C-subalgebra $\langle x_i \mid i \in T \rangle$ $(0 \in T)$ is said to be normal if (10.3) holds, replacing A_i and A_α by x_i and x_α. If (10.4) holds, then the quotient C-algebra $\mathfrak{A}/\langle x_i \mid i \in T \rangle$ can be defined to be $\langle \frac{1}{p} x_{T_i} \mid 0 \leq i \leq t \rangle$ with $p = \sum_{\alpha \in T} k_\alpha$ and $x_{T_i} = \sum_{\alpha \in T_i} x_\alpha$. Conversely, if $\langle \frac{1}{p} x_{T_i} \mid 0 \leq i \leq t \rangle$ is a (non-commutative) C-algebra, then (10.4) must hold. For commutative C-algebras, (10.3) automatically holds and the conditions $p_{ij}^k \geq 0$ and $q_{ij}^k \geq 0$ imply (10.4) (Theorem 9.9), but unfortunately so far we cannot find a nice condition like the Krein condition $q_{ij}^k \geq 0$ even for the adjacency algebra $\langle A_i \mid 0 \leq i \leq d \rangle$ of a non-commutative association scheme. The trouble comes from the fact that the primitive idempotents of $\langle A_i \mid 0 \leq i \leq d \rangle$ are no longer determined uniquely ($\langle A_i \mid 0 \leq i \leq d \rangle$ is semi-simple) nor are non-negative definite Hermitian matrices. So we ask the following question:

QUESTIONS. (2) For non-commutative C-algebras, does there exist any condition (like the Krein condition) which guarantees the existence of quotients by normal C-subalgebras?

(3) Is there any more reasonable definition of non-commutative C-algebras such that quotients always exist for normal C-subalgebras?

2.11 SPHERICAL FUNCTIONS OF MULTIPLICITY-FREE PERMUTATION REPRESENTATIONS

Let G be a transitive permutation group on a finite set Ω ($|\Omega| = n$) and let H be the stabilizer of α ($\alpha \in \Omega$). In this section, we always assume the permutation character 1_H^G is multiplicity free:

$$1_H^G = \chi_0 + \chi_1 + \cdots + \chi_d, \tag{11.1}$$

where $1_G = \chi_0$ and $\chi_0, \chi_1, \ldots, \chi_d$ are distinct irreducible characters of G. Let $C(H\backslash G)$ be the set of left H-invariant functions from G to the complex number field $\mathbb{C}$:

$$C(H\backslash G) = \{F\colon G \to \mathbb{C} \mid f(ax) = f(x) \quad \text{for all } a \in H\}. \tag{11.2}$$

Identifying the cosets $H\backslash G$ with Ω, $C(H\backslash G)$ is the set of functions from Ω to $\mathbb{C}$. $C(H\backslash G)$ has a vector space structure in the natural way and becomes a Hermitian space with the form

$$\langle f,g\rangle = \frac{1}{|G|}\sum_{x \in G} f(x)\overline{g(x)}. \tag{11.3}$$

G acts on $C(H\backslash G)$ by

$$f^a(x) = f(xa^{-1}). \tag{11.4}$$

This G-action preserves the Hermitian form:

$$\langle f^a, g^a \rangle = \langle f, g \rangle. \tag{11.5}$$

For a subset S of G, φ_S denotes the characteristic function of S:

$$\varphi_S(x) = \begin{cases} 1 & \text{if } x \in S, \\ 0 & \text{if } x \in G - S. \end{cases} \tag{11.6}$$

$\{\sqrt{n}\,\varphi_{Ha}\}$ is an orthonormal basis of $C(H\backslash G)$ and G permutes $\{\sqrt{n}\,\varphi_{Ha}\}$ transitively.

Since $\varphi_{Hx}^a = \varphi_{Hxa}$, the permutations of G on $H\backslash G$ and $C(H\backslash G)$ are isomorphic by the correspondence of Hx to φ_{Hx}. 1_H^G is the character of G afforded by $C(H\backslash G)$. $C(H\backslash G)$ is decomposed into a direct sum of mutually orthogonal irreducible G-spaces V_i:

$$C(H\backslash G) = V_0 + V_1 + \cdots + V_d, \tag{11.7}$$

$$V_i \perp V_j \qquad (i \neq j),$$

where V_i affords the character χ_i of G. The elements of V_i are called <u>spherical functions</u> belonging to χ_i.

PROPOSITION 11.1. For each χ_i, there exists a unique non-zero spherical function f belonging to χ_i up to scalar multiple such that $f^a = f$ for all $a \in H$. Such a function f is called a <u>zonal spherical function</u> belonging to χ_i.

Proof. Decompose V_i into a sum of irreducible H-spaces. Then the identity representation of H appears only once because $(1_H,\chi_i)_H = (1_H^G,\chi_i) = 1$ by the Frobenius reciprocity. □

Let $\{f_1,f_2,\ldots,f_m\}$ be an orthonormal basis of V_i where f_1 is a zonal spherical function. Let $\mathbb{F}$: $a \to F(a)$ be the unitary matrix representation of G with respect to the basis $\{f_j \mid 1 \le j \le m\}$ and $F_{ij}(a)$ the (i,j) entry of $F(a)$:

$$F_{ij}(a) = \langle f_i^a, f_j \rangle . \tag{11.8}$$

(The action of G on $C(H\backslash G)$ is from the right.)

THEOREM 11.2. Let $\mathbb{F}$: $a \to (F_{ij}(a))$ be the representation of G afforded by V_i with respect to an orthonormal basis $f_1 \ldots f_m$ of V_i, where $f_1^a = f_1$ for $a \in H$. Then $\{\sqrt{m}\ \overline{F_{1j}} \mid 1 \le j \le m\}$ is an orthonormal basis of V_i and $\overline{F}_{11} = \lambda f_1$ with $|\lambda| = 1/\sqrt{m}$.

Proof. Let Φ: $f \to \Phi_f$ be a linear map from V_i to $C(H\backslash G)$ given by

$$\Phi_f(x) = \langle f, f_1^x \rangle . \tag{11.9}$$

Then Φ is injective because $\{f_1^x \mid x \in G\}$ spans V_i by the irreducibility of V_i. Since Φ commutes with the action of G, Φ is a scalar map by Theorem 3.2 of Chapter I (a corollary to Schur's lemma): $\Phi_f = \lambda f$. Therefore, by (11.8) and (11.9), we have $\overline{F}_{1j} = \lambda f_j$ and

$$\lambda\overline{\lambda} = \langle F_{1j}, F_{1j} \rangle = \frac{1}{|G|} \sum_{x \in G} F_{1j}(x)\overline{F_{1j}(x)} .$$

So

$$m\lambda\bar{\lambda} = \frac{1}{|G|} \sum_{x \in G} \sum_{j=1}^{m} F_{1j}(x)\overline{F_{1j}(x)} = \frac{1}{|G|} \sum_{x \in G} 1 = 1,$$

because $F(x)$ is a unitary matrix. Thus $|\lambda| = \frac{1}{\sqrt{m}}$. □

Let $\omega = \bar{F}_{11} = \lambda f_1$, i.e.,

$$\omega(x) = \langle f_1, f_1^x \rangle, \tag{11.10}$$

where $f_1 \in V_i$, $\langle f_1, f_1 \rangle = 1$ and $f_1^a = f_1$ for all $a \in H$. Then ω is a spherical function belonging to χ_i characterized by

$$\omega(1) = 1 \quad (1 \text{ is the identity of } G) \tag{11.11}$$

$$\omega^a = \omega \quad \text{for all } a \in H.$$

ω is called <u>the zonal spherical function</u> belonging to χ_i. Let ω_i, ω_j be the zonal spherical functions belonging to χ_i and χ_j. Since V_i and V_j are perpendicular, we have

$$\langle \omega_i, \omega_j \rangle = \frac{1}{m_i} \delta_{ij}, \tag{11.12}$$

where $m_i = \dim V_i$. Let $C(H\backslash G/H)$ be the set of left and right H-invariant functions from G to $\mathbb{C}$:

$$C(H\backslash G/H) = \{f: G \to \mathbb{C} \mid f(axb) = f(x) \text{ for all } a,b \in H\}. \tag{11.13}$$

Then the zonal spherical functions $\omega_0, \omega_1, \ldots, \omega_d$ form an orthogonal basis of $C(H\backslash G/H)$. $C(H\backslash G/H)$ can be regarded as the set of functions from the double cosets $H\backslash G/H$ to $\mathbb{C}$.

THEOREM 11.3. Let ω_i be the zonal spherical function belonging to χ_i. Then for $x \in G$,

$$\omega_i(x) = \frac{1}{|H|} \sum_{a \in H} \chi_i(ax^{-1}).$$

Proof. Let $\{f_1, f_2, \ldots, f_m\}$ be an orthonormal basis of V_i, where f_1 is a zonal spherical function. For $a \in H$,

$$\chi_i(ax^{-1}) = \sum_{j=1}^{m} \langle f_j^{ax^{-1}}, f_j \rangle$$

by (11.8). Since $f_j^a = \sum_{k=1}^{m} \langle f_j^a, f_k \rangle f_k$,

$$\chi_i(ax^{-1}) = \sum_{j,k=1}^{m} \langle f_j^a, f_k \rangle \langle f_k^{x^{-1}}, f_j \rangle. \tag{11.14}$$

We claim

$$\frac{1}{|H|} \sum_{a \in H} \langle f_j^a, f_k \rangle = \begin{cases} 1 & \text{if } j = k = 1, \\ 0 & \text{otherwise.} \end{cases} \tag{11.15}$$

Since $\sum_{a \in H} f_j^a$ is H-invariant, $\sum_{a \in H} f_j^a = \lambda f_1$ for some $\lambda \in \mathbb{C}$ by Proposition 11.1. If $j = 1$, then $\lambda = |H|$, so (11.15) is true for this case. If $j \neq 1$, then $\langle f_j^a, f_1 \rangle = 0$, because the subspace of V_i orthogonal to f_1 is preserved by the H-action. So if $j \neq 1$, then $\lambda = 0$. Thus the claim (11.15) is proved.

If we sum up (11.14) over $a \in H$, then by (11.15), only the terms for $j = k = 1$ contribute to the sum, and so

$$\frac{1}{|H|}\sum_{a\in H} \chi_i(ax^{-1}) = \langle f_1^{x^{-1}}, f_1\rangle$$

$$= \omega_i(x) \qquad \text{(by (11.10))}. \qquad \square$$

Let $C(G)$ be the set of functions from G to $\mathbb{C}$. For $f,g \in C(G)$, define the <u>convolution product</u> $f * g$ of f and g by

$$(f * g)(x) = \sum_{y\in G} f(xy^{-1})g(y) . \tag{11.16}$$

Verify the associative law holds for the convolution, i.e.,

$$(f * g) * h = f * (g * h).$$

PROPOSITION 11.4. $C(G)$ becomes an algebra with respect to the convolution product. This algebra $C(G)$ is isomorphic to the group ring of G over $\mathbb{C}$ by the correspondence of φ_x to $x \in G$, where

$$\varphi_x(y) = \begin{cases} 1 & \text{if } x = y, \\ 0 & \text{otherwise.} \end{cases}$$

<u>Proof</u>. Check $\varphi_x * \varphi_y = \varphi_{xy}$. $\square$

$C(H\backslash G)$ and $C(H\backslash G/H)$ are closed under the convolution product. $C(H\backslash G/H)$ is called the <u>Hecke ring</u> (the Hecke algebra). The identity of $C(H\backslash G/H)$ is $\frac{1}{|H|}\varphi_H$.

For $\varphi \in C(H\backslash G/H)$, define a linear map I_φ from $C(H\backslash G)$ to $C(H\backslash G)$ by

$$I_\varphi(f) = \varphi * f . \tag{11.17}$$

By the associativity of the convolution,

$$I_{\varphi * \psi} = I_\varphi I_\psi . \tag{11.18}$$

Furthermore I_φ commutes with the G-action, i.e.,

$$I_\varphi(f^a) = I_\varphi(f)^a \quad \text{for} \quad a \in G. \tag{11.19}$$

THEOREM 11.5. (i) Let $\mathscr{I} = \{I_\varphi \mid \varphi \in C(H\backslash G/H)\} \subseteq \mathrm{Hom}(C(H\backslash G), C(H\backslash G))$. Then the Hecke ring $C(H\backslash G/H)$ is isomorphic to $\mathscr{I}$ as algebras by the correspondence of φ to I_φ. In particular, let $G = Ha_0H \cup Ha_1H \cup \ldots \cup Ha_dH$ with $a_0 = 1$ and let $\varphi_i = \frac{1}{|H|} \varphi_{Ha_iH}$. Then $\{I_{\varphi_i} \mid 0 \leq i \leq d\}$ is a basis of $\mathscr{I}$.

(ii) Consider the permutation representation of G on $\{\varphi_{Ha} \mid a \in G\}$ and let A_i be the ith adjacency matrix corresponding to the H-orbit $\{\varphi_{Ha} \mid Ha \subseteq Ha_iH\}$. Then the matrix expression of I_{φ_i} with respect to the basis $\{\varphi_{Ha}\}$ is the i'^{th} adjacency matrix. So $\mathscr{I}$ is the adjacency algebra and $\mathscr{I}$ is the set of all linear maps from $C(H\backslash G)$ to $C(H\backslash G)$ which commute with the action of G, i.e., $\mathscr{I} = \mathrm{Hom}_G(C(H\backslash G), C(H\backslash G))$.

(iii) Spherical functions are simultaneous eigenfunctions of $\mathscr{I}$. Keeping the notation of (11.7), let $I_\varphi(f) = \Delta_i(I_\varphi)f$ for $0 \neq f \in V_i$. Then $\Delta_0, \Delta_1, \ldots, \Delta_d$ are the set of non-isomorphic linear representations of $\mathscr{I}$.

<u>Proof</u>. (i) The correspondence is an algebra isomorphism by (11.18), and it is injective because $I_\varphi(\frac{1}{|H|}\varphi_H) = \varphi * \frac{1}{|H|}\varphi_H = \varphi \neq 0$ for $\varphi \neq 0$.

(ii) By Theorem 1.3 and (11.19), I_{φ_i} belongs to the adjacency algebra of G on $\{\varphi_{Ha}\}$. Since $I_{\varphi_i}(\varphi_H) = \varphi_{Ha_iH}$, the first column of I_{φ_i} coincides with that of the i'^{th} adjacency matrix in the matrix expression with respect to the basis $\{\varphi_{Ha}\}$. Since the elements of the adjacency algebra are determined by the first columns, I_{φ_i} is the i'^{th} adjacency matrix.

(iii) By Schur's lemma, each element of $\mathcal{A}$ is a scalar map on V_i. The latter half of the claim follows from Proposition 5.4. □

THEOREM 11.6. Let ω be the zonal spherical function belonging to χ_i. Then for $\varphi \in C(H\backslash G/H)$,

$$\varphi * \omega = \omega * \varphi = |G|\hat{\omega}(\varphi)\omega,$$

where $\hat{\omega}(\varphi) = \frac{1}{|G|} \sum_{x \in G} \varphi(x^{-1})\omega(x) = \langle\varphi,\omega\rangle$.

<u>Proof</u>. By Theorem 11.5(iii), $I_\varphi(\omega) = \lambda\omega$ for some $\lambda \in C$. By (11.11), $\omega(1) = 1$ and hence $\lambda = I_\varphi(\omega)(1) = (\varphi * \omega)(1) = \sum_{x \in G} \varphi(x)\omega(x^{-1})$. By (11.10),

$$\omega(x^{-1}) = \overline{\omega(x)}. \qquad (11.20)$$

So $\lambda = |G|\langle\varphi,\omega\rangle$. The Hecke ring $C(H\backslash G/H)$ is isomorphic to the adjacency algebra $\mathcal{A}$, which is commutative. So $\varphi * \omega = \omega * \varphi$. □

COROLLARY 11.7. (i) Let ω_i be the zonal spherical function belonging to χ_i. Then $\frac{m_i}{|G|}\omega_i$ $(0 \leq i \leq d)$ are the primitive idempotents of the Hecek ring $C(H\backslash G/H)$, where m_i is the degree of χ_i.

(ii) Let $(p_i(j))$ be the first eigenmatrix of G on $\{\varphi_{Ha}\}$. Then

$$p_{i'}(j) = \frac{1}{|H|} \sum_{x \in Ha_iH} \chi_j(x).$$

Proof. (i) By (11.12) and Theorem 11.6, this is trivial.

(ii) We keep the notation of Theorem 11.5. By Proposition 5.4, $p_{i'}(j) = \Delta_j(I_{\varphi_i})$. Since $I_{\varphi_i}(\omega_j) = \varphi_i * \omega_j = |G|\langle\varphi_i,\omega_j\rangle\omega_j$, the eigenvalue $\Delta_j(I_{\varphi_i})$ of I_{φ_i} on V_j is $|G|\langle\varphi_i,\omega_j\rangle$. Thus,

$$\begin{aligned} p_{i'}(j) &= |G|\langle\varphi_i,\omega_j\rangle \\ &= \sum_{x \in G} \varphi_i(x)\omega_j(x^{-1}) \\ &= \sum_{x \in Ha_iH} \frac{1}{|H|}\omega_j(x^{-1}) \\ &= \frac{1}{|H|} \sum_{\substack{x \in Ha_iH \\ a \in H}} \frac{1}{|H|}\chi_j(ax) \qquad \text{(by Theorem 11.3)} \\ &= \frac{1}{|H|} \sum_{x \in Ha_iH} \chi_j(x). \qquad \square \end{aligned}$$

THEOREM 11.8. (The Addition Formula) Let ω be the zonal spherical function belonging to χ_i. Then

$$\frac{1}{|H|} \sum_{a \in H} \omega(xay) = \omega(x)\omega(y)$$

for $x,y \in G$.

Proof. For a fixed $y \in G$, let f_y be an element of $C(H\backslash G/H)$ given by

$$f_y(x) = \frac{1}{|H|} \sum_{a \in H} \omega(xay).$$

For each zonal spherical function φ,

$$\begin{aligned}
\langle f_y, \varphi\rangle &= \frac{1}{|G|} \sum_{x \in G} f_y(x)\varphi(x^{-1}) \\
&= \frac{1}{|G||H|} \sum_{\substack{x \in G \\ a \in H}} \omega(xay)\varphi(x^{-1}) \\
&= \frac{1}{|G||H|} \sum_{\substack{z \in G \\ a \in H}} \omega(zy)\varphi(az^{-1}) \qquad (z = xa) \\
&= \frac{1}{|G|} \sum_{z \in G} \omega(zy)\varphi(z^{-1}) \\
&= \frac{1}{|G|} \sum_{t \in G} \varphi(yt^{-1})\omega(t) \qquad (t = zy) \\
&= \frac{1}{|G|} (\varphi * \omega)(y) \\
&= \langle\varphi,\omega\rangle\omega(y) \qquad \text{(by Theorem 11.6)}.
\end{aligned}$$

By (11.12), $\langle f_y,\varphi\rangle = 0$ if $\varphi \neq \omega$. Thus $f_y = \lambda\omega$ for some $\lambda \in \mathbb{C}$. Since $\omega(1) = 1$,

$$\begin{aligned}
\lambda &= f_y(1) \\
&= \frac{1}{|H|} \sum_{a \in H} \omega(ay) \\
&= \omega(y).
\end{aligned}$$

So $f_y(x) = \omega(y)\omega(x)$, which is the desired result. □

Identifying Ω with $H\backslash G = \{Hx_0, Hx_1, \ldots, Hx_{n-1}\}$ $(x_0 = 1)$, let

$$f(\alpha) = f(x) \qquad (\alpha = Hx \in \Omega) \tag{11.21}$$

for $f \in C(H\backslash G)$. Let $\Lambda_0, \Lambda_1, \ldots, \Lambda_d$ be the orbits of G on $\Omega \times \Omega$:

$$\begin{aligned} \Lambda_i(\alpha_0) &= \{\alpha \in \Omega \mid (\alpha_0, \alpha) \in \Lambda_i\} \\ &= \{\alpha \in \Omega \mid \alpha = Hx \subseteq Ha_iH\}, \end{aligned}$$

where $\alpha_0 = H$ and $H\backslash G/H = \{Ha_0H, Ha_1H, \ldots, Ha_dH\}$ $(a_0 = 1)$.

LEMMA 11.9. Let $\varphi_i = \frac{1}{|H|}\varphi_{Ha_iH}$. Then for $f \in C(H\backslash G)$ and $\alpha \in \Omega$, $(\varphi_i * f)(\alpha) = \sum_{(\beta,\alpha) \in \Lambda_i} f(\beta)$.

Proof. Let $\alpha = Hx$. Then

$$\begin{aligned} (\varphi_i * f)(\alpha) &= \sum_{y \in G} \varphi_i(xy^{-1}) f(y) \\ &= \sum_{xy^{-1} \in Ha_iH} \frac{1}{|H|} f(y) \\ &= \frac{1}{|H|} \sum_{y:\, (Hy,Hx) \in \Lambda_i} f(y) \\ &= \sum_{(\beta,\alpha) \in \Lambda_i} f(\beta) \qquad (\beta = Hy). \qquad \square \end{aligned}$$

THEOREM 11.10. Let ω_i be the zonal spherical function belonging to χ_i. For $x \in Ha_jH$, let $\omega_i(j) = \omega_i(x)$ and $\mathfrak{w}_i = (\overline{\omega_i(0)}, \overline{\omega_i(1)}, \ldots, \overline{\omega_i(d)})$. Then

$$p_j(i)\mathbf{u}_i = \mathbf{u}_i B_j,$$

where $B_j = (p_{jt}^s)$ is the jth intersection matrix and $P = (p_j(i))$ is the first eigenmatrix. In particular

$$\omega_i(j) = \frac{1}{k_j} p_j(i),$$

and

$$p_j(i)\,{}^t\mathbf{v}_i = B_j\,{}^t\mathbf{v}_i,$$

where

$$\mathbf{v}_i = (k_0\omega_i(0), k_1\omega_i(1), \ldots, k_d\omega_i(d)).$$

<u>Proof</u>. Since ω_i is an eigenfunction of the adjacency matrix I_{φ_j} belonging to the eigenvalue $p_{j'}(i)$, we have $\varphi_j * \omega_i = p_{j'}(i)\omega_i$ and so for $\alpha = Hx \subseteq Ha_sH$,

$$p_{j'}(i)\omega_i(s) = p_{j'}(i)\omega_i(\alpha) = (\varphi_j * \omega_i)(\alpha)$$

$$= \sum_{(\beta,\alpha)\in\Lambda_j} \omega_i(\beta) \qquad \text{(by Lemma 11.9)}.$$

Let $\alpha_0 = H$. Then $\omega_i(\beta) = \omega_i(t)$ for β such that $(\alpha_0,\beta) \in \Lambda_t$ and $(\beta,\alpha) \in \Lambda_j$, and the number of such β is p_{tj}^s. So $p_{j'}(i)\omega_i(s) = \sum_{t=0}^{d} \omega_i(t)p_{tj}^s$. Taking the complex conjugate,

$$p_j(i)\overline{\omega_i(s)} = \sum_{t=0}^{d} \overline{\omega_i(t)}p_{jt}^s$$

because $p_{tj}^s = p_{jt}^s$. This implies $p_j(i)\mathbf{u}_i = \mathbf{u}_i B_j$. By

Theorem 4.1, ω_i is unique up to scalar multiple. Since $\omega_i(0) = 1$, we get

$$\omega_i(j) = \frac{1}{k_j} p_j(i) .$$

The rest follows from Theorem 4.1. □

For spherical functions on a compact symmetric homogeneous space, see Cartan (1929), Selberg (1956), Godement (1952), etc.

2.12 NOTES TO CHAPTER II

It seems that the present definition and the name of association schemes (for symmetric association schemes) were first given by Bose and Shimamoto (1952) in 1952 and its algebraic properties were subsequently extensively studied by Bose and Mesner (1959). (Readers are referred to Dembowski (1968, section 7.1) for a further discussion of the history of association schemes.) However, essentially the same concept and the same idea were, quite independently and sometimes earlier, originated in many other branches of mathematics; in particular, in the work of Schur (1933), in the work of Hoheisel (1939) and Kawada (1942) and in the study of "hypergroups" (cf. Ross (1977) for earlier references for hypergroups).

We think that Delsarte's thesis (1973b) still serves as an excellent and thorough standard text to study association schemes. Several introductory books and articles are also available, e.g. Delsarte (1974), Cameron and van Lint (1975), Sloane (1975). In the finite permutation group theory, the concept is studied under the names of centralizer rings (algebras), Schur rings (algebras) or Hecke rings (algebras). Section 2.1 is just an introduction to the centralizer rings of finite transitive permutation groups and the treatment here is based on the book of Wielandt (1964). We recommend the reader further reading of his book, which gives an excellent and thorough treatment of this topic. The methods of Schur rings and centralizer rings are, for example, applied to show certain primitive permutation groups are doubly transitive, or to study permutation groups of certain specific degrees, notably as in the work of Wielandt (1964, Chapter V) on groups of degree 2p.

Also rank 3 permutation groups and strongly regular graphs are extensively studied by these methods, encouraged by the discovery of several sporadic simple groups as rank 3 permutation groups (cf. D. G. Higman (1964), Sims (1967, 1968), Hubaut (1975), etc.).

The foundation of these methods in permutation groups was axiomatized as coherent configurations by D. G. Higman (1975). The article of Higman gives a very thorough investigation of the algebraic aspect of the subject, and Chapter II of these lecture notes overlaps it in many places. A <u>coherent configuration</u> is the object $\mathfrak{X} = (X, \{R_i\}_{0 \leq i \leq d})$ with the properties (i), (ii), (iii), (iv) in (2.1) and without (v), where the property (i) is weakened as follows:

(i)' The diagonal set $\Delta = \{(x,x) \mid x \in X\}$ is a (disjoint) union of some of $R_0, R_1, \ldots, R_d$.

A coherent configuration with the property (i) $\Delta = R_0$ is called a <u>homogeneous</u> coherent configuration. So we have three classes of association schemes:

(a) symmetric association schemes (by Bose et al),

(b) commutative association schemes (by Delsarte),

(c) coherent configurations (by D. G. Higman),

with the inclusion (a) $\subset$ (b) $\subset$ (c).

The Krein condition $q_{ij}^k \geq 0$ (Theorem 3.8) was first discovered by Dunkl and Scott (cf. Scott (1973) (in the work of Krein in harmonic analysis) and independently by Delsarte (1973b). The present proof which uses Lemma 3.9 (due to Schur) is that of Delsarte (1973b) and

Biggs (1976).

Theorem 4.1(i) is mentioned in Biggs (1976). The formula in Ott (1981, page 88) is exactly this theorem. In these lecture notes, we only used the theory of normal matrices to develop the algebraic theory of association schemes, but we can also use the theory of semi-simple algebras systematically (cf. Curtis-Reiner (1962)) for this purpose (cf. D. G. Higman (1975)). Theorem 4.2 and Theorem 4.3 are taken from Wielandt (1964). Theorem 4.6 is from Cameron-Goethals-Seidel (1978b). The treatment of Lemma 4.7, Theorem 4.8 and Theorem 4.9 are based on Neumaier (1981b). The absolute bound condition will be treated in Part II again, where another understanding is given to it from the viewpoint of spherical designs.

It seems that the paper by Kawada (1942) was unnoticed until recently by those who were working on association schemes and permutation groups. The paper was given a motivation by Hoheisel (1939), in which Hoheisal attached commutative association schemes to finite groups G as in Example 2.1(2) and calculated all the relations between the parameters, in particular the orthogonality relation of the character table of G along this line. Kawada (1942) distilled from Hoheisel's work the abstract concept of a commutative association scheme, which he called a C-algebra (character algebra), and he carried out the calculations for the parameter relations of C-algebra. Kawada (1942) also considered C-subalgebras and quotient C-algebras. Theorem 9.9 is due to him. Rao, Ray-Chaudhuri and Singhi (to appear) used the terminology "parametric scheme" for C-algebras which come from association schemes and considered composition series and composition factors, of them.

Example 2.1(2) is slightly generalized by weakening the group G to be a quasigroup Q. Namely, let Q be a quasigroup, that is Q has a binary operation satisfying left and right cancellation laws (no identity nor associative law in general). Let $\mathrm{Mult}(Q)$ be the permutation group on Q generated by left and right multiplications $L(a)$, $R(a)$ $(a \in Q)$ in the symmetric group on Q, where

$L(a)\colon x \to ax \qquad (x \in Q)$

$R(a)\colon x \to xa \qquad (x \in Q).$

Then Mult(Q) acts transitively on Q and the association scheme constructed as in Example 2.1(1) becomes commutative. If Q is a group, then this association scheme amounts to Example 2.1(2). For the details, see Johnson (1976) and J. D. H. Smith (1979).

S-rings (Schur rings) found their original source in Schur (1933) and were developed by Wielandt (1964, Chapter IV). Originally S-rings were used to study primitive permutation groups which have regular subgroups but most of the results there are also true for primitive association schemes which have regular groups acting on them. The foundation of S-rings was studied by Tamaschke in a series of his papers (see Tamaschke (1970)). This theory overlaps the study of coherent configurations by D. G. Higman (1975) in many calculations. We recommend to the readers Chapter IV of Wielandt (1964).

McKay (1980) made the following very interesting observation on the character tables of finite groups (originally announced at the Santa Cruz AMS Summer Institute on Finite Groups, 1979).

MCKAY'S OBSERVATION.

(1) The representation graphs for certain finite subgroups G of $SU_2(C)$ (with respect to the representation of degree 2) are given by the Euclidean type Dynkin diagram $\tilde{A}_n$, $\tilde{D}_n$, $\tilde{E}_6$, $\tilde{E}_7$, $\tilde{E}_8$ (which are all the Euclidean type Dynkin diagrams with only single bonds).

(2) For each above group G, each of the columns of the character table of G becomes an eigenvector of the incidence matrix of the graph.

(This is one of McKay's many important and exciting observations on different topics.) Some of the generalizations and explanations are given by Happel-Preiser-Ringel (1980), Steinberg (manuscript), etc. In section 2.7, we explained the mechanism of (2), i.e., we showed the columns of the character table of a finite group G are uniquely determined as the set of standardized common eigenvectors of all the representation graphs. It is clear that the character table of G determines the decomposition of $\chi_i\chi_j$. Conversely, our explanation of (2) claims that the knowledge of the decomposition of $\chi_i\chi_j$ determines the character table uniquely, which is a well-known fact in finite group theory (cf. Curtis-Reiner (1962)). The Kawada-Delsarte duality is nothing but this mechanism for commutative association schemes. It seems that the appearance of extended Dynkin diagrams here has not yet been explained intrincically, though there are many factual explanations. It is an interesting problem to determine the types of the graphs which appear as representation graphs of finite groups. The paper by Gallagher (1979) may be related to McKay's observation (2).

It is a famous theorem of Brauer (1945, 1947) that $Q(\sqrt[e]{1})$ is a splitting field for a finite group G, i.e., all the irreducible representations of G over $\mathbb{C}$ are realizable in the field, where $e = \mathrm{Min}\{i \in \mathbb{N} \mid a^i = 1$ for all $a \in G\}$, the exponent of G. (It is clear that the values of the character table of G are in $Q(\sqrt[e]{1})$.) The fact that they are in a cyclotomic field was known before Brauer (1945) (cf. Curtis and Reiner (1962, section 41A)). Many examples suggest that the values of the eigenmatrices P and Q of association schemes are all in a cyclotomic field, but we do not know whether this is always true. This problem was posed by Norton at Oberwolfach, May, 1980. The affirmative answer would imply that the theory of association scheme could be very similar to the group theory.

The contents of section 2.8 and section 2.9 are taken from Cameron-Goethals-Seidel (1978b). Several materials are added.

PROBLEM. Let G be a transitive primitive permutation group on Ω and let $\theta = 1 + \chi_1 + \cdots + \chi_d$ be the permutation character, which is multiplicity free, i.e., $\chi_0 = 1, \chi_1, \ldots, \chi_d$ are distinct irreducible characters of G. Determine the groups G which satisfy the property that χ_i $(i = 0, 1, \ldots, d)$ appears in χ_1^i but not in χ_1^j, $j < i$ (or χ_i $(i = 0, 1, \ldots, d)$ appears in $\mathrm{Sym}^i\chi_1$ but not in $\mathrm{Sym}^j\chi_1$, $j < i$).

As is immediately seen from Theorem 8.1(i) (or (ii)), this problem is a special case (the group theory version) of the classification of Q-polynomial association schemes (Chapter III). This can be regarded as a dual version of the classification problem of distance-transitive graphs (cf. Saxl (1981)). The Norton algebras (section 2.8) are relevant to recent works by S. Smith (1977), Griess (1981), etc. on the study and construction of some finite simple groups as automorphism groups of <u>nonassociative</u> algebras.

The content of section 2.10 is from Bannai (1982). We admit that the content of this section is not profound but formal; the original Tannaka-Krein duality for compact groups and thereafter for locally compact groups has difficulty in handling the topological complexities. Our motivation was the desire that there should be a general duality theorem for non-commutative association schemes which includes the Kawada-Delsarte duality for commutative association schemes, i.e., the duality between the ordinary product and Hadamard product in $\mathfrak{A}$.

In section 2.11, we explain the relation between the Hecke algebra (defined in the space $C(H\backslash G)$ of left H-invariant functions of G) and the adjacency algebra of the multiplicity free permutation representations of G on $\Omega = H\backslash G$. This theory was developed in the representation theory of topological groups. Since we neglect the topological aspect in these lecture notes, the theory becomes rather trivial, nevertheless we believe that this section will help the reader to read Dunkel (1976, 1977, 1978), Stanton (1980, 1981a), etc. Theorem 11.2 is known as Peter-Weyl's theorem for compact topological groups. The treatment here is based on Takeuchi (1975), which is a good textbook in this area for those who can read the Japanese language. Cartan (1929), Selberg (1956), Godement (1952) seem to be fundamental papers in this area. We will return to spherical functions of compact symmetric spaces of rank 1 in Part II.

CHAPTER III. DISTANCE-REGULAR GRAPHS AND (P AND Q)-POLYNOMIAL ASSOCIATION SCHEMES

This chapter is the core of Part I of these lecture notes. In the previous chapter, we have discussed the properties of association schemes in a general situation and shown rich mathematical content in them. But as far as we are considering the general class of association schemes, there is certainly a limitation to the extent of the theory owing to the general structure. In fact, we have too many examples of association schemes to hope to classify them. Thus the following question naturally arises:

What are the important classes of association schemes?

The most natural answer, in particular from a combinatorial viewpoint, seems to be the class of P-polynomial association schemes. This terminology was first introduced by Delsarte (1973b). But this concept is older and is equivalent to that of distance-regular graphs, and there have been plenty of studies on them from plenty of viewpoints. Another important class seems to be Q-polynomial association schemes, which were first defined by Delsarte (1973b). The concept of Q-polynomial association schemes is an algebraic one and cannot be exactly understood from a combinatorial viewpoint. Nonetheless it is no less important than the concept of P-polynomial association schemes, notably when developing design theory. The purpose of this chapter is to study association schemes which have the P-polynomial and/or Q-polynomial property, and to present a program for classifying them.

The answer to the question of why the P- or Q-polynomial property is important may not be a matter of logic but taste and philosophy. Nonetheless this chapter as well as the whole of Part I was written in the hope that we could convince the reader of the importance of these properties. Firstly, we possess so nice a list of known examples of association schemes with the P- and/or Q-polynomial property that we can dream of completely classifying such association schemes with large diameters. Secondly, P- or Q-polynomial association schemes seem as appealing to us as semi-simple Lie groups in the Lie group theory and finite simple groups in the finite group theory (no matter how one likes the present existing proofs). Thirdly, there exist incredibly important connections with many other branches of mathematics, in particular orthogonal polynomials (cf. Introduction). The connection with orthogonal polynomials will culminate in the theorem of D. Leonard (section 3.5, Theorem 5.1) that orthogonal polynomials with the (P and Q)-property are Askey-Wilson polynomials, including their certain limiting cases.

We should like to comment on the arrangement of sections in Chapter III. At first glance, the reader may have an impression that the sections are not mutually connected. This is partly true because there is tremendous diversity in the techniques used in each section. However, the connecting line of each section is explained as follows. Section 3.1 is the introduction to the concepts of P-polynomial and Q-polynomial properties. Section 3.2 is the discussion of two important fundamental examples. Section 3.3 is the discussion of what we can say if we have just one P-polynomial structure (without assuming the Q-polynomial property). Section 3.4 is the discussion of what happens if there are two P-polynomial structures. Section 3.5 (and afterwards) is the discussion of association schemes which have the (P and Q)-property. (We should like to, but so far are unable to, discuss association schemes with two Q-polynomial structures, and so in section 3.7, we discuss (P and Q)-polynomial schemes with another Q-polynomial structure.) More detailed contents of each section will be given in the following.

In section 3.1, we define distance-regular graphs and we show that they are equivalent to the P-polynomial schemes defined by Delsarte. Two most important examples, namely Hamming schemes $H(n,q)$ and Johnson schemes $J(v,k)$, are discussed in the next section. The parameters and orthogonal polynomials attached to these association schemes are described.

In section 3.3, we discuss the nonexistence of certain distance-regular graphs (corresponding to certain extremal cases), starting with Moore graphs. Many updated current researches on this topic will be explained, and many on-going approaches to this topic will be reviewed.

In section 3.4, we study association schemes which have at least two P-polynomial structures. It is shown that such association schemes are of very special types.

In section 3.5, we study (P and Q)-polynomial schemes, i.e., association schemes which have both a P-polynomial structure and a Q-polynomial structure. The deepest result here is the recent theorem of Leonard (Theorem 5.1) that the orthogonal polynomials which satisfy the (P and Q)-polynomial property do coincide exactly with the Askey-Wilson polynomials (including their certain limiting cases). Askey-Wilson polynomials are a class of discrete orthogonal polynomials which include "all" classical type orthogonal polynomials (some of them as limiting cases). There are lots of definitions of "classical" orthogonal polynomials. We suggest that the "classical orthogonal polynomials" should be defined as the ones which satisfy the conclusion of Leonard's theorem (= Askey-Wilson polynomials, including their limiting cases). This also indicates that (P and Q)-polynomial schemes are association schemes of "classical" type and these are very important classes of association schemes. These association schemes are very close to those well-known important association schemes which are obtained from classical forms and groups. Theorem 5.1 will be used later to study the <u>feasibility problem</u>, i.e., to determine which (P and Q)-polynomial orthogonal polynomials possibly correspond to association schemes (in section 3.7). We will include the most updated results on this topic.

In section 3.6, we give a list of known (P and Q)-polynomial association schemes. We hope and expect that this list covers (almost) all (P and Q)-polynomial association schemes if the diameter d is sufficiently large.

In section 3.7, under the assumption that d is sufficiently large, we will determine (P and Q)-polynomial schemes which have two Q-polynomial structures. As a consequence of this result we prove that all the eigenvalues θ_i and θ_i^* and all the parameters q_{ij}^k (and of course p_{ij}^k) of (P and Q)-polynomial association schemes which are not ordinary n-gons and which have sufficiently large d are rational numbers. We believe that this information is very helpful when we try to solve the feasibility problem of (P and Q)-polynomial schemes. Several conjectures will be mentioned concerning possible further researches in this direction.

In section 3.8, we review current researches on the characterization of known (P and Q)-polynomial association schemes by their parameters. We have enough reason to believe that the final classification of (P and Q)-polynomial schemes is not far off, at least if the diameter d is sufficiently large. Many possible directions for further researches will be discussed.

3.1 DISTANCE-REGULAR GRAPHS AND P-POLYNOMIAL ASSOCIATION SCHEMES

Let Γ be a connected undirected graph on X ($|X| = n$). A <u>path</u> of length r from x to y $(x,y \in X)$ is a sequence of vertices $x_0 = x, x_1, \ldots, x_{r-1}, x_r = y$ such that each (x_i, x_{i+1}) is an edge of Γ. The <u>distance</u> of x and y is the minimum length of paths from x to y and denoted by $\partial(x,y)$ ($\partial(x,x)$ is defined to be zero). It is easy to check that $\partial(x,y)$ satisfies the axioms of distance. The <u>diameter</u> of Γ is the maximum distance of two vertices and denoted by d. Let R_i be the relation on X defined by $(x,y) \in R_i$ if and only if $\partial(x,y) = i$. If $\mathfrak{X} = (X, \{R_i\}_{0 \le i \le d})$ becomes an association scheme, Γ is called <u>distance-regular</u>. Then obviously, $\mathfrak{X}$ is always symmetric. If the automorphism group of Γ is transitive on each R_i $(0 \leq i \leq d)$, Γ is called <u>distance-transitive</u>. It is clear that if Γ is distance-transitive, then Γ is distance-regular. The association schemes in Example 2.1(3), (4) of Chapter II come from distance-transitive graphs.

A symmetric association scheme $\mathcal{X} = (X, \{R_i\}_{0 \leq i \leq d})$ is called a <u>P-polynomial scheme</u> with respect to the ordering $R_0, R_1, \ldots, R_d$, if there exist some complex coefficient polynomials $v_i(x)$ of degree i $(0 \leq i \leq d)$ such that $A_i = v_i(A_1)$, where A_i is the adjacency matrix with respect to R_i. (Later, it is shown in (1.2) that $v_i(x)$ have real (more strongly rational) coefficients.)

We use the following notation: a tridiagonal matrix

$$B = \begin{pmatrix} a_0 & c_1 & & & 0 \\ b_0 & a_1 & \ddots & c_{d-1} & \\ & b_1 & \ddots & a_{d-1} & c_d \\ 0 & & \ddots & b_{d-1} & a_d \end{pmatrix}$$

is denoted by

$$\left\{\begin{matrix} * & c_1 & \cdots & c_{d-1} & c_d \\ a_0 & a_1 & \cdots & a_{d-1} & a_d \\ b_0 & b_1 & \cdots & b_{d-1} & * \end{matrix}\right\}.$$

PROPOSITION 1.1. Let $\mathcal{X} = (X, \{R_i\}_{0 \leq i \leq d})$ be a symmetric association scheme. Let $A_i, B_1 = (p^j_{1i})$ and $P = (p_i(j))$

be the adjacency matrix with respect to R_i, the intersection matrix with respect to R_1 and the first eigenmatrix, respectively. Then the following (i), (ii), (iii) and (iv) are equivalent to each other.

(i) Let $\Gamma = (X, R_1)$ be the graph whose vertex and edge sets are X and R_1. Let $\partial(x,y)$ be the distance of x and y in Γ. Then $\partial(x,y) = i$ if and only if $(x,y) \in R_i$, i.e., Γ is distance-regular.

(ii) B_1 is a tridiagonal matrix with non-zero off-diagonal entries:

$$B_1 = \begin{Bmatrix} * & 1 & c_2 & \cdots & c_{d-1} & c_d \\ 0 & a_1 & a_2 & \cdots & a_{d-1} & a_d \\ k & b_1 & b_2 & \cdots & b_{d-1} & * \end{Bmatrix}$$

$$(b_i \neq 0,\ c_i \neq 0).$$

(iii) $\mathfrak{X}$ is a P-polynomial scheme with respect to the ordering $R_0, R_1, \ldots, R_d$, i.e.,

$$A_i = v_i(A_1) \qquad (i = 0, 1, \ldots, d)$$

for some polynomials $v_i(x)$ of degree i.

(iv) P satisfies

$$p_i(j) = v_i(\theta_j) \qquad (i,j = 0, 1, \ldots, d)$$

for some polynomials $v_i(x)$ of degree i, where

$$\theta_j = p_1(j).$$

(The polynomials $v_i(x)$ $(i = 0, 1, \ldots, d)$ in (iii) and

(iv) coincide with each other. B_1 is called the *intersection matrix of the distance-regular graph* Γ, $\theta_0 = k, \theta_1, \ldots, \theta_d$ are called the *eigenvalues of* Γ, and k is called the *valency* of Γ.)

Proof. (i) ⇒ (ii): Suppose (i) holds. By the triangle inequality, $p^j_{1i} = 0$ if $|i - j| \geq 2$, and by the connectivity of Γ, $p^j_{1i} \neq 0$ if $|i - j| = 1$. So $B_1 = (p^j_{1i})$ is a tridiagonal matrix with non-zero off-diagonal entries.

(ii) ⇒ (i): Suppose (ii) holds. We use induction on i. Let $R_i(x) = \{y|\ (x,y) \in R_i\}$. Then for $y \in R_i(x)$, $R_{i-1}(x) \cap R_1(y) \neq \emptyset$ and so $\partial(x,y) \leq i$. For $y \in R_i(x)$, $R_j(x) \cap R_1(y) = \emptyset$ $(j \leq i-2)$ and so $\partial(x,y) \geq i$.

(ii) ⇒ (iii): Suppose (ii) holds. Then since $A_1A_i = \sum_j p^j_{1i}A_j$,

$$A_1A_i = b_{i-1}A_{i-1} + a_iA_i + c_{i+1}A_{i+1} \qquad (1.1)$$

$$(0 \leq i \leq d-1),$$

where $b_{-1} = 0$, $b_0 = k$, $a_0 = 0$, $c_1 = 1$. Let $v_0(x) = 1$, $v_1(x) = x$ and $v_i(x)$ the polynomial of degree i defined by the recurrence

$$xv_i(x) = b_{i-1}v_{i-1}(x) + a_iv_i(x) + c_{i+1}v_{i+1}(x) \qquad (1.2)$$

$$(0 \leq i \leq d-1).$$

Then $v_i(A_1) = A_i$.

(iii) ⇒ (ii): Suppose (iii) holds. Since the polynomial $xv_i(x)$ is a linear combination of

$v_{i+1}(x), v_i(x), \ldots, v_0(x)$, $A_1 A_i$ is a linear combination of $A_{i+1}, A_i, \ldots, A_0$, in which the coefficient of A_{i+1} is non-zero, i.e., $p^j_{1i} = 0$ if $j \geq i + 2$, and $p^j_{1i} \neq 0$ if $j = i + 1$. Since $\mathfrak{X}$ is symmetric, $p^j_{1i} = 0$ holds if and only if $p^i_{1j} = 0$. So we have $p^j_{1i} = 0$ if $|i - j| \geq 2$, and $p^j_{1i} \neq 0$ if $|i - j| = 1$.

(iii) ⇔ (iv): Let $E_0, E_1, \ldots, E_d$ be the primitive idempotents of the adjacency algebra of $\mathfrak{X}$. Set $\theta_j = p_1(j)$. Then $A_1 = \sum_j \theta_j E_j$. For any polynomial $v(x)$, $v(A_1) = \sum_j v(\theta_j)E_j$. Suppose (iii) holds. Then $A_i = v_i(A_1) = \sum_j v_i(\theta_j)E_j$ and so $p_i(j) = v_i(\theta_j)$. Suppose (iv) holds. Then $A_i = \sum_j p_i(j)E_j = \sum_j v_i(\theta_j)E_j = v_i(A_1)$. □

REMARK. (1) A connected undirected graph Γ on X is usually said to be distance-regular (see Biggs (1974)) if:

> For any non-negative integers i and j, the number p^j_{1i} of vertices z with $\partial(x,z) = 1$, $\partial(z,y) = i$ depends only on i and j, but not on the choice of x, y with $\partial(x,y) = j$. (DR)

This definition of distance regularity seems weaker than the previous one, but they are actually the same.

Suppose Γ satisfies (DR). Let R_i be the relation on X defined by $(x,y) \in R_i$ if and only if $\partial(x,y) = i$. Let A_i be the incidence matrix with respect to R_i. Then (DR) implies that (1.1) holds for $A_1 A_j$ and $A_1 A_d = b_{d-1} A_{d-1} + a_d A_d$. So $A_i = v_i(A_1)$ and $v_{d+1}(A_1) = 0$, where $v_{d+1}(x) = (x - a_d)v_d(x) - b_{d-1}v_{d-1}(x)$. A_1 spans an algebra of degree $d + 1$

with a linear basis $A_0, A_1, \ldots, A_d$. Thus, the conditions (2.1)' of Chapter II are satisfied.

Let $\mathfrak{X} = (X, \{R_i\}_{0 \leq i \leq d})$ be a symmetric association scheme and let E_i $(i = 0, 1, \ldots, d)$ be the primitive idempotents of the adjacency algebra $\mathfrak{A}$. $\mathfrak{X}$ is called a <u>Q-polynomial</u> <u>scheme</u> with respect to the ordering $E_0, E_1, \ldots, E_d$, if there exists some complex coefficient polynomials $v_i^*(x)$ such that $E_i = v_i^*(E_1)$ and $v_i^*(x)$ has degree i, where the multiplication of $\mathfrak{A}$ is under the Hadamard product. (Later, it is shown in $(1.2)^*$ that $v_i^*(x)$ have real coefficients.) Let q_{ij}^k be the Krein parameters of $\mathfrak{X}$ and $\hat{B}_1 = (q_{1i}^j)$. Let $Q = (q_i(j))$ be the second eigenmatrix of $\mathfrak{X}$. As in Proposition 1.1, the following (ii), (iii), (iv) are equivalent.

(ii) $\hat{B}_1$ is a tridiagonal matrix with non-zero off-diagonal entries:

$$\hat{B}_1 = \begin{Bmatrix} * & 1 & c_2^* & \cdots & c_{d-1}^* & c_d^* \\ 0 & a_1^* & a_2^* & \cdots & a_{d-1}^* & a_d^* \\ m & b_1^* & b_2^* & \cdots & b_{d-1}^* & * \end{Bmatrix}$$

(iii) $\mathfrak{X}$ is a Q-polynomial scheme with respect to the ordering $E_0, E_1, \ldots, E_d$, i.e.,

$$E_i = v_i^*(E_1) \qquad (i = 0, 1, \ldots, d)$$

for some polynomials $v_i^*(x)$ of degree i under the Hadamard product.

(iv) Q satisfies

$$q_i(j) = v_i^*(\theta_j^*) \qquad (i,j = 0, 1, \ldots, d)$$

for some polynomials $v_i^*(x)$ of degree i, where

$$\theta_j^* = q_1(j).$$

The polynomials $v_i^*(x)$ in (iii) and (iv) coincide with each other, and satisfy the recurrence

$$xv_i^*(x) = b_{i-1}^* v_{i-1}^*(x) + a_i^* v_i^*(x) + c_{i+1}^* v_{i+1}^*(x) \quad (1.2)^*$$

$$(0 \leq i \leq d-1)$$

with $v_0^*(x) = 1$ and $v_1^*(x) = x$.

In Part II, we shall define t-designs in Q-polynomial schemes and show that these generally defined t-designs are orthogonal arrays of strength t and ordinary t-designs, when the underlying Q-polynomial schemes are $H(n,q)$ and $J(v,k)$ respectively.

QUESTIONS. (1) Can one find a combinatorial definition of Q-polynomial schemes?

(2) Can one find a group theoretical definition of Q-polynomial schemes that come from generously transitive groups?

PROPOSITION 1.2. Let $\mathfrak{X} = (X, \{R_i\}_{0 \leq i \leq d})$ be a P-polynomial scheme with respect to the ordering $R_0, R_1, \ldots, R_d$. Let

$$B_1 = (p_{1i}^j) = \begin{Bmatrix} * & 1 & c_2 & \cdots & c_d \\ 0 & a_1 & a_2 & \cdots & a_d \\ k & b_1 & b_2 & \cdots & * \end{Bmatrix}$$

be the intersection matrix with respect to R_1, k_i the valency of R_i and $k = k_1$. Then

(i) $a_i + b_i + c_i = k \quad (0 \le i \le d)$ with $c_0 = b_d = 0$,

(ii) $k_i = kb_1b_2 \ldots b_{i-1}/c_2c_3 \ldots c_i \quad (2 \le i \le d)$,

(iii) $k \ge b_1 \ge b_2 \ge \ldots \ge b_{d-1}$,

(iv) $1 \le c_2 \le c_3 \le \ldots \le c_d$.

Proof. (i) By Proposition 2.2(v) of Chapter I, (i) holds.

(ii) By Proposition 2.2(vi) of Chapter I, $k_i p^i_{1j} = k_j p^j_{1i}$ and so $k_i b_i = k_{i+1} c_{i+1}$. By induction on i, (ii) holds.

(iii) Let Γ be the graph on X with the edge set R_1 and let $\partial(x,y)$ be the distance of x, y in Γ. For x, y with $\partial(x,y) = i$, $b_i = p^i_{1i+1}$ is the number of w such that $\partial(x,w) = 1$ and $\partial(w,y) = i + 1$. Take $z \in X$ such that $\partial(x,z) = i - 1$ and $\partial(z,y) = 1$. Then $\partial(z,w) \le i$ becuase $\partial(x,w) = 1$, and $\partial(z,w) \ge i$ because $\partial(w,y) = i + 1$, so $\partial(w,z) = i$. Thus b_i is less than or equal to b_{i-1} which is the number of w such that $\partial(x,w) = 1$, $\partial(w,z) = i$.

(iv) For x, y with $\partial(x,y) = i$, $c_i = p^i_{1i-1}$ $(i \le d-1)$ is the number of w such that $\partial(x,w) = 1$ and $\partial(w,y) = i - 1$. Take $z \in X$ such that $\partial(x,z) = i + 1$ and $\partial(y,z) = 1$. Then $\partial(w,z) \le i$ because $\partial(w,y) = i - 1$, and $\partial(w,z) \ge i$ because $\partial(x,w) = 1$, so $\partial(w,z) = i$. Thus c_i is less than or equal tc c_{i+1} which is the number of w such that $\partial(x,w) = 1$, $\partial(w,z) = i$. □

QUESTIONS. (3) Let

$$B = \left\{\begin{matrix} * & \overbrace{1 \quad 1 \quad \cdots \quad 1}^{s_1} & \overbrace{\alpha \cdots \alpha}^{s_2} & \overbrace{\beta \cdots \beta}^{s_3} & \cdots \\ 0 & a_1 \quad a_2 \quad \cdots & \cdots & \cdots & \cdots \\ k & b_1 \quad b_2 \quad \cdots & \cdots & \cdots & \cdots \end{matrix}\right\}$$

be the intersection matrix of a distance regular graph. Is there any relation between $s_1, s_2, s_3, \ldots$? For example, is it true that $s_1 \geq s_2 \geq s_3 \geq \ldots$? Is there any relation between $\alpha - 1, \beta - \alpha, \ldots$? For example, is it true that $\alpha - 1 \leq \beta - \alpha \leq \ldots$? Consider the same type of question about the sequence $k, b_1, b_2, \ldots$. (See Gardiner (1978) and Terwilliger (1982a, 1983).)

(4) Let $\mathfrak{X} = (X, \{R_i\}_{0 \leq i \leq d})$ be a Q-polynomial scheme with respect to the ordering $E_0, E_1, \ldots, E_d$, and let q_{ij}^k be the Krein parameters. Let

$$\hat{B} = (q_{1j}^k) = \left\{\begin{matrix} * & 1 & c_2^* & \cdots & c_{d-1}^* & c_d^* \\ 0 & a_1^* & a_2^* & \cdots & a_{d-1}^* & a_d^* \\ m & b_1^* & b_2^* & \cdots & b_{d-1}^* & * \end{matrix}\right\}.$$

When do the inequalities $m \geq b_1^* \geq b_2^* \geq \cdots \geq b_{d-1}^*$ and $1 \leq c_2^* \leq c_3^* \leq \cdots \leq c_d^*$ hold? If these inequalities hold, the multiplicities m_i have the unimodal property by $m_i = m b_1^* b_2^* \cdots b_{i-1}^* / c_2^* c_3^* \cdots c_i^*$, i.e., $m = m_0 \leq m_1 \leq \cdots \leq m_{i_0}$ and $m_{i_0} \geq m_{i_0+1} \geq \cdots \geq m_d$. (Actually, the stronger condition of log-concavity holds.)

We have counterexamples for the inequalities of c_i^* (see Remark (2) of section 3.2), but even in these counterexamples, the unimodal property of m_i holds. Perhaps the unimodal property holds for the multiplicities m_i of every Q-polynomial scheme with sufficiently large class d (see Conjecture (1) of section 3.1).

THEOREM 1.3. Let $\mathfrak{X} = (X, \{R_i\}_{0 \leq i \leq d})$ be a P-polynomial scheme, k_i $(0 \leq i \leq d)$ the valencies, $k = k_1$, and m_i $(0 \leq i \leq d)$ the multiplicities. Then, keeping the notation of Proposition 1.1, the following hold:

(i) $\theta_0 = k, \theta_1, \ldots, \theta_d$ are distinct real numbers and the eigenvalues of B_1. Furthermore $k \geq \theta_i \geq -k$ $(0 \leq i \leq d)$.

(ii) $c_i v_i(x) = (x - a_{i-1})v_{i-1}(x) - b_{i-2}v_{i-2}(x)$ for $2 \leq i \leq d$ and $v_0(x) = 1$, $v_1(x) = x$.

(iii) Let $v_{d+1}(x)$ be a polynomial of degree $d + 1$ defined by

$$v_{d+1}(x) = (x - a_d)v_d(x) - b_{d-1}v_{d-1}(x).$$

Then
$$v_{d+1}(x) = \frac{1}{c_2 c_3 \cdots c_d}(x - \theta_0)(x - \theta_1) \cdots (x - \theta_d)$$
$$= (x - \theta_0) \sum_{i=0}^{d} v_i(x).$$

(iv) $$\sum_{\nu=0}^{d} v_i(\theta_\nu)v_j(\theta_\nu)m_\nu = nk_i\delta_{ij}.$$

(v) (Christoffel-Darboux formula)

$$\sum_{i=0}^{\ell} \frac{1}{k_i} v_i(x)v_i(y) = \frac{c_{\ell+1}}{k_\ell} \frac{v_{\ell+1}(x)v_\ell(y) - v_\ell(x)v_{\ell+1}(y)}{x - y}$$

for $\ell = 0, 1, \ldots, d$, where $c_{d+1} = 1$.

(vi) $m_i = \dfrac{nk_d}{v'_{d+1}(\theta_i)v_d(\theta_i)}$ $(0 \leq i \leq d)$,

where $n = |X|$ and $v'_{d+1}(x)$ is the derivative of the polynomial $v_{d+1}(x)$.

Proof. (i) By Proposition 1.1(iv), $P = (v_i(\theta_j))$. Since P is non-singular, $\theta_0, \theta_1, \ldots, \theta_d$ are distinct. Since ${}^tA_1 = A_1$, $1' = 1$ and so $\overline{\theta}_j = \theta_j$ by (3.7) of Chapter II. By (5.3) of Chapter II, $\theta_i = p_1(i)$ are eigenvalues of B_1.

By Proposition 1.1(iv), $k_i = v_i(k)$. By (1.2), ${}^t(k_0, k_1, \ldots, k_d)$ is a right eigenvector of B_1 belonging to the eigenvalue k. By Perron-Frobenius' Theorem (cf. any advanced book in linear algebra), the absolute values of eigenvalues of B_1 are no more than k. For the convenience of the reader, we shall give a proof of this assertion. Let $B_1\mathbf{x} = \theta\mathbf{x}$. Suppose every entry of $\mathbf{x}$ is non-negative. Then obviously $\theta \geq 0$. Let $\mathbf{x}_0 = {}^t(k_0, k_1, \ldots, k_d)$ and take $c > 0$ maximal such that every entry of $\mathbf{x}_0 - c\mathbf{x}$ is non-negative. Then $B_1(\mathbf{x}_0 - c\mathbf{x})$ has entries all non-negative. But $B_1(\mathbf{x}_0 - c\mathbf{x}) = k(\mathbf{x}_0 - \frac{\theta}{k} c\mathbf{x})$ and so $\frac{\theta}{k} \leq 1$. Suppose $\mathbf{x}$ has both positive and negative entries. Take $|c|$ maximal such that every entry of $\mathbf{x}_0 - c\mathbf{x}$ is non-negative. Then $B_1(\mathbf{x}_0 - c\mathbf{x})$ has entries all non-negative. But $B_1(\mathbf{x}_0 - c\mathbf{x}) = k(\mathbf{x}_0 - \frac{\theta}{k} c\mathbf{x})$ and so $|\frac{\theta}{k}| \leq 1$.

(ii) This comes from (1.2).

(iii) Let θ be an eigenvalue of B_1 and ${}^t(x_0, x_1, \ldots, x_d)$ a right eigenvector of B_1 belonging to θ. We may assume $x_0 = 1$ because if $x_0 = 0$, then $x_0 = x_1 = \ldots = x_d = 0$, which is a contradiction. By

(1.2), $x_i = v_i(\theta)$ $(0 \leq i \leq d)$ and $\theta x_d = b_{d-1}x_{d-1} + a_dx_d$, i.e., $v_{d+1}(\theta) = (\theta - a_d)x_d - b_{d-1}x_{d-1} = 0$. Hence $\theta_0, \theta_1, \ldots, \theta_d$ are the roots of $v_{d+1}(x)$. Since the leading coefficient of $v_{d+1}(x)$ is $1/c_2c_3 \ldots c_d$, we get

$$v_{d+1}(x) = \frac{1}{c_2c_3 \ldots c_d}(x - \theta_0)(x - \theta_1) \ldots (x - \theta_d).$$

By taking the sum of $\theta v_i(\theta) = b_{i-1}v_{i-1}(\theta) + a_iv_i(\theta) + c_{i+1}v_{i+1}(\theta)$ $(0 \leq i \leq d - 1)$ and $\theta v_d(\theta) = b_{d-1}v_{d-1}(\theta) + a_dv_d(\theta)$, we get

$$\theta \sum_{i=0}^{d} v_i(\theta) = k \sum_{i=0}^{d} v_i(\theta),$$

because $a_i + b_i + c_i = k$. If $\theta \neq k$, θ is a root of $\sum_{i=0}^{d} v_i(x)$, and hence

$$\sum_{i=0}^{d} v_i(x) = \frac{1}{c_2c_3 \ldots c_d}(x - \theta_1)(x - \theta_2) \ldots (x - \theta_d).$$

(iv) This comes from the second orthogonality relation (Theorem 3.5(iii) of Chapter II).

(v) $v_{i+1}(x)v_i(y) - v_i(x)v_{i+1}(y)$

$$= \frac{1}{c_{i+1}}\{(x - a_i)v_i(x) - b_{i-1}v_{i-1}(x)\}v_i(y)$$

$$- \frac{1}{c_{i+1}}\{(y - a_i)v_i(y) - b_{i-1}v_{i-1}(y)\}v_i(x)$$

$$= \frac{1}{c_{i+1}}(x - y)v_i(x)v_i(y)$$

$$+ \frac{b_{i-1}}{c_{i+1}}\{v_i(x)v_{i-1}(y) - v_{i-1}(x)v_i(y)\}.$$

Since $\dfrac{b_{i-1}}{k_i} = \dfrac{c_i}{k_{i-1}}$ by Proposition 1.2(ii), we get

$$\frac{1}{k_i} v_i(x)v_i(y) = \frac{c_{i+1}}{k_i} \frac{v_{i+1}(x)v_i(y) - v_i(x)v_{i+1}(y)}{x - y}$$

$$- \frac{c_i}{k_{i-1}} \frac{v_i(x)v_{i-1}(y) - v_{i-1}(x)v_i(y)}{x - y}$$

for $0 \le i \le d$, where $v_{-1}(x) = 0$ and $c_{d+1} = 1$. By taking the sum over $0 \le i \le \ell$, we get the desired result.

(vi) The limit of (v) with y tending to x is

$$\sum_{i=0}^{\ell} \frac{1}{k_i} v_i(x)^2 \tag{1.3}$$

$$= \frac{c_{\ell+1}}{k_\ell}\{v'_{\ell+1}(x)v_\ell(x) - v'_\ell(x)v_{\ell+1}(x)\},$$

because

$$v_{\ell+1}(x)v_\ell(y) - v_\ell(x)v_{\ell+1}(y)$$

$$= (v_{\ell+1}(x) - v_{\ell+1}(y))v_\ell(y)$$

$$- (v_\ell(x) - v_\ell(y))v_{\ell+1}(y) .$$

By Theorem 4.1 of Chapter II,

$$m_i = n/\sum_{\nu=0}^{d} \frac{1}{k_\nu} |p_\nu(i)|^2 = n/\sum_{\nu=0}^{d} \frac{1}{k_\nu} v_\nu(\theta_i)^2 .$$

By (1.3),

$$\sum_{\nu=0}^{d} \frac{1}{k_\nu} v_\nu(\theta_i)^2 = \frac{1}{k_d}\{v'_{d+1}(\theta_i)v_d(\theta_i) - v'_d(\theta_i)v_{d+1}(\theta_i)\}$$

$$= \frac{1}{k_d} v'_{d+1}(\theta_i)v_d(\theta_i) \qquad \text{(by (iii))},$$

and hence we get the desired result. □

REMARKS. (2) The m_i in Theorem 1.3(vi) is the multiplicity of θ_i as an eigenvalue of the first adjacency matrix A_1.

(3) Take a, b such that $a < \theta_i < b$ $(0 \leq i \leq d)$ and define a weight function $w(x)$ by

$$w(x) = \begin{cases} m_i & \text{if } x = \theta_i \quad (0 \leq i \leq d), \\ 0 & \text{otherwise.} \end{cases}$$

For real-valued functions $f(x)$, $g(x)$ on $[a,b]$, let

$$(f,g) = \int_a^b f(x)g(x)w(x)dx = \sum_{i=0}^{d} f(\theta_i)g(\theta_i)m_i.$$

Then by Theorem 1.3(iv), $(v_i,v_j) = nk_i\delta_{ij}$, and so $\frac{1}{\sqrt{nk_i}} v_i(x)$ are the orthogonal polynomials associated with the weight distribution $w(x)\,dx$. The numbers m_i are called the Christoffel numbers. (See Szegő (1939), Erdély; et al. (1953), etc.)

(4) Set

$$F_i(x) = c_2c_3 \cdots c_i(v_0(x) + v_1(x) + \cdots + v_i(x))$$

$$(i = 2, \ldots, d)$$

and $F_0(x) = 1$, $F_1(x) = x + 1$. Then $F_i(x)$ satisfies the recurrence

$$F_i(x) = (x - k + b_{i-1} + c_i)F_{i-1}(x) - b_{i-1}c_{i-1}F_{i-2}(x) \qquad (i = 2, \ldots, d). \tag{1.4}$$

By the argument used to prove Theorem 1.3(iii), $(x - k)F_i(x)$ is the minimal polynomial of the tridiagonal matrix

$$\left\{\begin{matrix} * & 1 & \cdots & c_{i-1} & c_i \\ 0 & a_1 & \cdots & a_{i-1} & a_i + b_i \\ k & b_1 & \cdots & b_{i-1} & * \end{matrix}\right\}.$$

In particular, $(x - k)F_d(x) = c_2c_3 \cdots c_dv_{d+1}(x)$.

Since θ_i $(i \neq 0)$ is a root of $\sum_{i=0}^{d} v_i(x)$ by Theorem 1.3(iii), we have

$$v_d(\theta_i) = -F_{d-1}(\theta_i)/c_2 \cdots c_{d-1} \qquad (i \neq 0).$$

So the formula for m_i in Theorem 1.3(vi) is written in terms of $F_d'(\theta_i)$ and $F_{d-1}(\theta_i)$:

$$m_i = \frac{nkb_1 \cdots b_{d-1}c_2 \cdots c_{d-1}}{(k - \theta_i)F_d'(\theta_i)F_{d-1}(\theta_i)} \qquad (i \neq 0), \tag{1.5}$$

$$m_0 = 1.$$

The identity $m_0 = 1$ is directly calculated by using Theorem 1.3(vi), $v_d(k) = k_d$ and

$$v'_{d+1}(k) = (k - \theta_1) \ldots (k - \theta_d)$$

$$= \sum_{i=0}^{d} v_i(k) = \sum_{i=0}^{d} k_i = n.$$

(5) Let $f_i(x)$ be a polynomial of degree i with real coefficients defined by the recurrence:

$$f_i(x) = (A_i x + B_i) f_{i-1}(x) - C_i f_{i-2}(x)$$

$$(A_i > 0,\ C_i > 0),$$

where $f_0(x) = C > 0$ and $f_1(x) = Ax + B$ with $A > 0$. Then $f_i(x)$ has roots all real and distinct. Let $\alpha_1 > \alpha_2 > \ldots > \alpha_i$ be the roots of $f_i(x)$. Then $f_{i-1}(x)$ has a root in the open interval (α_j, α_{j-1}) $(2 \leq j \leq i)$. (This can be easily proved by induction on i.) In particular, $v_i(x)$, $F_i(x)$ have roots all real and distinct.

(6) Let

$$B_1 = \begin{Bmatrix} * & 1 & \cdots & c_{d-1} & c_d \\ 0 & a_1 & \cdots & a_{d-1} & a_d \\ k & b_1 & \cdots & b_{d-1} & * \end{Bmatrix}$$

be an arbitrary tridiagonal matrix with non-negative diagonal, positive non-zero off-diagonal entries and a constant column sum k. (We do not assume the existence of an association scheme.) Let $v_i(x)$ be the polynomials of degree i defined as in Theorem 1.3(ii).

$(x - k)F_d(x)$ is the minimal polynomial of B_1 and

has roots all real and distinct (Remarks (4), (5)). $v_i(B_1)$ has the form

$$\begin{pmatrix} 0 & \cdots & 0 & 1 & 0 & \cdots & 0 \\ \vdots & & & & & & \\ 0 & & & & & & \\ k_i & & & * & & & \\ 0 & & & & & & \\ \vdots & & & & & & \\ 0 & & & & & & \end{pmatrix} \tag{1.6}$$

(with 1 in the ith place of the first row and k_i in the ith place of the first column) with $k_i = v_i(k)$. The algebra spanned by B_1 is a C-algebra with the basis $v_0(B_1), v_1(B_1), \ldots, v_d(B_1)$. Thus Theorem 1.3 formally holds for B_1.

(7) Let $u_i(x) = \frac{1}{k_i} v_i(x)$. Then for the eigenvalues θ_j of B_1, the row vectors $(u_0(\theta_j), u_1(\theta_j), \ldots, u_d(\theta_j))$ are the left eigenvectors of B_1 belonging to θ_j by Theorem 4.1 of Chapter II. In particular, $u_i(x)$ satisfies the recurrence

$$xu_i(x) = c_i u_{i-1}(x) + a_i u_i(x) + b_i u_{i+1}(x) \tag{1.7}$$

$$(1 \leq i \leq d-1)$$

with $u_0(x) = 1$, $u_1(x) = \frac{1}{k_1} x$.

QUESTION. (5) In Remark (6), let B_1 be the intersection matrix of a P-polynomial scheme. Then it holds that

$$v_i(B_1) \text{ has entries all non-negative} \tag{1.8}$$

for $i = 0, 1, \ldots, d$. What is the condition on the entries a_i, b_i, c_i under which (1.8) holds for B_1 in Remark (6)?

PROPOSITION 1.4. The valencies k_i $(0 \leq i \leq d)$ of distance-regular graphs have the unimodal property, i.e., there exists i_0 such that $k_0 \leq k_1 \leq \ldots \leq k_{i_0}$, $k_{i_0+1} \geq k_{i_0+2} \geq \ldots \geq k_d$.

Proof. Since $k_i = \dfrac{c_{i+1}}{b_i} k_{i+1} = \dfrac{b_{i-1}}{c_i} k_{i-1}$ and $c_i \leq c_{i+1}$, $b_{i-1} \geq b_i$ by Proposition 1.2, the log-concavity holds for $\{k_i\}$, i.e., $k_i^2 \geq k_{i+1}k_{i-1}$. So the unimodal property holds. □

CONJECTURES. (1) The multiplicities m_i $(0 \leq i \leq d)$ of a Q-polynomial scheme have the unimodal property.

(2) The multiplicities m_i $(0 \leq i \leq d)$ of a distance-regular graph have the unimodal property, if the diameter d is large enough (relative to the valency), where the ordering of m_i is such that $\theta_0 > \theta_1 > \theta_2 > \ldots > \theta_d$.

If the inequalities $b^*_{i-1} \geq b^*_i$ and $c^*_i \leq c^*_{i+1}$ should hold for Q-polynomial schemes, the conjecture (1) is true. If d is small, there are counterexamples for the conjecture (2), but regarding m_i as the Christoffel numbers in the light of Theorem 1.3, we have some grounds for the conjecture (2), namely the Christoffel numbers of classical orthogonal polynomials have the unimodal property.

There is a conjecture that P-polynomial schemes are Q-polynomial schemes if d is large enough (see section 3.6), except when the P-polynomial schemes are $2 \cdot O_k$ (the double covering of the odd graph O_k) or its q-analogue (see section 3.7). If this conjecture is true, the validity of the conjecture (1) implies that of the conjecture (2).

PROPOSITION 1.5. Keeping the notation of Theorem 1.3, let $\mathfrak{X} = (X, \{R_i\}_{0 \le i \le d})$ be a P-polynomial scheme. Then $m_i = m_j$ whenever θ_i and θ_j are algebraically conjugate over the rational field. In particular, for each i, $\#\{m_j \mid m_j = m_i\} \ge \#\{\theta_j \mid \theta_j$ is algebraically conjugate to θ_i over the rational field$\}$.

Proof. Since the coefficients of $v_{d+1}(x)$ and $v_d(x)$ are all rational, Theorem 1.3(vi) implies that $m_j = m_i$, whenever θ_j is algebraically conjugate to θ_i over the rational field. □

For example, suppose the multiplicities m_i have the strict unimodal property: $m_0 < m_1 < \ldots < m_k$ and $m_{k+1} > m_{k+2} > \ldots > m_d$ for some k. Then for each m_i, there is at most one m_j such that $i \ne j$ and $m_i = m_j$. By Proposition 1.5, $v_{d+1}(x)$ must be decomposed into the product of polynomials of degree at most 2 over the rational field, which is a very strong condition on the intersection matrix B_1.

THEOREM 1.6. Let G be a distance-transitive group on X, i.e., G is transitive on X and $\mathfrak{X} = (X, \{R_i\}_{0 \le i \le d})$ is a P-polynomial scheme, where R_i $(0 \le i \le d)$ are the orbits of G on $X \times X$. Let $\chi_0 + \chi_1 + \ldots + \chi_d$ be the permutation character with χ_i irreducible characters of G. Then $\chi_i \ne \chi_j$ $(i \ne j)$. Let ω_i be the zonal spherical function on X belonging to χ_i and set $\omega_i(j) = \omega_i(x)$ for $(x_0, x) \in R_j$, where x_0 is a fixed base point. Then

$$\omega_i(j) = \frac{1}{k_j} v_j(\theta_i),$$

where $\theta_i = p_1(i)$ and $v_j(x)$ are the polynomials defined in Theorem 1.3.

Proof. By Theorem 11.10 of Chapter II and Proposition 1.1(iv), this is trivial. □

3.2 THE HAMMING SCHEMES H(n,q) AND THE JOHNSON SCHEMES J(v,k)

HAMMING SCHEME $H(n,q)$. Let F be a set of cardinality q $(q \geq 2)$ and $X = F^n$, the nth Cartesian power of F. The Hamming distance between two points $x = (x_1, \ldots, x_n)$ and $y = (y_1, \ldots, y_n)$ of X is

$$\partial(x,y) = \#\{i \mid x_i \neq y_i, \quad 1 \leq i \leq n\}. \tag{2.1}$$

Let R_i be the ith distance relation on X, i.e.,

$$R_i = \{(x,y) \in X \times X \mid \partial(x,y) = i\}. \tag{2.2}$$

Then $\mathfrak{X} = (X, \{R_i\}_{0 \leq i \leq n})$ is a symmetric association scheme and is called the Hamming scheme $H(n,q)$ of length n over F. Let G be the wreath product of S_q by S_n (of order $(q!)^n n!$). Then G acts on X distance-transitively, i.e., G is transitive on R_i $(0 \leq i \leq n)$. The intersection matrix B_1 is

$$B_1 = \left\{\begin{matrix} * & 1 & \cdots & i & \cdots & n \\ 0 & q-2 & \cdots & i(q-2) & \cdots & n(q-2) \\ n(q-1) & (n-1)(q-1) & \cdots & (n-i)(q-1) & \cdots & * \end{matrix}\right\}. \tag{2.3}$$

We want to find the eigenmatrices P, Q and show $P = Q$. Provide F with the structure of an Abelian group, say the cyclic group of order q. Let F^* be the character group of F. Then $\eta: a \to \zeta$ generates F^* and F^* is a cyclic group of order q, where a is a generator of F and ζ is a primitive q^{th} root of unity.

$X = \underbrace{F \times \cdots \times F}_{n}$ is an abelian regular subgroup of G. Let

$$X_i = \{x = (x_1,\ldots,x_n) \in X \mid \partial(x_0,x) = i\}, \tag{2.4}$$

where $x_0 = (0,0,\ldots,0)$. Then $\mathfrak{X}_0, \mathfrak{X}_1, \ldots, \mathfrak{X}_n$ generate an S-ring $\mathfrak{S}$ over X, where $\mathfrak{X}_i = \sum_{x \in X_i} x$. We shall show the dual S-ring $\mathfrak{S}^*$ is isomorphic to $\mathfrak{S}$.

Let Y be the character group of X. Then

$$Y = \underbrace{F^* \times F^* \times \cdots \times F^*}_{n}.$$

LEMMA 2.1. Let η_α $(1 \le \alpha \le j)$ be a non-identity element of F^*. Then

$$\sum \eta_1(x_1)\eta_2(x_2) \cdots \eta_j(x_j) = (-1)^j,$$

where the summation is over $x_1 \in F - \{0\}, \ldots, x_j \in F - \{0\}$.

<u>Proof</u>. By the orthogonality relation of characters of F,

$$\sum_{\substack{x_\alpha \in F-\{0\} \\ 1 \le \alpha \le j}} \eta_1(x_1)\eta_2(x_2) \cdots \eta_j(x_j) = \prod_{\alpha=1}^{j} \Big(\sum_{x \in F-\{0\}} \eta_\alpha(x) \Big) = (-1)^j . \qquad \square$$

PROPOSITION 2.2. For $\Delta = (\eta_1, \eta_2, \ldots, \eta_n) \in Y$, let i be the number of α such that $\eta_\alpha \neq$ identity. Then

$$\Delta(\mathfrak{X}_k) = \sum_{j=0}^{k} (-1)^j (q-1)^{k-j} \binom{i}{j}\binom{n-i}{k-j} .$$

Proof. Let $S = \{\alpha \mid \eta_\alpha \neq \text{identity}\}$ $(|S| = i)$. Apply Lemma 2.1, arranging the elements $x = (x_1, \ldots, x_n) \in X_k$ according to the number j of α such that $\alpha \in S$ and $x_\alpha \neq 0$. $\square$

Let Y_i be the set of $\Delta = (\eta_1, \ldots, \eta_n) \in Y$ such that

$$\#\{\alpha \mid \eta_\alpha \neq \text{identity},\ 1 \le \alpha \le n\} = i. \tag{2.5}$$

Then by Theorem 6.3 of Chapter II, $\mathfrak{Y}_0, \mathfrak{Y}_1, \ldots, \mathfrak{Y}_n$ span the dual S-ring $\mathfrak{S}^*$, where $\mathfrak{Y}_i = \sum_{\Delta \in Y_i} \Delta$. So comparing (2.4) and (2.5), the correspondence of $\mathfrak{X}_i$ to $\mathfrak{Y}_i$ gives an isomorphism of $\mathfrak{S}$ to $\mathfrak{S}^*$. By Proposition 5.4 of Chapter II,

$$\Delta(\mathfrak{X}_k) = p_k(i) \qquad (\Delta \in Y_i). \tag{2.6}$$

We introduce the Krawtchouk polynomials (cf. Szegö (1939)) as follows:

$$K_k(u) = \sum_{j=0}^{k} (-1)^j (q-1)^{k-j} \binom{u}{j}\binom{n-u}{k-j} . \tag{2.7}$$

This is a polynomial of degree k in the variable u.

THEOREM 2.3. Let $P = (p_k(i))$ and $Q = (q_k(i))$ be the eigenmatrices of the Hamming scheme $H(n,q)$. Then

$$p_k(i) = q_k(i) = K_k(i).$$

Proof. By Proposition 2.2, (2.6) and the fact $\mathfrak{S} = \mathfrak{S}^*$, the assertion is trivial. □

Verify that the Krawtchouk polynomials can be expressed as follows (cf. (2.10)):

$$K_k(u) = \sum_{i=0}^{k} (-q)^i (q-1)^{k-i} \binom{n-i}{k-i}\binom{u}{i}. \tag{2.8}$$

The polynomials $v_i(x)$ in Theorem 1.3 differ from $K_i(u)$ just by an affine transformation, namely

$$v_i(x) = K_i(u) \qquad (x = K_1(u) = (q-1)n - qu). \tag{2.9}$$

The generating function of $K_k(u)$ is

$$\sum_{k=0}^{\infty} K_k(u) t^k = (1 + (q-1)t)^{n-u}(1-t)^u. \tag{2.10}$$

Let ${}_rF_s$ be a hypergeometric series:

$${}_rF_s\left(\begin{matrix} a_1, \ldots, a_r \\ & ;x \\ b_1, \ldots, b_s \end{matrix}\right) = \sum_{i=0}^{\infty} \frac{(a_1)_i \cdots (a_r)_i}{(b_1)_i \cdots (b_s)_i} \frac{x^i}{i!}, \tag{2.11}$$

where

$$(a)_i = \begin{cases} a(a+1) \ldots (a+i-1) & i = 1, 2, \ldots, \\ 1 & i = 0. \end{cases} \tag{2.12}$$

Then

$$K_k(u) = \binom{n}{k} (q-1)^k {}_2F_1\left(\begin{matrix}-k, -u\\ -n\end{matrix}; \frac{q}{q-1}\right). \tag{2.13}$$

To see this, put $\binom{u}{i} = (-1)^i \frac{(-u)_i}{i!}$ and $\binom{n-i}{k-i} = \binom{n}{k}\frac{(-k)_i}{(-n)_i}$ into (2.8).

By Theorem 1.6 and Theorem 2.3, ${}_2F_1\left(\begin{matrix}-k, -u\\ -n\end{matrix}; \frac{q}{q-1}\right)$ is the zonal spherical function $\omega_u(k)$ of the permutation group $G = S_q \text{ wr } S_n$ on X.

JOHNSON SCHEME $J(v,k)$. Let S be a set of cardinality v and $X = \{T \subset S \mid |T| = k\}$ $(k \leq v/2)$. The distance between $T_1, T_2 \in X$ is

$$k - |T_1 \cap T_2|. \tag{2.14}$$

Let R_i be the ith distance relation on X, i.e.,

$$R_i = \{(T_1, T_2) \mid |T_1 \cap T_2| = k - i\}. \tag{2.15}$$

Then $\mathfrak{X} = (X, \{R_i\}_{0 \leq i \leq k})$ is a symmetric association scheme and called the <u>Johnson scheme</u> $J(v,k)$. The intersection matrix B_1 is

$$B_1 = \begin{Bmatrix} * & 1 & \ldots & i^2 & \ldots & k^2 \\ 0 & v-2 & \ldots & i(v-2i) & \ldots & k(v-2k) \\ k(v-k) & (k-1)(v-k-1) & \ldots & (k-i)(v-k-i) & \ldots & * \end{Bmatrix}. \tag{2.16}$$

Let G be a symmetric group on S. Then G acts on X naturally and this action is distance-transitive.

We want to find the eigenmatrices P, Q, but the situation is not so simple as in the case of Hamming schemes because G does not have a regular subgroup. Let $\mathcal{P}_i(S) = \{T \subset S \mid |T| = i\}$. Then G acts on $\mathcal{P}_i(S)$ transitively in the natural way. Let π_i be the permutation character of G on $\mathcal{P}_i(S)$. In particular, $X = \mathcal{P}_k(S)$ and π_k is the character of G on X. Notice $\pi_i(a)$ is the number of elements of $\mathcal{P}_i(S)$ fixed by a.

LEMMA 2.4.

$$(\pi_i, \pi_j)_G = i + 1 \qquad (0 \leq i \leq j \leq v/2).$$

<u>Proof</u>. Since $\pi_i\pi_j$ is the permutation character of G on $\mathcal{P}_i(S) \times \mathcal{P}_j(S)$,

$$\begin{aligned}(\pi_i, \pi_j)_G &= (1, \pi_i\pi_j)_G \\ &= \text{the number of orbits of } G \text{ on } \mathcal{P}_i(S) \times \mathcal{P}_j(S) \\ &= i + 1. \quad \square\end{aligned}$$

THEOREM 2.5. Let $\chi_0 = 1_G$, the identity character, and $\chi_i = \pi_i - \pi_{i-1}$ $(1 \leq i \leq v/2)$. Then

(i) $\chi_0, \chi_1, \ldots, \chi_{[v/2]}$ are distinct irreducible characters of G,

(ii) $\pi_i = \chi_0 + \chi_1 + \cdots + \chi_i \qquad (0 \leq i \leq v/2)$.

<u>Proof</u>. Assume that $\chi_0, \chi_1, \ldots, \chi_{i-1}$ are distinct irreducible characters of G and

$$\pi_{i-1} = \chi_0 + \chi_1 + \cdots + \chi_{i-1}.$$

For $1 \leq j \leq i - 1$,

$$(\pi_i, \chi_j) = (\pi_i, \pi_j) - (\pi_i, \pi_{j-1})$$

$$= (j + 1) - j \qquad \text{(by Lemma 2.4)}$$

$$= 1.$$

Thus π_i contains χ_j with multiplicity one for $1 \le j \le i-1$. Since G is transitive on $\mathcal{P}_i(S)$, we also have $(\pi_i, \chi_0) = 1$. Since $(\pi_i, \pi_i) = i + 1$, $\pi_i - \pi_{i-1}$ must be an irreducible character different from each χ_j $(0 \le j \le i-1)$. So the proof is completed by induction. □

χ_i is the irreducible character corresponding to the Young diagram

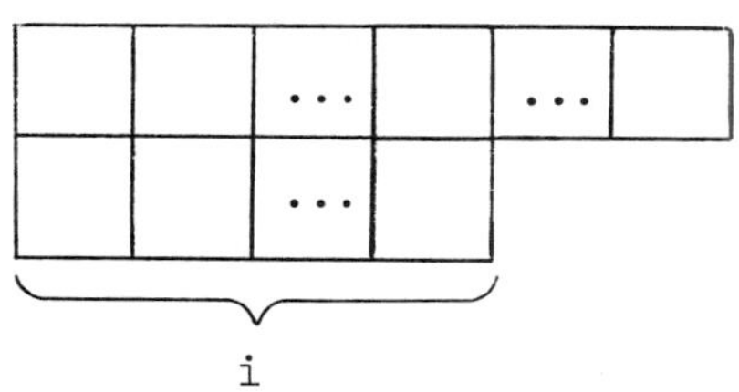

For the complete determination of irreducible characters of G, see James and Kerber (1981). (Orginally due to Frobenius, Young, et al.)

THEOREM 2.6. For $1 \le i \le v/2$, $1 \le j \le v/2$,

$$(\chi_1, \chi_i\chi_j) = \begin{cases} 1 & \text{if } |i-j| \le 1 \text{ and unless } i = j = v/2, \\ 0 & \text{if } |i-j| > 1 \text{ or } i = j = v/2. \end{cases}$$

Proof. We may assume $i \le j$. By counting the number of G-orbits on $\mathcal{P}_1(S) \times \mathcal{P}_i(S) \times \mathcal{P}_j(S)$, we get

$$(\pi_1, \pi_i\pi_j) = \begin{cases} 4i + 2 & \text{if } 0 \le i < j \le v/2, \\ 4i + 1 & \text{if } 0 \le i = j < v/2, \\ 4i & \text{if } 1 \le i = j = v/2. \end{cases} \tag{2.17}$$

Since $(\pi_0, \pi_i\pi_j) = i + 1 \quad (0 \le i \le j \le v/2)$,

$$(\chi_1, \pi_i\pi_j) = \begin{cases} 3i + 1 & \text{if } 0 \le i < j \le v/2, \\ 3i & \text{if } 0 \le i = j < v/2, \\ 3i - 1 & \text{if } 1 \le i = j = v/2. \end{cases} \tag{2.18}$$

Apply (2.18) to $\chi_i\chi_j = \pi_i\pi_j + \pi_{i-1}\pi_{j-1} - \pi_{i-1}\pi_j - \pi_i\pi_{j-1}$ to get the desired result. □

COROLLARY TO THEOREM 2.6. Let q_{ij}^k be the Krein parameters. Then $q_{1i}^j = 0$ for $|i - j| > 1$ or $i = j = v/2$. In particular, $J(v,k)$ is a Q-polynomial scheme.

<u>Proof</u>. By Theorem 8.1 of Chapter II, this is trivial. □

One possible way to find $p_i(j)$ is to use the formula of Corollary 11.7 in Chapter II:

$$p_i(j) = \frac{1}{|H|} \sum_{x \in Ha_iH} \chi_j(x),$$

where H is the stabilizer of $T_0 \in \mathcal{P}_k(S)$ and $Ha_iH = \{x \in G \mid |T_0 \cap T_0^x| = k - i\}$. Actually this method works by counting in two ways the number of

$$\{(x,U) \in Ha_iH \times \mathcal{P}_j(S) \mid U^x = U\},$$

giving $\sum_{x \in Ha_iH} \pi_j(x)$ in one way, but we end up with a somewhat complicated sum of binomial coefficients. Here we adopt a different approach, namely the use of spherical functions. (See Dunkl (1979b).)

Let x_i be the ith coordinate function, i.e., for $T \subseteq S = \{1,2,\ldots,v\}$,

$$x_i(T) = \begin{cases} 1 & \text{if } i \in T, \\ 0 & \text{otherwise.} \end{cases} \qquad (2.19)$$

Regarding x_i as an element of $C(\mathcal{P}_j(S))$, the set of functions from $\mathcal{P}_j(S)$ to $\mathbb{C}$, $\prod_{i \in T} x_i$ is the characteristic function of T for $T \in \mathcal{P}_j(S)$. So $\{\prod_{i \in T} x_i \mid T \in \mathcal{P}_j(S)\}$ is a basis of $C(\mathcal{P}_j(S))$. G acts on $C(\mathcal{P}_j(S))$ by $f^a(T) = f(T^{a^{-1}})$ for $f \in C(\mathcal{P}_j(S))$ and $T \in \mathcal{P}_j(S)$. This action on the basis $\{\prod_{i \in T} x_i\}$ is isomorphic to that on $\mathcal{P}_j(S)$ and so has the same permutation character π_j.

Since the coordinate functions x_i are elements of $C(\mathcal{P}_k(S))$, each $C(\mathcal{P}_j(S))$ is embedded in $C(\mathcal{P}_k(S))$ naturally $(0 \le j \le k)$. From now on, the $C(\mathcal{P}_j(S))$ $(0 \le j \le k)$ are regarded as subspaces of $C(\mathcal{P}_k(S))$ by this embedding.

Let D be the formal differential operator of $C(\mathcal{P}_k(S))$ given by

$$D = \sum_{i=0}^{d} \frac{\partial}{\partial x_i}. \qquad (2.20)$$

Let

$$V_j = \ker D \cap C(\mathcal{P}_j(S)). \qquad (2.21)$$

PROPOSITION 2.7. V_j is the space of spherical functions

belonging to χ_j.

<u>Proof</u>. D maps $C(\mathcal{P}_j(S))$ into $C(\mathcal{P}_{j-1}(S))$. We want to show this is an onto map. For $T \subseteq S$, let x_T denote $\prod_{i \in T} x_i$. Fix $U_0 \in \mathcal{P}_{j-1}(S)$. Set $y_i = \sum_{\substack{|T \cap U_0| = i \\ T \in \mathcal{P}_j(S)}} x_T$ and $z_i = \sum_{\substack{|U \cap U_0| = i \\ U \in \mathcal{P}_{j-1}(S)}} x_U$. Then

$$D(y_i) = (v - 2j + 2 + i)z_i + (j - i)z_{i-1}.$$

So,

$$D\left(\sum_{t=0}^{j-1} \frac{t!}{(-v+j)_t} y_{j-1-t}\right) = (v - j + 1)z_{j-1}$$
$$= (v - j + 1)x_{U_0} .$$

Since D commutes with the action of G, $C(\mathcal{P}_j(S))/V_j$ is isomorphic to $C(\mathcal{P}_{j-i}(S))$ as G-spaces. So, the character of G afforded by V_j is $\pi_j - \pi_{j-1} = \chi_j$. □

For $T \subseteq S$, let $\sigma_i(T)$ be the fundamental symmetric function of degree i in the variables x_j, $j \in T$.

LEMMA 2.8. (i) $D\sigma_i(T) = (|T| - i + 1)\sigma_{i-1}(T)$.

(ii) For disjoint subsets T, U of S, let

$$f = \sum_{t=0}^{j} (-1)^t(|U| - j + 1)_t(|T| - j + 1)_{j-t}\sigma_t(T)\sigma_{j-t}(U)$$

for $j \leq \min\{|T|, |U|\}$. Then

$Df = 0.$

<u>Proof</u>. (i) Check directly.

(ii) By (i),

$$Df = \sum_{t=0}^{j} (-1)^t (|U| - j + 1)_t (|T| - j + 1)_{j-t+1} \sigma_{t-1}(T) \cdot$$

$$\sigma_{j-t}(U) + \sum_{t=0}^{j} (-1)^t (|U| - j + 1)_{t+1} \cdot$$

$$(|T| - j + 1)_{j-t} \sigma_t(T) \sigma_{j-t-1}(U)$$

$$= \sum_{s=0}^{j-1} (-1)^{s+1} (|U| - j + 1)_{s+1} (|T| - j + 1)_{j-s} \cdot$$

$$\sigma_s(T) \sigma_{j-s-1}(U) + \sum_{t=0}^{j-1} (-1)^t (|U| - j + 1)_{t+1} \cdot$$

$$(|T| - j + 1)_{j-t} \sigma_t(T) \sigma_{j-t-1}(U)$$

$$= 0. \qquad \square$$

Let T_0 be a fixed base point of $\mathcal{P}_k(S)$ and H the stabilizer of T_0 in G. H has order $k!(v-k)!$. Put $T = T_0$, $U = S - T_0$ in Lemma 2.8(ii). Then f is H-invariant and so a zonal spherical function belonging to χ_j.

THEOREM 2.9. Let ω_j be the zonal spherical function belonging to χ_j and let $\omega_j(i)$ be the value of ω_j at $T \in \mathcal{P}_k(S)$ with $|T_0 \cap T| = k - i$. Then

$$\omega_j(i) = {}_3F_2\left(\begin{matrix}-i\,,\,-j\,,\,-v-1+j\\ -k\,,\,-v+k\end{matrix};1\right).$$

(This is called the <u>Hahn polynomial</u> $Q_j(i\,;-v+k-1\,,-k-1\,,k)$, where

$$Q_j(y) = Q_j(y\,;\alpha\,,\beta\,,N) = {}_3F_2\left(\begin{matrix}-y\,,\,-j\,,\,j+\alpha+\beta+1\\ -N\,,\,\alpha+1\end{matrix};1\right).)$$

<u>Proof</u>. Put $T = T_0$, $U = S - T_0$ in Lemma 2.8(ii). Then the value of $\sigma_t(T_0)\sigma_{j-t}(S - T_0)$ at $T \in \mathcal{P}_k(S)$ with $|T_0 \cap T| = k - i$ is $\binom{k-i}{t}\binom{i}{j-t}$. Changing t to $j-t$,

$$f(T) = \sum_{t=0}^{j} (-1)^{j-t}(v-k-j+1)_{j-t}(k-j+1)_t \cdot \binom{k-i}{j-t}\binom{i}{t},$$

and

$$f(T_0) = (-1)^j(v-k-j+1)_j\binom{k}{j}.$$

Thus

$$\begin{aligned}\omega_j(i) &= \omega_j(T) = f(T)/f(T_0)\\ &= \frac{1}{\binom{k}{j}} \sum_{t=0}^{j} (-1)^t \frac{(v-k-j+1)_{j-t}}{(v-k-j+1)_j}(k-j+1)_t \cdot \binom{k-i}{j-t}\binom{i}{t}.\end{aligned}$$

Since $\dfrac{(v-k-j+1)_{j-t}}{(v-k-j+1)_j} = \dfrac{(-1)^t}{(-v+k)_t}$, $\dbinom{k-i}{j-t} =$
$\dfrac{(-1)^{j-t}(-k+i)_{j-t}}{(j-t)!} = (-1)^{j-t}\dfrac{(-j)_t(-k+i)_j}{j!(k-i-j+1)_t}$, and since
$\dbinom{i}{t} = (-1)^t\dfrac{(-i)_t}{t!}$, we have

$$\omega_j(i) = \frac{(-k+i)_j}{(-k)_j}\, {}_3F_2\left(\begin{matrix}-i\,,\,-j\,,\,k-j+1\\ -v+k\,,\,k-i-j+1\end{matrix};1\right). \tag{2.22}$$

We apply the following formula to (2.22) with $a = -v-1+j$, $b = -i$, $c = -v+k$, $d = -k$, $n = j$:

$$ {}_3F_2\left(\begin{matrix}-n\,,\,a\,,\,b\\ c\,,\,d\end{matrix};1\right) \tag{2.23}$$

$$= \frac{(d-b)_n}{(d)_n}\, {}_3F_2\left(\begin{matrix}-n\,,\,c-a\,,\,b\\ c\,,\,1+b-d-n\end{matrix};1\right).$$

(See Gasper (1974, p. 187 (2.7)), Bailey (1935, p. 22).)
Then we get the desired result. □

COROLLARY TO THEOREM 2.9. Let $P = (p_i(j))$, $Q = (q_j(i))$ be the eigenmatrices of $J(v,k)$ and let k_i and m_j be the valencies and multiplicities. Then

(1) $k_i = \dbinom{k}{i}\dbinom{v-k}{i}$,

$m_j = \dfrac{v-2j+1}{v-j+1}\dbinom{v}{j}$,

and

$$(2)\quad \frac{p_i(j)}{k_i} = \frac{q_j(i)}{m_j} = {}_3F_2\left(\begin{matrix}-i\,,\,-j\,,\,-v-1+j\\ -k\,,\,-v+k\end{matrix};1\right).$$

Proof. (1) The valency k_i is the number of $T \in \mathcal{P}_k(S)$ such that $|T_0 \cap T| = k - i$. The multiplicity m_j is the degree of χ_j.

(2) By Theorem 11.10 of Chapter II, $p_i(j)/k_i = \omega_j(i)$. By Theorem 3.5 of Chapter II, $q_j(i)/m_j = p_i(j)/k_i$. □

THEOREM 2.10. (1) Let $E_i(u)$ be the polynomial of degree $2i$ in u given by

$$E_i(u) = \sum_{t=0}^{i} (-1)^{i-t}\binom{k-t}{i-t}\binom{k-u}{t}\binom{v-k+t-u}{t}$$

$$= (-1)^i\binom{k}{i}{}_3F_2\left(\begin{matrix}-i\,,\,-k+u\,,\,v-k-u+1\\ -k\,,\,1\end{matrix};1\right).$$

(This is called the Eberlein polynomial (or dual Hahn polynomial), see Delsarte (1973b, p. 48).) Then the eigenmatrix $P = (p_i(j))$ is given by

$$p_i(j) = E_i(j).$$

Another expression for $E_i(u)$ is

$$E_i(u) = \sum_{t=0}^{i} (-1)^t \binom{u}{t}\binom{k-u}{i-t}\binom{v-k-u}{i-t}$$

$$= \binom{k-u}{i}\binom{v-k-u}{i}\cdot$$

$${}_3F_2\left(\begin{matrix}-i\,,\,-i\,,\,-u\\ k-i-u+1\,,\,v-i-u-k+1\end{matrix};1\right).$$

(2) The polynomials $v_i(x)$ in Theorem 1.3 are given by

$$v_i(x) = E_i(u) \qquad (x = u^2 - (v+1)u + k(v-k)).$$

Proof. (1) Apply (2.23) with $n = i$, $a = -v - 1 + j$, $b = -j$, $c = -k$, $d = -v + k$ to get

$$p_i(j) = k_i \; {}_3F_2\left(\begin{matrix} -i, -v-1+j, -j \\ -k, -v+k \end{matrix}; 1\right) \tag{2.24}$$

$$= k_i \frac{(-v+k+j)_i}{(-v+k)_i} \cdot {}_3F_2\left(\begin{matrix} -i, -j, v+1-j-k \\ -k, v+1-i-j-k \end{matrix}; 1\right).$$

Apply (2.23) to (2.24) with $n = i$, $a = -j$, $b = v + 1 - j - k$, $c = -k$, $d = v + 1 - i - j - k$ to get

$$p_i(j) = k_i \frac{(-v+k+j)_i}{(-v+k)_i} \frac{(-i)_i}{(v-i-j-k+1)_i} \cdot$$

$${}_3F_2\left(\begin{matrix} -i, -k+j, v+1-j-k \\ -k, 1 \end{matrix}; 1\right).$$

Since

$$k_i(-v+k+j)_i(-1)_i/(-v+k)_i(v-i-j-k+1)_i = (-1)^i\binom{k}{i},$$

we get $p_i(j) = E_i(j)$.

By (2.22),

$$p_i(j) = k_i \frac{(-k+i)_j}{(-k)_j} \; {}_3F_2\left(\begin{matrix} -i, k-j+1, -j \\ k-i-j+1, -v+k \end{matrix}; 1\right).$$

Apply (2.23) with $n = i$, $a = k - j + 1$, $b = -j$, $c = k - i - j + 1$, $d = -v + k$. Then we get the second expression for $E_i(u)$.

(2) By the first expression for $E_i(u)$, $E_i(u)$ is a polynomial in $(k-u)(v-k-u+1)$ and hence in $x = (k-u)(v-k-u+1)-k$. Let $E_i(u) = F_i(x)$. Then $F_i(x)$ is a polynomial of degree i and $F_i(\theta_j) = p_i(j) = v_i(\theta_j)$ $(0 \le i \le d)$, where $\theta_j = E_1(j) = (k-j)(v-k-j+1)-k$ are the eigenvalues of B_1 $(0 \le j \le d)$. Since the polynomial $F_i(x) - v_i(x)$ of degree at most d has at least $d+1$ roots, we get $F_i(x) = v_i(x)$. □

Let q_{ij}^h be the Krein parameter of $J(v,k)$ and $\hat{B}_1 = (q_{1i}^j)$ the dual intersection matrix. By Corollary to Theorem 2.6, $\hat{B}_1$ is a tridiagonal matrix:

$$\hat{B}_1 = \begin{Bmatrix} * & c_1^* & \cdots & c_i^* & \cdots & c_d^* \\ 0 & a_1^* & \cdots & a_i^* & \cdots & a_d^* \\ m & b_1^* & \cdots & b_i^* & \cdots & * \end{Bmatrix}, \tag{2.25}$$

where $m = m_1 = v-1$, $d = k$. We want to calculate a_j^*, c_j^*, b_j^*. Let $u_j^*(x)$ be the polynomial of degree j given by

$$xu_j^*(x) = c_j^* u_{j-1}^*(x) + a_j^* u_j^*(x) + b_j^* u_{j+1}^*(x) \tag{2.26}$$

for $1 \le i \le d-1$, and $u_0^*(x) = 1$, $u_1^*(x) = \frac{1}{m}x$. Then by the Q-polynomial scheme version of Proposition 1.1(iv) and Remark (7) of section 3.1, we have $u_j^*(\theta_i^*) = \frac{1}{m_j} q_j(i)$, where the $\theta_i^* = q_1(i)$ are the eigenvalues of $\hat{B}_1$. Since $\frac{1}{m_j} q_j(i) = \frac{1}{k_i} p_i(j) = Q_j(i)$ by Corollary to Theorem 2.9,

$$u_j^*(\theta_i^*) = Q_j(i),$$

where $Q_j(y) = Q_j(y\colon -v+k-1\,,-k-1\,,k)$ is the Hahn polynomial and has degree j in y. Let

$$x = (v-1)(1 - \frac{vy}{k(v-k)}) \,. \tag{2.27}$$

Then

$$u_j^*(x) = Q_j(y) \qquad (0 \le i \le d)\,, \tag{2.28}$$

because $\theta_i^* = (v-1)(1 - \frac{vi}{k(v-k)})$ and $u_j^*(\theta_i^*) = Q_j(i)$ for $0 \le i \le d$. It is known that in general, the Hahn polynomials $Q_j(y) = Q_j(y\,;\alpha\,,\beta\,,N)$ satisfy the recurrence

$$\begin{aligned} -yQ_j(y) &= s_jQ_{j-1}(y) - (s_j + t_j)Q_j(y) \\ &\qquad + t_jQ_{j+1}(y), \end{aligned} \tag{2.29}$$

for all y if $0 \le j \le N-1$, and for $y = 0, 1, 2, \ldots, N$ if $j = N$, where

$$s_j = \frac{j(j+\beta)(j+\alpha+\beta+N+1)}{(2j+\alpha+\beta)(2j+\alpha+\beta+1)}$$

and

$$t_j = \frac{(j+\alpha+\beta+1)(j+\alpha+1)(N-j)}{(2j+\alpha+\beta+1)(2j+\alpha+\beta+2)} \,.$$

(See Karlin and McGregor (1961, p. 34).) In our case, we can directly check (2.29) because we know the $Q_j(y)$'s satisfy a 3-term recurrence by (2.26), (2.28), and we can determine the coefficients of the recurrence by putting $y = 0, 1, 2$. Rewriting (2.29) in terms of x and $u_j^*(x)$ with $\alpha = -v+k-1$, $\beta = -k-1$, $N = k$, we get the following theorem:

THEOREM 2.11. Let

$$\hat{B}_1 = \begin{Bmatrix} * & c_1^* & \cdots & c_d^* \\ 0 & a_1^* & \cdots & a_d^* \\ m & b_1^* & \cdots & * \end{Bmatrix}$$

be the dual intersection matrix (q_{1i}^j) of $J(v,k)$. Then

$$c_j^* = \frac{v(v-1)}{k(v-k)} s_j,$$

$$a_j^* = (v-1) - \frac{v(v-1)}{k(v-k)}(s_j + t_j)$$

and

$$b_j^* = \frac{v(v-1)}{k(v-k)} t_j,$$

where

$$s_j = \frac{j(k-j+1)(v-k-j+1)}{(v-2j+2)(v-2j+1)}$$

and

$$t_j = \frac{(v-j+1)(v-k-j)(k-j)}{(v-2j+1)(v-2j)} .$$

REMARKS. (1) $a_j^* = 0$ $(j = 0, 1, \ldots, d)$ if $v = 2k$.

(2) In $J(v,k)$, $b_j^* > b_{j+1}^*$, but the inequality $c_j^* \leq c_{j+1}^*$ <u>does</u> <u>not</u> hold in general.

3.3 MOORE GRAPHS AND RELATED EXTREMAL CASES

The purpose of this section is to study P-polynomial schemes, i.e., distance-regular graphs. If, in addition to a P-polynomial structure, we assume another P-polynomial structure or a Q-polynomial structure, we can say very strong things about the parameters (cf. section 3.4 and section 3.5). But with only one P-polynomial structure, the general classification of them is very difficult. So we treat here very limited, "extremal" but important classes and classify some of them. We hope that the method here will be applied to the further study of P-polynomial association schemes.

Let Γ be a distance-regular graph with valency k, diameter d and n vertices. The <u>girth</u> of Γ is the minimum length of circuits. Then

$$n \leq 1 + k + k(k-1) + k(k-1)^2 + \dots + k(k-1)^{d-1}, \qquad (3.1)$$

and

$$g \leq 2d + 1. \qquad (3.2)$$

REMARK. The bound (3.1) is of order k^d, while the absolute bound condition (Theorem 4.9 of Chapter II) is of order $m^d/d!$.

Let B be the intersection matrix of Γ. Then it is easy to see the following conditions are equivalent:

(i) The equality holds in (3.1),

(ii) The equality holds in (3.2),

(iii)
$$B = \begin{Bmatrix} * & 1 & 1 & \dots & 1 & 1 \\ 0 & 0 & 0 & \dots & 0 & k-1 \\ k & k-1 & k-1 & \dots & k-1 & * \end{Bmatrix}.$$

If a distance-regular graph Γ satisfies (one of) the above three equivalent conditions, Γ is called a <u>Moore</u>

graph.

THEOREM 3.1. Let Γ be a Moore graph with valency k and diameter d. Then either (i) $k = 2$ or (ii) $d = 2$ and $k = 2, 3, 7, 57$.

If $k = 2$, then Γ is a $(2d+1)$-gon. If $d = 2$ and $k = 3$, Γ is Petersen's graph, i.e., O_3 $(n = 10)$. If $d = 2$ and $k = 7$, Γ is the Hoffman-Singleton graph obtained by the permutation group $U_3(5)$ on $U_3(5)/A_7$ $(n = 50)$. If $d = 2$ and $k = 57$, the existence and uniqueness problem of Γ is still open $(n = 3250)$, but if such Γ exists, Γ is not distance-transitive (Aschbacher (1971)) and the full automorphism group of Γ contains a subgroup of odd order whose index is at most 2 (G. Higman, unpublished). Theorem 3.1 is due to Hoffman-Singleton (1960) for the case $d = 2, 3$ and Bannai-Ito (1973), Damerell (1973) for $d \geq 3$.

Proof of Theorem 3.1. The proof is divided into several steps. In steps 1, 2 and 3, we use (1.4), (1.5) to calculate the eigenvalues and the multiplicities. In our case, they become

$$F_i(x) = xF_{i-1}(x) - (k-1)F_{i-2}(x) \qquad (2 \leq i \leq d) \tag{3.3}$$

with $F_0(x) = 1$, $F_1(x) = x + 1$, and

$$m_i = \frac{nk(k-1)^{d-1}}{(k-\theta_i)F'_d(\theta_i)F_{d-1}(\theta_i)} \quad (\theta_i \neq k). \tag{3.4}$$

Step 1. Let

$$T = \begin{pmatrix} x & 1 \\ -(k-1) & 0 \end{pmatrix}.$$

Then the recurrence (3.3) can be written

$$(F_i(x), F_{i-1}(x)) = (F_{i-1}(x), F_{i-2}(x))T \tag{3.5}$$

$$= (F_1(x), F_0(x))T^{i-1}$$

$$= (1,0)T^i + (1,0)T^{i-1}$$

$(2 \le i \le d)$.

Let λ, μ be the eigenvalues of T. Then

$$\lambda + \mu = x, \quad \lambda\mu = s^2 \qquad (s = \sqrt{k-1}). \tag{3.6}$$

The eigenvectors of T are $(\lambda,1)$ and $(\mu,1)$. Since $(1,0) = \frac{1}{\lambda - \mu}(\lambda,1) - \frac{1}{\lambda - \mu}(\mu,1)$, the identity (3.5) becomes

$$F_i(x) = \frac{\lambda^i(\lambda + 1) - \mu^i(\mu + 1)}{\lambda - \mu} \qquad (0 \le i \le d). \tag{3.7}$$

Let $\lambda = se^{\sqrt{-1}\,\varphi}$ and $\mu = se^{-\sqrt{-1}\,\varphi}$. Then

$$x = \lambda + \mu = 2s\cos\varphi, \tag{3.8}$$

and by (3.7),

$$F_d(x) = \frac{s^{d-1}}{\sin\varphi}\{s\sin(d+1)\varphi + \sin d\varphi\}. \tag{3.9}$$

<u>Step 2</u>. Let $\theta_1 > \theta_2 > \dots > \theta_d$ be the roots of $F_d(x)$ and $\theta_i = 2s\cos\varphi_i$. We want to show

$$\frac{i\pi}{d+1} < \varphi_i < \frac{i\pi}{d}. \tag{3.10}$$

The signs of $F_d(x)$ are $(-1)^{i-1}$ and $(-1)^i$ for $x = 2s\cos\frac{i\pi}{d+1}$ and $2s\cos\frac{i\pi}{d}$ because $s\sin(d+1)\varphi + \sin d\varphi = (-1)^{i-1}\sin\frac{i\pi}{d+1}$ if $\varphi = \frac{i\pi}{d+1}$, $(-1)^i s\sin\frac{i\pi}{d}$ if $\varphi = \frac{i\pi}{d}$, and because

$$\lim_{\varphi\to\pi} \frac{1}{\sin\varphi}\{s\sin(d+1)\varphi + \sin d\varphi\} = (-1)^d s(d+1) + (-1)^{d-1}d.$$

By the intermediate value theorem, (3.10) holds.

<u>Step 3</u>. Let θ be a root of $F_d(x)$. By (3.10), we may assume $\theta = \lambda + \mu = 2s\cos\varphi$ with $0 < \varphi < \pi$. Then $\lambda \neq \mu$ and by putting $i = d$ in (3.7), $\lambda^d(\lambda+1) - \mu^d(\mu+1) = 0$, i.e.,

$$\left(\frac{\lambda}{\mu}\right)^d = \frac{\mu+1}{\lambda+1}. \tag{3.11}$$

So

$$\begin{aligned} F_{d-1}(\theta) &= \frac{\mu^{d-1}}{\lambda-\mu}\left\{\left(\frac{\lambda}{\mu}\right)^{d-1}(\lambda+1) - (\mu+1)\right\} \\ &= \frac{\mu^{d-1}}{\lambda-\mu}\left\{\frac{\mu}{\lambda}(\mu+1) - (\mu+1)\right\} \\ &= -\frac{1}{\lambda}\mu^{d-1}(\mu+1). \end{aligned} \tag{3.12}$$

Take the derivative of (3.6) with respect to x to get $\lambda'\mu + \lambda\mu' = 0$, $\lambda' + \mu' = 1$, and so

$$\lambda' = \frac{\lambda}{\lambda-\mu}, \quad \mu' = -\frac{\mu}{\lambda-\mu}. \tag{3.13}$$

Since $F_d(\theta) = 0$, the value of the derivative of (3.7) $(i = d)$ at θ is

$$F_d'(\theta) = \frac{1}{\lambda - \mu}\{(d+1)\lambda^d\lambda' + d\lambda^{d-1}\lambda'$$

$$- (d+1)\mu^d\mu' - d\mu^{d-1}\mu'\}$$

$$= \frac{1}{(\lambda - \mu)^2}\{(d+1)\lambda^{d+1} + d\lambda^d + (d+1)\mu^{d+1}$$

$$+ d\mu^d\} \qquad \text{(by (3.13))}$$

$$= \frac{\lambda^d}{(\lambda - \mu)^2}\{(d+1)\lambda + d + (d+1)\mu\,\frac{\lambda+1}{\mu+1}$$

$$+ d\,\frac{\lambda+1}{\mu+1}\} \qquad \text{(by (3.11))}$$

$$= \frac{\lambda^d}{(\lambda - \mu)^2(\mu+1)}\{(d+1)(2\lambda\mu + \lambda + \mu)$$

$$+ d(\lambda + \mu + 2)\}$$

$$= \frac{\lambda^d}{(\theta^2 - 4s^2)(\mu+1)}\{(2d+1)(\theta + k) + k - 2\} \qquad (3.14)$$

(by (3.6)).

By (3.12) and (3.14),

$$F_d'(\theta)F_{d-1}(\theta) = \frac{(\lambda\mu)^{d-1}}{4s^2 - \theta^2}\{(2d+1)(\theta + k) + k - 2\}.$$

So the multiplicity of θ in A (the adjacency matrix of Γ) is, by (3.4),

$$m(\theta) = \frac{nk}{g}\,\frac{4s^2 - \theta^2}{(k - \theta)(\theta + k + \frac{k-2}{g})} \qquad (g = 2d + 1). \qquad (3.15)$$

<u>Step 4</u>. We want to show that if $d \geq 3$ and $k \geq 3$, then

$$
\begin{aligned}
&m(\theta_1) < m(\theta_i) \qquad (i = 2, 3, \ldots, d-1),\\
&m(\theta_d) < m(\theta_i) \qquad (i = 2, 3, \ldots, d-1).
\end{aligned} \tag{3.16}
$$

We can regard $m(\theta)$ in (3.15) as a continuous function of θ in the interval $[-2s, 2s]$. For any real number y, the equation $m(\theta) = y$ has at most two solutions. So the graph of $y = m(\theta)$ is unimodal in $[-2s, 2s]$. Thus in order to prove (3.16), it suffices to show

$$
\begin{aligned}
&m(\theta_1) < m(\theta_{d-1}),\\
&m(\theta_d) < m(\theta_2).
\end{aligned} \tag{3.17}
$$

The inequality $m(\theta_d) < m(\theta_2)$ is trivial, because $m(\theta) = \frac{nk}{g} a(\theta)b(\theta)$ and $a(\theta_d) < a(\theta_2)$, $b(\theta_d) < b(\theta_2)$, where

$$a(\theta) = (4s^2 - \theta^2)/(k^2 - \theta^2),$$

$$b(\theta) = 1/\left(1 + \frac{k-2}{g(k+\theta)}\right).$$

To prove $m(\theta_1) < m(\theta_{d-1})$, we need to know more precise locations of θ_1 and θ_{d-1} than (3.10), namely

$$
\begin{aligned}
&\frac{\pi}{d+1} < \varphi_1 < \frac{2\pi}{2d+1},\\
&\frac{3\pi}{2d+1} < \pi - \varphi_{d-1} < \frac{2\pi}{d+1}.
\end{aligned} \tag{3.18}
$$

We can easily verify (3.18) by the intermediate value theorem. Since $m(\theta)$ is unimodal and

$$-2s \cos \frac{3\pi}{2d+1} < \theta_{d-1} < 2s \cos \frac{2\pi}{2d+1} < \theta_1,$$

in order to prove $m(\theta_1) < m(\theta_{d-1})$, it suffices to show

$$m(2s \cos 2\alpha) < m(-2s \cos 3\alpha) \quad (\alpha = \pi/(2d+1)). \tag{3.19}$$

The sign of $m(-2s \cos 3\alpha) - m(2s \cos 2\alpha)$ is the same as that of

$$\sin^2 3\alpha\{k^2 - 4s^2\cos^2 2\alpha + \frac{k-2}{g}(k - 2s \cos 2\alpha)\} \tag{3.20}$$
$$-\sin^2 2\alpha\{k^2 - 4s^2\cos^2 3\alpha + \frac{k-2}{g}(k + 2s \cos 3\alpha)\}.$$

Since

$$\sin^2 3\alpha - \sin^2 2\alpha = \sin^2 3\alpha \cos^2 2\alpha - \sin^2 2\alpha \cos^2 3\alpha$$
$$= \sin \alpha \sin 5\alpha,$$

and since

$$\sin^2 3\alpha \cos 2\alpha - \sin^2 2\alpha \cos 3\alpha = \sin \alpha \sin 5\alpha$$
$$+ 8 \sin^2 \frac{3\alpha}{2} \sin^2\alpha \sin \frac{5\alpha}{2} \sin \frac{\alpha}{2},$$

(3.2) is equal to

$$\sin \alpha \sin 5\alpha\{k^2 + \frac{(k-2)k}{g} - 4s^2 \tag{3.21}$$
$$- \frac{2s(k-2)}{g}(1 + \frac{4 \sin^2 \frac{3\alpha}{2} \sin \alpha \sin \frac{\alpha}{2}}{\cos \frac{5\alpha}{2}})\}.$$

Since $\alpha \leqq \frac{\pi}{7}$,

$$4 \sin^2 \frac{3\alpha}{2} \sin \alpha \sin \frac{\alpha}{2}/\cos \frac{5\alpha}{2} < 0.35$$

and so (3.21) is greater than

$$\sin \alpha \sin 5\alpha\{k^2 + \frac{(k - 2)k}{g} - 4s^2 - \frac{2.7s(k - 2)}{g}\},$$

which is positive because $k \geq 3$ and $g \geq 7$. Thus the inequalities (3.16) are proved.

Step 5. Suppose $d \geq 3$ and $k \geq 3$. By Proposition 1.5 and (3.16), we get that $\theta_1 + \theta_d$ is a rational number. Since $F_d(x)$ is a monic polynomial, θ_i $(0 \leq i \leq d)$ are algebraic integers. Hence,

$$\theta_1 + \theta_d \text{ is an integer.} \tag{3.22}$$

By the recurrence (3.3),

$$F_d(x) = x^d + x^{d-1} + \text{the terms in } x^i \quad (0 \leq i \leq d-2)$$

and so

$$\theta_1 + \theta_2 + \dots + \theta_d = -1. \tag{3.23}$$

By (3.10),

$$\theta_i + \theta_{d-i} < 0. \tag{3.24}$$

By (3.23) and (3.24), we get $-1 < \theta_i + \theta_{d-i} < 0$ $(0 \leq i \leq d)$, which contradicts (3.22). Therefore we get $d = 2$ or $k = 2$.

Step 6. Suppose $d = 2$ and $k \neq 2$. Then by (3.3),

$$F_2(x) = x^2 + x - (k - 1). \tag{3.25}$$

Since $n = 1 + k^2$ and $g = 5$, (3.15) becomes

$$m(\theta) = \frac{k(k^2 + 1)(4k - 4 - \theta^2)}{(k - \theta)(6k - 2 + 5\theta)} . \tag{3.26}$$

Since $\theta_2 < -\theta_1$, we get $a(\theta_2) < a(\theta_1)$, $b(\theta_2) < b(\theta_1)$ and so $m(\theta_2) < m(\theta_1)$, where $m(\theta) = \frac{kn}{5} a(\theta)b(\theta)$ with

$$a(\theta) = (4k - 4 - \theta^2)/(k^2 - \theta^2),$$

$$b(\theta) = 1/(1 + \frac{k - 2}{5(k + \theta)}).$$

Hence the roots of $F_2(x)$ are rational and so $1 + 4(k - 1)$ is a square. Let $t^2 = 1 + 4(k - 1)$. Then $k = (t^2 + 3)/4$ and $\theta = (t - 1)/2$ is a root of $F_2(x)$. The identity (3.26) becomes

$$t^5 + t^4 + 6t^3 - 2t^2 + (9 - 32m)t - 15 = 0, \tag{3.27}$$

where $m = m(\theta)$ is an integer. Thus t divides 15 and we get $t = 1, 3, 5, 15$, giving $k = 1, 3, 7, 57$. This completes the proof of Theorem 3.1. □

Another extremal case is the distance-regular graph with intersection matrix

$$B = \begin{Bmatrix} * & 1 & 1 & \cdots & 1 & k \\ 0 & 0 & 0 & \cdots & 0 & 0 \\ k & k-1 & k-1 & \cdots & k-1 & * \end{Bmatrix}.$$

This is a special case of generalized d-gons with $s = t = k - 1$. An incidence system (P,L,I) is called a <u>generalized</u> <u>d-gon</u> if

(i) each element of P is incident to exactly $t + 1$ elements of L,

(ii) each element of L is incident to exactly

s + 1 elements of P, and

(iii) the associated bipartite graph on $P \cup L$ has diameter d and girth 2d.

Given a BN-pair whose Weyl group is a dihedral group of order 2d, the coset geometry becomes a generalized d-gon. (Feit and Higman (1964).)

THEOREM. (Feit-Higman (1964)) Let (P,L,I) be a generalized d-gon with parameters s, t. Then (P,L,I) is an ordinary polygon or $d = 2, 3, 4, 6, 8$ or 12. If furthermore $s > 1$ and $t > 1$, then $d = 2, 3, 4, 6$ or 8, and if $d = 3, 6, 8$, then the square-free part of st is 1, 1, 2, respectively.

In particular, if $s = t > 1$, then $d = 2, 3, 4$ or 6. (see also Singleton (1966), Biggs (1974).)

Let Γ be a distance-regular graph with intersection matrix

$$B = \begin{Bmatrix} * & 1 & 1 & \cdots & 1 & c \\ 0 & 0 & 0 & \cdots & 0 & k-c \\ k & k-1 & k-1 & \cdots & k-1 & * \end{Bmatrix}.$$

Γ is called a <u>generalized Moore graph</u>. If $c = 1$, Γ is a Moore graph. If $c = k$, Γ is a generalized d-gon. Let $k = \theta_0, \theta_1, \ldots, \theta_d$ $(\theta_i > \theta_{i+1})$ be the eigenvalues of B and $m(\theta_i)$ the multiplicity of θ_i in the incidence matrix A. Then it holds that

$$m(\theta_d) < m(\theta_1) \leq m(\theta_{d-1}) < m(\theta_2) \leq m(\theta_{d-2}) \cdots$$

$$m(\theta_i) \leq m(\theta_{d-i}) < m(\theta_{i+1}) \leq \cdots,$$

the equality $m(\theta_i) = m(\theta_{d-i})$ holding if and only if Γ is either a generalized d-gon or the odd graph O_4. (Bannai and Ito (1981).) This is a stronger version of the unimodal property and in particular, all θ_i are rational unless Γ is a generalized d-gon (because the eigenvalues of O_4 are rational). Using this fact, R. M. Damerell and M. A. Georgiacodis showed that the diameter of such a graph must be at most 5 (Damerell and Georgiacodis (1981a)).

An incidence system (P,L,I) is called a Moore geometry of diameter d if

(i) each element of P is incident to exactly $t+1$ elements of L,

(ii) each element of L is incident to exactly $s+1$ elements of P and

(iii) any two distinct elements of P are joined by a unique irreducible path of length at most d,

where an irreducible path of length r joining $p,q \in P$ means a sequence $p = p_0, \ell_1, p_1, \ell_2, p_2, \ldots, p_{r-1}, \ell_r, p_r = q$ such that $p_i \in P$ and $\ell_i \in L$ are all distinct and ℓ_i is incident both to p_{i-1} and p_i. A Moore geometry (P,L,I) induces a distance-regular graph on P with intersection matrix

$$B = \left\{ \begin{matrix} * & 1 & 1 & \cdots & 1 & 1 \\ 0 & s-1 & s-1 & \cdots & s-1 & (t+1)s-1 \\ (t+1)s & st & st & \cdots & st & * \end{matrix} \right\}.$$

Such a distance-regular graph must have diameter at most 2 unless it is an ordinary $(2d+1)$-gon. (Fuglister (1977) for $d = 3$, Damerell and Georgiacodis (1981b) for $d > 12$, Damerell (1981a) for $12 \geq d > 4$ and Fuglister (1983) for $d = 4$.)

THEOREM. (Fuglister-Damerell-Georgiacodis) Let (P,L,I) be a Moore geometry of diameter d. Then either (P,L,I) is an ordinary $(2d+1)$-gon or $d = 1, 2$.

The methods used to prove this theorem are quite number theoretical and involved.

A generalized Moore geometry of diameter d is an incidence system (P,L,I) with the properties (i), (ii), (iii)', where (i) and (ii) are the same as in Moore geometries and

(iii)' in the induced graph on P, any two distinct points with distance less than d are joined by a unique shortest path, and any two distinct points with distance d are joined by exactly c distinct shortest paths.

The induced graph on P is distance-regular with intersection-matrix

$$B = \left\{ \begin{matrix} * & 1 & 1 & \dots & 1 & c \\ 0 & s-1 & s-1 & \dots & s-1 & (t+1)s-c \\ (t+1)s & st & st & \dots & st & * \end{matrix} \right\}.$$

Generalized Moore geometries include, as special cases, Moore geometries (c = 1) and generalized n-gons ($d = \frac{n}{2}$ and $c = t+1$ if n is even, $d = \frac{n-1}{2}$ and $c = 1$ if n is odd). The induced graph on P of a generalized Moore geometry with s = 1 is a generalized Moore graph. The classification of generalized Moore geometries seems possible for diameters large enough, but so far only very special cases have been finished (cf. Roos and van Zanten (to appear) for c = s + 1).

PROBLEM. (1) Classify generalized Moore geometries.

Let Γ be a distance-regular graph with intersection matrix

$$B = \left\{ \begin{matrix} * & 1 & 2 & 3 & \dots & d-1 & c \\ 0 & 0 & 0 & 0 & \dots & 0 & k-c \\ k & k-1 & k-2 & k-3 & \dots & k-d+1 & * \end{matrix} \right\}.$$

The Hamming scheme $H(d,2)$ gives an example with $c = k = d$. Enomoto (1973a) and Egawa (1981) characterized $H(d,2)$ by the parameters, so Γ with $c = k = d$ is isomorphic to $H(d,2)$. There are two more known examples with $(k,d,c) = (23,3,3)$, $(22,3,6)$ which come from the Mathieu groups M_{23}, M_{22}.

PROBLEM. (2) Classify distance-regular graphs with the intersection matrix B given above.

If $c = d$, this problem is perhaps not so difficult.

The classification problem of distance-regular graphs with small valency k is interesting in view of the following conjectures.

CONJECTURES. (1) Let k be an arbitrarily given integer $(k \geq 3)$. If d is sufficiently large, say $d \geq d_0(k)$, then there exist no distance-regular graphs of diameter d and valency k. In particular, there are only finitely many distance-regular graphs of valency k.

(2) Let Γ be a distance-regular graph with valency k, diameter d and girth g. Then $d \leq c_0 kg$, where c_0 is an absolute constant. (Terwilliger (1982a) proved that $d \leq \frac{1}{2}(k-1)(g-1) + 1$ for <u>bipartite</u> distance-regular graphs. There are some subsequent related works by Terwilliger on this topic. See also section 3.9.)

There are only 12 distance-transitive graphs of valency 3 (Tutte (1947), Biggs and Smith (1971), Sims (1967, 1968), Wong (1967)), and also distance-transitive graphs of valency 4 are determined (Sims (1967, 1968), Quirin (1971), D. H. Smith (1973, 1974a, 1974b)). Recently Cameron (1982b) showed the conjecture (1) is true for distance-transitive graphs by using Cameron, Praeger, Saxl and Seitz (to appear), which has recourse to the classification of finite simple groups. But it is not known whether there are finitely many distance-regular graphs even of valency 3. (There are exactly 8 bipartite distance-regular graphs of valency three. Cf. T. Ito (1982).)

Let Γ be a distance-regular graph of diameter d with intersection matrix

$$B = \left\{\begin{matrix} * & \overbrace{1 & 1 & \cdots & 1}^{t} & c_{t+1} & \cdots & c_d \\ 0 & 0 & 0 & \cdots & 0 & a_{t+1} & \cdots & a_d \\ k & k-1 & k-1 & \cdots & k-1 & b_{t+1} & \cdots & * \end{matrix}\right\}.$$

QUESTION. Let d, k be arbitrarily given diameter and valency. How large can t be?

In connection with the above condition, consider a graph whose automorphism group acts transitively on paths of length t. Such a graph is called <u>t-arc</u> <u>transitive</u> (Biggs (1974)). (Here we do not assume the distance-transitivity.) If a t-arc transitive graph has valency 3, then $t \leq 5$ (Tutte (1947)). Much work has been done on t-arc transitive graphs (Gardiner (1973, 1974, 1976), Weiss (1976), etc.). Recently, Weiss (1981) proved that there are no t-arc transitive graphs with $t \geq 8$, having recourse to the classification of finite simple groups.

3.4 MULTIPLE P-POLYNOMIAL STRUCTURES

In this section, we study association schemes which have at least two P-polynomial structures, and show that their parameters are strongly restricted.

Let $\mathfrak{X} = (X, \{R_i\}_{0 \leq i \leq d})$ be a P-polynomial scheme with respect to the ordering $R_0, R_1, R_2, \ldots, R_d$. The graph Γ on X with respect to R_1 is distance-regular, i.e., $(x,y) \in R_i$ if and only if $\partial(x,y) = i$. In some cases,

$\mathfrak{X}$ becomes a P-polynomial scheme with respect to another ordering $R_0, R_{s_1}, R_{s_2}, \ldots, R_{s_d}$.

EXAMPLES 4.1. (1) The association scheme of an ordinary n-gon is a P-polynomial scheme with respect to $R_0, R_s, R_{2s}, \ldots$ for $(n,s) = 1$.

(2) Let $\mathfrak{X} = (X, \{R_i\}_{0 \le i \le k-1})$ be the Johnson scheme $J(2k-1, k-1)$, where R_i is the usual i^{th} relation of the Johnson scheme. The graph with respect to R_{k-1} is distance-transitive and consequently distance-regular. This graph is called the <u>odd graph</u> O_k. O_k has the ordering $R_0, R_{k-1}, R_1, R_{k-2}, R_2, R_{k-3}, \ldots,$ and the intersection matrix

$$\begin{Bmatrix} * & 1 & 1 & 2 & 2 & 3 & 3 & \cdots \\ 0 & 0 & 0 & 0 & 0 & 0 & 0 & \cdots \\ k & k-1 & k-1 & k-2 & k-2 & k-3 & k-3 & \cdots \end{Bmatrix}.$$

(3) Let $\mathfrak{X} = (X, \{R_i\}_{0 \le i \le n})$ be the Hamming scheme $H(n,2)$, where R_i is the usual i^{th} relation defined by (2.2). In addition to the natural ordering, $\mathfrak{X}$ becomes a P-polynomial scheme with respect to $R_0, R_{n-1}, R_2, R_{n-3}, R_4, R_{n-5}, \ldots, R_{n-4}, R_3, R_{n-2}, R_1, R_n$ if n is even. The graph with respect to R_{n-1} has the intersection matrix

$$\begin{Bmatrix} * & 1 & 2 & 3 & \cdots & n \\ 0 & 0 & 0 & 0 & \cdots & 0 \\ n & n-1 & n-2 & n-3 & \cdots & * \end{Bmatrix}.$$

(If n is odd, the graph with respect to R_{n-1} is not connected.)

(4) Let $\mathfrak{X} = (X, \{R_i\}_{0 \le i \le d})$ be a P-polynomial scheme with respect to $R_0, R_1, R_2, \ldots$. Suppose the intersection matrix with respect to R_1 is

$$B_1 = \begin{Bmatrix} * & 1 & c_2 & \cdots & c_{d-1} & c_d \\ 0 & 0 & 0 & \cdots & 0 & a_d \\ k & b_1 & b_2 & \cdots & b_{d-1} & * \end{Bmatrix} \qquad (a_d > 0).$$

Let Γ be the distance-regular graph on X with respect to R_1. We construct an <u>antipodal double covering</u> $\tilde{\Gamma}$ of of Γ as follows. Let $\tilde{\Gamma}$ be a graph with the vertex set

$$\tilde{X} = \{0,1\} \times X$$

and the edge set

$$\{((i,x),(j,y)) \in \tilde{X} \times \tilde{X} \mid i \neq j \text{ and } (x,y) \in R_1\}.$$

This graph $\tilde{\Gamma}$ becomes distance-regular, bipartite and of diameter $2d + 1$. The intersection matrix of $\tilde{\Gamma}$ is

$$\begin{Bmatrix} * & 1 & \cdots & c_{d-1} & c_d & a_d & b_{d-1} & \cdots & b_1 & k \\ 0 & 0 & \cdots & 0 & 0 & 0 & 0 & \cdots & 0 & 0 \\ k & b_1 & \cdots & b_{d-1} & a_d & c_d & c_{d-1} & \cdots & 1 & * \end{Bmatrix}.$$

Each vertex of $\tilde{\Gamma}$ has a unique vertex at distance $2d + 1$. So $\tilde{\Gamma}$ is imprimitive and the quotient becomes Γ.

Let $\tilde{\Gamma} = \tilde{\Gamma}_0 \cup \tilde{\Gamma}_1$ be the bipartition, i.e., $\tilde{\Gamma}_i$ $(i = 0, 1)$ has the vertex set $\{(i,x) \mid x \in X\}$ and two vertices of $\tilde{\Gamma}_i$ are adjacent if and only if they have distance 2 in $\tilde{\Gamma}$. $\tilde{\Gamma}_i$ becomes distance-regular and is called the <u>halved graph</u>. Identifying the vertices of $\tilde{\Gamma}_0$ and Γ in the natural way, we get the ordering $R_0, R_2, R_4, R_6, \ldots, R_5, R_3, R_1$.

If we apply this method to O_k, we get the natural ordering of the Johnson scheme $J(2k-1, k-1)$.

THEOREM 4.2. Let $\mathcal{X} = (X, \{R_i\}_{0 \le i \le d})$ be a P-polynomial scheme with respect to the ordering $R_0, R_1, R_2, \ldots$. Let p_{ij}^k be the intersection numbers of $\mathcal{X}$. Suppose the graph on X with respect to R_1 is not the ordinary n-gon.

(1) Suppose $\mathcal{X}$ is a P-polynomial scheme with respect to another ordering. Then the new ordering is one of the following:

(I) $R_0, R_2, R_4, R_6, \ldots, R_5, R_3, R_1$.

(II) $R_0, R_d, R_1, R_{d-1}, R_2, R_{d-2}, R_3, R_{d-3}, \ldots$

(III) $R_0, R_d, R_2, R_{d-2}, R_4, R_{d-4}, \ldots, R_{d-5}, R_5, R_{d-3}, R_3, R_{d-1}, R_1$.

(IV) $R_0, R_{d-1}, R_2, R_{d-3}, R_4, R_{d-5}, \ldots, R_5, R_{d-4}, R_3, R_{d-2}, R_1, R_d$.

(2) $\mathcal{X}$ has at most two P-polynomial structures.

(3) Let Γ be the graph on X with respect to R_1. Let B be the intersection matrix of Γ. Then,

(I) holds if and only if

$$B = \begin{Bmatrix} * & 1 & c_2 & \cdots & c_{d-1} & c_d \\ 0 & 0 & 0 & \cdots & 0 & a_d \\ k & b_1 & b_2 & \cdots & b_{d-1} & * \end{Bmatrix} \quad (a_d > 0).$$

(II) holds if and only if $p^d_{jd} = 0$ for $j = 2, 3, \ldots, d$.

(III) holds if and only if

$p^3_{13} = 0$ and if $d = 3$ then the graph with respect to R_3 is connected,

$p^4_{14} = 0$ and $p^3_{14}(p^1_{11} + p^2_{12} - p^3_{12}) = p^2_{12}$ and if $d = 4$ then the graph with respect to R_4 is connected,

$$B = \begin{Bmatrix} * & 1 & c_2 & \cdots & c_\ell & c_{\ell+1} & c_{\ell+2} & \cdots & c_{d-1} & k \\ 0 & 0 & 0 & \cdots & 0 & a_{\ell+1} & 0 & \cdots & 0 & 0 \\ k & b_1 & b_2 & \cdots & b_\ell & b_{\ell+1} & b_{\ell+2} & \cdots & b_{d-1} & * \end{Bmatrix}$$

with $a_{\ell+1}$ arbitrary and $p^d_{jd} = 0$ $(j \neq 0, 2)$ for $d = 2\ell + 1 \geq 5$, and

$$B = \begin{Bmatrix} * & 1 & c_2 & \cdots & c_{\ell-1} & c_\ell & c_{\ell+1} & c_{\ell+2} & \cdots & c_{d-1} & k \\ 0 & 0 & 0 & \cdots & 0 & a_\ell & a_{\ell+1} & 0 & \cdots & 0 & 0 \\ k & b_1 & b_2 & \cdots & b_{\ell-1} & b_\ell & b_{\ell+1} & b_{\ell+2} & \cdots & b_{d-1} & * \end{Bmatrix}$$

with $a_\ell > 0$, $a_{\ell+1} > 0$ and $p^d_{jd} = 0$ $(j \neq 0, 2)$ for $d = 2\ell \geq 6$.

(IV) holds if and only if

$$B = \begin{Bmatrix} * & 1 & c_2 & \cdots & c_{\ell-1} & c_\ell & b_{\ell-1} & \cdots & b_1 & k \\ 0 & 0 & 0 & \cdots & 0 & a_\ell & 0 & \cdots & 0 & 0 \\ k & b_1 & b_2 & \cdots & b_{\ell-1} & b_\ell & c_{\ell-1} & \cdots & 1 & * \end{Bmatrix}$$

with a_ℓ arbitrary for $d = 2\ell$, and

$$B = \begin{Bmatrix} * & 1 & c_2 & \cdots & c_{\ell-1} & c_\ell & b_\ell & b_{\ell-1} & \cdots & b_1 & k \\ 0 & 0 & 0 & \cdots & 0 & a_\ell & a_\ell & 0 & \cdots & 0 & 0 \\ k & b_1 & b_2 & \cdots & b_{\ell-1} & b_\ell & c_\ell & c_{\ell-1} & \cdots & 1 & * \end{Bmatrix}.$$

with $a_\ell > 0$ for $d = 2\ell + 1$.

REMARKS. (1) The orderings (I) and (II) are dual to each other, i.e., exchanging the roles of the first and second orderings, the first is of type (II) (respectively (I)) in terms of the second if the second is of type (I) (respectively (II)) in terms of the first. The orderings (III), (IV) are self-dual.

(2) The ordering (I) is obtained by the antipodal double covering method (Example 4.1(4)).

(3) If $\mathfrak{X}$ has a P-polynomial structure of type (IV), then the graph Γ with respect to R_1 is an antipodal double covering of a distance-regular graph whose intersection matrix is

$$\begin{Bmatrix} * & 1 & c_2 & \cdots & c_{\ell-1} & c_\ell + b_\ell \\ 0 & 0 & 0 & \cdots & 0 & a_\ell \\ k & b_1 & b_2 & \cdots & b_{\ell-1} & * \end{Bmatrix} \quad \text{for } d = 2\ell$$

and

$$\begin{Bmatrix} * & 1 & c_2 & \cdots & c_{\ell-1} & c_\ell \\ 0 & 0 & 0 & \cdots & 0 & a_\ell + b_\ell \\ k & b_1 & b_2 & \cdots & b_{\ell-1} & * \end{Bmatrix}$$

for $d = 2\ell + 1$.

(4) The condition $p^d_{jd} = 0$ $(j = 2, 3, \ldots, d)$ in Theorem 4.2(3)(II) can be replaced by $p^d_{jd} = 0$ $(j = 2, 3, \ldots, d-1)$. Similarly, the condition $p^d_{jd} = 0$ $(j \neq 0, 2)$ in (3)(III) can be replaced by $p^d_{jd} = 0$ $(j \neq 0, 2, d)$.

We prepare some lemmas for the proof of Theorem 4.2. In what follows, $\mathfrak{X} = (X, \{R_i\}_{0 \leq i \leq d})$ is a P-polynomial scheme with respect to $R_0, R_1, R_2, \ldots$.

LEMMA 4.3. If $k_i < k_{i+1}$, then $k_j < k_{j+1}$ for $j \leq i$. If $k_i > k_{i+1}$, then $k_j > k_{j+1}$ for $j \geq i$.

Proof. Apply Proposition 1.2(iii), (iv) to $\dfrac{k_{i+1}}{k_i} = \dfrac{b_i}{c_{i+1}}$.
□

LEMMA 4.4. Let

$$B_1 = \begin{Bmatrix} * & 1 & c_2 & \cdots & c_d \\ 0 & a_1 & a_2 & \cdots & a_d \\ k & b_1 & b_2 & \cdots & * \end{Bmatrix}.$$

Then

(1) $c_i \leq b_j$ for $i + j \leq d$,

and

(2) $k_i \leq k_j$ for $i \leq j \leq d - i$.

Proof. (1) Choose x, y, z such that $(x,y) \in R_i$, $(y,z) \in R_j$ and $(x,z) \in R_{i+j}$. Then

$$R_{i-1}(x) \cap R_1(y) \subsetneq R_{j+1}(z) \cap R_1(y),$$

where

$$R_k(x) = \{y \in X \mid (x,y) \in R_k\}.$$

So

$$p^i_{1i-1} \leq p^j_{1j+1}, \quad \text{i.e.,} \quad c_i \leq b_j .$$

$$(2) \quad k_j = k_i \frac{b_i}{c_{i+1}} \frac{b_{i+1}}{c_{i+2}} \cdots \frac{b_{j-1}}{c_j}$$

$$= k_i \frac{b_i}{c_j} \frac{b_{i+1}}{c_{j-1}} \cdots \frac{b_{j-1}}{c_{i+1}}$$

$$\geq k_i \qquad \text{(by (1))}. \qquad \square$$

LEMMA 4.5. (Taylor-Levingston (1978)) If $k_1 \geq k_2$, then either $d \leq 3$ or $k_1 = 2$.

Proof. By Lemma 4.3 and Lemma 4.4,

$$k_0 < k_1 < k_2 < \cdots < k_{i_0} = \cdots \tag{4.1}$$

$$= k_{j_0} > \cdots > k_{d-1} > k_d$$

with $j_0 \geq d - i_0$. Since $i_0 = 1$,

$$k_1 = k_2 = \cdots = k_{d-1}. \tag{4.2}$$

So $b_i = c_{i+1}$ for $i = 1, \ldots, d-2$. These identities together with the inequalities $b_j \geq b_{j+1}$, $c_j \leq c_{j+1}$ imply

$$b_1 = b_2 = \cdots = b_{d-2} = c_2 = c_3 = \cdots = c_{d-1}. \tag{4.3}$$

Let $b_1 = b$ and $k_1 = k$.

Suppose $d \geq 4$. We shall show b divides k. For $x \in X$, choose y, z such that $y \in R_1(x)$, $z \in R_3(x)$ and $(y,z) \in R_2$. Then $R_1(y) \cap R_1(z) \subseteq R_1(y) \cap R_2(x)$. But $|R_1(y) \cap R_1(z)| = c_2 = b_1 = |R_1(y) \cap R_2(x)|$. So the inclusion must be identity:

$$R_1(y) \cap R_1(z) = R_1(y) \cap R_2(x). \tag{4.4}$$

Similarly

$$R_1(y) \cap R_1(z) = R_1(z) \cap R_2(x). \tag{4.5}$$

The identities (4.4), (4.5) imply that for $x \in X$, the set

$$\{R_1(y) \cap R_1(z) \mid (y,z) \in R_1(x) \times R_3(x) \cap R_2\}$$

is a partition of $R_2(x)$. Hence b divides $k = |R_2(x)|$.

Next we shall show $k \leq 3b-1$. For $x \in X$, choose $y \in R_1(x)$ and $z \in R_2(x)$ such that $(y,z) \in R_1$. Then $R_1(y) \cap R_1(z)$ is contained in the union of $R_1(y) \cap R_2(x) - \{z\}$ and $R_1(z) \cap R_1(x) - \{y\}$. So $a_1 \leq b_1 - 1 + c_2 - 1$, i.e., $k - b - 1 \leq 2b - 2$.

Since b divides k and $k \leq 3b - 1$, we get $k = 2b$ and so $a_2 = k - b_2 - c_2 = 0$. By the following lemma, $a_1 = 0$. But $a_1 = k - b - 1 = b - 1$ and hence $b = 1$, $k = 2$. □

LEMMA 4.6. (Gardiner (1978)) With the notation in Lemma 4.4, let $a_1 \neq 0$. Then $a_i \neq 0$ for $1 \leq i \leq d-1$.

Proof. For $x \in X$, choose $y \in R_i(x)$ and $z \in R_{i+1}(x)$ such that $(y,z) \in R_1$. $R_1(y) \cap R_1(z)$ is not empty by the assumption $a_1 \neq 0$. Choose $w \in R_1(y) \cap R_1(z)$. Then either $w \in R_i(x)$ or $w \in R_{i+1}(x)$. If $w \in R_i(x)$, the triple (x,y,w) gives $a_i \neq 0$. If $w \in R_{i+1}(x)$, then take $u \in R_1(x)$ such that $(u,y) \in R_{i-1}$. The triple (u,z,w) gives $a_i \neq 0$. □

Proof of Theorem 4.2(1). Let $R_0, R'_1, R'_2, \ldots, R'_d$ be the second ordering and $R'_i = R_{s_i}$ $(i = 1, 2, \ldots, d)$. If $d \leq 3$, the assertion is trivial. We may assume $d \geq 4$.

Step 1. We shall show the following:

$$R'_1 = R_2,\ R_{d-1} \ \text{or}\ R_d. \tag{4.6}$$

$$R_1 = R'_d \ \text{if}\ R'_1 = R_2, \ \text{and} \tag{4.7}$$

$$R_d = R'_d \ \text{if}\ R'_1 = R_{d-1}.$$

By Lemma 4.5, $k'_1 < k'_2$, i.e., $k_{s_1} < k_{s_2}$, where k'_i and k_i are the valencies of R'_i and R_i, respectively. By Lemma 4.3, $k_i < k_{s_1} < k_j$ for $i < s_1 < j < s_2$ if $s_1 < s_2$; $k_i < k_{s_1} < k_j$ for $i > s_1 > j > s_2$ if $s_1 > s_2$. But $k'_0 = 1$ and k'_d are the only possible valencies less than $k'_1 = k_{s_1}$ by Lemma 4.4(2). So we get $s_1 = 2, d-1$ or d, and $R_1 = R'_d$ if $s_1 = 2$, $R_d = R'_d$ if $s_1 = d-1$.

<u>Step 2</u>. Let $R_1' = R_2$. We shall show (I) occurs. We use the following notation:

$$R_i \circ R_j := \{(x,y) \in X \times X \mid \text{there exists } z \in X \text{ such that } (x,z) \in R_i \text{ and } (z,y) \in R_j\}. \quad (4.8)$$

By the triangle inequality,

$$R_i \circ R_j \subseteq \bigcup_{k=|j-i|}^{r} R_k \quad (r = \min\{i+j,d\}) \quad (4.9)$$

and

$$R_{|j-i|} \subseteq R_i \circ R_j,$$

$$R_{i+j} \subseteq R_i \circ R_j \quad \text{if} \quad i+j \leq d.$$

The same inclusions hold for $R_i' \circ R_j'$. We use these inclusion relations very often in our argument.

Since $R_{d-1}' \subseteq R_d' \circ R_1' = R_1 \circ R_2 \subseteq R_1 \cup R_2 \cup R_3 = R_d' \cup R_1' \cup R_3$, we get $R_{d-1}' = R_3$. Since $R_2' \subseteq R_1' \circ R_1' = R_2 \circ R_2 \subseteq R_0 \cup R_1 \cup R_2 \cup R_3 \cup R_4 = R_0 \cup R_d' \cup R_1' \cup R_{d-1}' \cup R_4$, we get $R_2' = R_4$. Repeat this argument in turn for $R_{d-1}' \circ R_1'$, $R_2' \circ R_1'$, $R_{d-2}' \circ R_1'$, $R_3' \circ R_1'$, ... to get the ordering $R_0, R_2, R_4, R_6, \ldots, R_5, R_3, R_1$.

<u>Step 3</u>. Let $R_1' = R_{d-1}$. By (4.7), $R_d' = R_d$. We shall show (IV) occurs.

The relations $R_{d-1}' = R_1$ and $R_{d-2}' = R_{d-2}$ follow from $R_1 \circ R_1' = R_1 \circ R_{d-1} \supseteq R_d \cup R_{d-2} = R_d' \cup R_{d-2}$. We want to show

$$R_2' = R_2. \quad (4.10)$$

Since $R_2 \circ R_1' = R_2 \circ R_{d-1} \subseteq R_{d-3} \cup R_{d-2} \cup R_{d-1} \cup R_d = R_{d-3} \cup R_{d-2}' \cup R_1' \cup R_d'$, we get $R_2 \circ R_1' \subseteq R_{d-3} \cup R_{d-2}' \cup R_1'$. So $(R_2 \circ R_1') \cap R_{d-2}' \neq \emptyset$ or $(R_2 \circ R_1') \cap R_1' \neq \emptyset$. This implies $R_2 = R_{d-3}'$ or R_2'. Suppose $R_2 = R_{d-3}' \neq R_2'$. Then $R_2' \subseteq R_{d-3}' \circ R_{d-1}' = R_2 \circ R_1 \subseteq R_1 \cup R_2 \cup R_3 = R_{d-1}' \cup R_{d-3}' \cup R_3$. So $R_2' = R_3$. Since we have already obtained the relation $R_1 = R_{d-1}'$, which is the dual of $R_1' = R_{d-1}$, exchange the roles of the first and second orderings and apply the same argument to get $R_2' = R_{d-3}$ or R_2. So if $R_2 \neq R_2'$, then $R_2' = R_3 = R_{d-3}$. Thus $d = 6$ and the ordering is

$$R_0, R_5, R_3, R_2, R_4, R_1, R_6. \tag{4.11}$$

We shall show (4.11) is impossible. Since $R_5 \circ R_5 = R_1' \circ R_1' \subseteq R_0 \cup R_5 \cup R_3$, we have $p_{15}^5 = p_{25}^5 = 0$. So $p_{11}^1 = 0$ by Lemma 4.6. If either $b_4 \geq 2$ or $c_6 \geq 2$ holds, then by $p_{11}^1 = 0$, there exist $x,y,z \in X$ such that $(x,y) \in R_5$, $(x,z) \in R_5$ and $(y,z) \in R_2$, which contradicts $p_{25}^5 = 0$. So $b_4 = c_6 = 1$. By Proposition 1.2, $b_5 = c_5 = 1$ and hence $k = a_5 + b_5 + c_5 = 2$, which is a contradiction. Thus we get (4.10).

Since $R_{d-3}' \subseteq R_{d-1}' \circ R_2' = R_1 \circ R_2 \subseteq R_1 \cup R_2 \cup R_3 = R_{d-1}' \cup R_2' \cup R_3$, we get $R_{d-3}' = R_3$. Since $R_3' \subseteq R_1' \circ R_2' = R_{d-1} \circ R_2 \subseteq R_{d-3} \cup R_{d-2} \cup R_{d-1} \cup R_d = R_{d-3} \cup R_{d-2}' \cup R_1' \cup R_d'$, we get $R_3' = R_{d-3}$. Repeat this argument in turn for $R_{d-2}' \circ R_2'$, $R_2' \circ R_2'$, $R_{d-3}' \circ R_2'$, $R_3' \circ R_2'$, ... to get the ordering $R_0, R_{d-1}, R_2, R_{d-3}, R_4, R_{d-5}, \ldots, R_5, R_{d-4}, R_3, R_{d-2}, R_1, R_d$.

<u>Step 4</u>. Let $R_1' = R_d$. If $R_2' = R_1$, exchange the roles

of R_i, R_i' in the step 2 to get the ordering $R_0, R_1, R_2, R_3, \ldots$ is $R_0, R_2', R_4', R_6', \ldots, R_5', R_3', R_1'$ in terms of R_i'. So the ordering $R_0, R_1', R_2', R_3', \ldots$ is $R_0, R_d, R_1, R_{d-1}, R_2, \ldots$ in terms of R_i. We may assume $R_2' \neq R_1$. We shall show (III) occurs.

Since $R_1 \circ R_1' = R_1 \circ R_d \subseteq R_{d-1} \cup R_d = R_{d-1} \cup R_1'$ and $R_2' \neq R_1$, $R_1 \circ R_1' \subseteq R_{d-1}$. Since $R_1' \circ R_i' \supseteq R_{i-1}' \cup R_{i+1}'$ unless $i = 0$ or d, this implies $R_1 = R_d'$ and $R_{d-1} = R_{d-1}'$. We want to show (4.10) $R_2' = R_2$ again.

Since $R_2 \circ R_1' = R_2 \circ R_d \subseteq R_{d-2} \cup R_{d-1} \cup R_d = R_{d-2} \cup R_{d-1}' \cup R_1'$, we get $R_2 = R_{d-2}'$ or R_2'. Suppose $R_2 = R_{d-2}' \neq R_2'$. Then $R_2' \subseteq R_{d-2}' \circ R_d' = R_2 \circ R_1 \subseteq R_1 \cup R_2 \cup R_3 = R_d' \cup R_{d-2}' \cup R_3$. So $R_2' = R_3$. By the duality ($R_1' = R_d$, $R_1 = R_d'$), $R_2' = R_{d-2}$ or R_2. So $R_2' = R_3 = R_{d-2}$ and we get that $d = 5$ and the ordering is

$$R_0,\ R_5,\ R_3,\ R_2,\ R_4,\ R_1. \tag{4.12}$$

We shall show (4.12) is impossible. Applying Lemma 4.4(2) to k_i and k_i', we have $k_2 = k_3$. Since $R_5 \circ R_5 = R_1' \circ R_1' \subseteq R_0 \cup R_5 \cup R_3$ and $R_3 \circ R_3 = R_2' \circ R_2' \subseteq R_0 \cup R_5 \cup R_3 \cup R_2 \cup R_4$, we have $p_{15}^5 = p_{25}^5 = p_{13}^3 = 0$. So $p_{11}^1 = 0$ by Lemma 4.6. This information about k_i and p_{ij}^k together with Proposition 1.2 gives

$$B_1 = \begin{Bmatrix} * & 1 & c_2 & b_2 & c_4 & k \\ 0 & 0 & a_2 & 0 & a_4 & 0 \\ k & k-1 & b_2 & b_3 & 1 & * \end{Bmatrix}.$$

The valencies are 1, k, $k(k-1)/c_2$, $k(k-1)/c_2$, $b_3k(k-1)/c_4c_2$ and $b_3(k-1)/c_4c_2$. Since $k_2 = k_3 \geq k_4$ by Lemma 4.3, $b_3/c_4 \leq 1$. So $k_5 = b_3(k-1)/c_4c_2 < k = k_1$. The ordering $R_0, R_1, R_2, R_3, R_4, R_5$ is $R_0, R_5', R_3', R_2', R_4', R_1'$ in terms of R_i', which is exactly the same as (4.12). So $k_5' < k_1'$ holds. This contradicts $k_5 < k_1$, because $k_5' = k_1$, $k_1' = k_5$. Thus we get (4.10) $R_2' = R_2$.

Since $R_{d-2}' \subseteq R_d' \circ R_2' = R_{-} \circ R_2 \subseteq R_1 \cup R_2 \cup R_3 = R_d' \cup R_2' \cup R_3$, we get $R_{d-2}' = R_3$. Since $R_3' \subseteq R_1' \circ R_2' = R_d \circ R_2 \subseteq R_{d-2} \cup R_{d-1} \cup R_d = R_{d-2} \cup R_{d-1}' \cup R_1'$, we get $R_3' = R_{d-2}$. Repeat this argument in turn for $R_{d-1}' \circ R_2'$, $R_2' \circ R_2'$, $R_{d-2}' \circ R_2'$, $R_3' \circ R_2'$, ... to get the ordering $R_0, R_d, R_2, R_{d-2}, R_4, R_{d-4}, \ldots, R_{d-5}, R_5, R_{d-3}, R_3, R_{d-1}, R_1$.

Proof of Theorem 4.2(2). Suppose $\mathfrak{X}$ has three P-polynomial structures $R_0, R_1, R_2, \ldots$; $R_0, R_1', R_2', \ldots$; $R_0, R_1'', R_2'', \ldots$.

Step 1. We treat the case $d = 3$ separately. Changing the roles of the orderings suitably, there are only two patterns of triple P-polynomial structures: one of the orderings is R_0, R_1, R_2, R_3 and the rest are

(A) R_0, R_2, R_3, R_1 and R_0, R_3, R_1, R_2 ,

or

(B) R_0, R_2, R_3, R_1 and R_0, R_3, R_2, R_1.

We may assume the first and second orderings are R_0, R_1, R_2, R_3 and R_0, R_2, R_3, R_1, respectively. Since $R_2 \circ R_2 \subseteq R_0 \cup R_2 \cup R_3$, we have $p_{12}^2 = 0$ and so $p_{11}^1 = 0$ by Lemma 4.6.

Suppose the case (A) holds. Then by Lemma 4.4(2), $k_1 = k_2 = k_3$. So we have

$$B_1 = \begin{Bmatrix} * & 1 & k-1 & 1 \\ 0 & 0 & 0 & k-1 \\ k & k-1 & 1 & * \end{Bmatrix}.$$

Thus $k = 2$ by Proposition 1.2(iv), and hence $\mathfrak{X}$ is an ordinary n-gon.

Suppose the case (B) holds. Since $R_3 \circ R_3 \subseteq R_0 \cup R_3 \cup R_2$, we have $p^3_{13} = 0$. By Lemma 4.4(2), $k_2 = k_3$ and hence $b_2 = c_3$. But since $p^3_{13} = 0$, $c_3 = k$ and so $b_2 = k$, which contradicts $c_2 > 0$.

In what follows, we assume $d \geq 4$.

Step 2. Let the second ordering $R_0, R'_1, R'_2, \ldots$ be $R_0, R_2, R_4, \ldots, R_5, R_3, R_1$ in terms of the first ordering. The third ordering satisfies $R''_1 = R'_2$, R'_{d-1} or R'_d in terms of the second ordering by Theorem 4.2(1). If $R''_1 = R'_2$, then $R''_1 = R_4$ in terms of the first ordering. So $d = 4$ or 5 and the third ordering is one of the following:

$$R_0,\ R_4,\ R_1,\ R_3,\ R_2\ ,$$

$$R_0,\ R_4,\ R_2,\ R_3,\ R_1,$$

$$R_0,\ R_4,\ R_2,\ R_3,\ R_1,\ R_5\ .$$

But in terms of the second ordering, it is $R_0, R'_2, R'_4, R'_3, R'_1$; $R_0, R'_2, R'_1, R'_3, R'_4$; $R_0, R'_2, R'_1, R'_4, R'_5, R'_3$. Thus only $d = 4$ is possible and the third ordering is R_0, R_4, R_1, R_3, R_2. By Lemma 4.5, $k_1 < k_2$, $k'_1 < k'_2$, $k''_1 < k''_2$, i.e., $k_1 < k_2 < k_4 < k_1$, which is a contradiction.

If $R''_1 = R'_{d-1}$, then $R''_1 = R_3$. So $d = 4$ and the third ordering is R_0, R_3, R_2, R_1, R_4. But in terms of the second ordering, it is $R_0, R'_3, R'_1, R'_4, R'_2$, which is a contradiction.

If $R''_1 = R'_d$, then $R''_1 = R_1$, which is a contradiction.

Step 3. Since the orderings of (I) and (II) are dual, the remaining case is that the second and third orderings are of type (III) and (IV) in terms of the first ordering, respectively. Since $R''_1 = R_{d-1} = R'_{d-1}$, the third ordering is also of type (IV) in terms of the second ordering. So $R_d = R''_d = R'_d = R_1$, which is a contradiction.

Proof of Theorem 4.2(3).

Step 1. Suppose $\mathfrak{X}$ has a P-polynomial structure of type (I). Then $R_i \circ R_2 \subseteq R_{i-2} \cup R_i \cup R_{i+2}$ $(i < d-1)$. This implies

$$p^{i+1}_{2\,i} = 0 \qquad (0 \leq i \leq d-2). \tag{4.13}$$

Since $B_2 = \frac{1}{c_2}(B_1^2 - a_1 B_1 - kB_0)$,

$$p^{i+1}_{2\,i} = \frac{c_{i+1}}{c_2}(a_{i+1} + a_i - a_1) \tag{4.14}$$

$$(0 \leq i \leq d-1).$$

By (4.13) and (4.14), $a_2 = a_4 = \ldots = 0$ and $a_1 = a_3 = \ldots$ except for a_d. By Lemma 4.6, $a_2 = 0$ implies $a_1 = 0$. Since $R_{d-1} \circ R_2 \supseteq R_d$, $p^d_{2\,d-1} \neq 0$. By (4.14), $a_d \neq 0$.

Conversely, suppose $a_i = 0$ $(0 \leq i \leq d-1)$ and $a_d > 0$. Then $\mathfrak{X}$ has a P-polynomial structure of type (I) by the antipodal double covering method given in Example 4.1(4).

Step 2. Suppose $\mathcal{X}$ has a P-polynomial structure of type (II). Since (I) and (II) are dual, $a_1' = 0$ for the intersection matrix with respect to $R_1' = R_d$. So $(R_d \circ R_d) \cap R_d = \emptyset$ and hence $R_d \circ R_d = R_0 \cup R_1$, i.e., $p_{jd}^d = 0 \quad (2 \leq j \leq d)$.

Conversely, suppose $R_d \circ R_d \subseteq R_0 \cup R_1 \cup R_d$, i.e., $p_{jd}^d = 0 \quad (2 \leq j \leq d-1)$. Since $(\bigcup_{i=0}^{j} R_i) \circ R_d \subseteq \bigcup_{i=0}^{j} R_{d-i}$ and $(\bigcup_{i=0}^{j} R_{d-i}) \circ R_d \subseteq \bigcup_{i=0}^{j} (R_i \circ R_d \circ R_d) \subseteq \bigcup_{i=0}^{j} R_i \circ (R_0 \cup R_1 \cup R_d) \subseteq (\bigcup_{i=0}^{j+1} R_i) \cup (\bigcup_{i=0}^{j} R_{d-i})$, we get

$$R_d \circ R_d \circ R_d \subseteq R_0 \cup R_1 \cup R_{d-1} \cup R_d,$$

$$R_d \circ R_d \circ R_d \circ R_d \subseteq R_0 \cup R_1 \cup R_2 \cup R_{d-1} \cup R_d,$$

$$R_d \circ R_d \circ R_d \circ R_d \circ R_d \subseteq R_0 \cup R_1 \cup R_2 \cup R_{d-2} \cup R_{d-1} \cup R_d,$$

and so on. Thus we need d products of R_d to cover $X \times X$, which means the graph (X, R_d) is distance-regular with the ordering $R_0, R_d, R_1, R_{d-1}, R_2, R_{d-2}, \ldots$

Step 3. Suppose $\mathcal{X}$ has a P-polynomial structure of type (III). Since $R_d \circ R_d \subseteq R_0 \cup R_d \cup R_2$,

$$p_{1d}^d = 0. \tag{4.15}$$

If $d = 3$, we have nothing more to prove. If $d = 4$, $p_{34}^4 = 0$ and

$$p_{34}^{4} = \frac{k}{c_2c_3}\{b_3(a_3 - a_2 - a_1) + a_2\}, \tag{4.16}$$

because $B_2 = \frac{1}{c_2}(B_1^2 - a_1B_1 - kB_0)$ and

$B_3 = \frac{1}{c_3}(B_1B_2 - a_2B_2 - b_1B_1)$. So $b_3(a_1 + a_2 - a_3) = a_2$.

Let $d > 4$. Since $R_2 \circ R_2 \subsetneq R_0 \cup R_d \cup R_2 \cup R_{d-2} \cup R_4$, $p_{12}^2 = 0$. By Lemma 4.6, $a_1 = 0$. By the triangle inequality for (X,R_1'), $(R_i \circ R_2) \cap R_{i+1} = \emptyset$ holds for $i \neq \ell, \ell + 1$ if $d = 2\ell + 1$ and for $i \neq \ell - 1, \ell, \ell + 1$ if $d = 2\ell$. Thus

$$p_{2\,i}^{i+1} = 0 \begin{cases} \text{for } i \neq \ell,\ \ell + 1 \text{ if } d = 2\ell + 1, \\ \\ \text{for } i \neq \ell,\ \ell \pm 1 \text{ if } d = 2\ell. \end{cases} \tag{4.13)'$$

By (4.13)', (4.14) and $a_1 = a_2 = a_d = 0$, we get $a_i = 0$ for $i \neq \ell + 1$ if $d = 2\ell + 1$ and for $i \neq \ell, \ell + 1$ if $d = 2\ell$. If $d = 2\ell$, $R_{\ell-1} \circ R_2 \supseteq R_\ell$ and $R_{\ell+1} \circ R_2 \supseteq R_{\ell+2}$. So $p_{2\,\ell-1}^{\ell} > 0$, $p_{2\,\ell+1}^{\ell+2} > 0$ and we get $a_\ell > 0$, $a_{\ell+1} > 0$. Since (III) is self-dual, $a_1' = 0$ for the intersection matrix with respect to $R_1' = R_d$. So $R_d \circ R_d = R_0 \cup R_2$.

Conversely, suppose $R_d \circ R_d \subsetneq R_0 \cup R_2 \cup R_d$. Also suppose $a_i = 0$ $(i \neq \ell + 1)$ if $d = 2\ell + 1 \geq 5$; $a_i = 0$ $(i \neq \ell, \ell + 1)$, $a_\ell > 0$, $a_{\ell+1} > 0$ if $d = 2\ell \geq 6$. By (4.14), (4.13)' holds, and $p_{2\,\ell-1}^{\ell} > 0$, $p_{2\,\ell+1}^{\ell+2} > 0$ if $d = 2\ell$. R_2 is the distance 2 relation in the graph (X,R_d), i.e., $(x,y) \in R_2$ if and only if x and y have distance 2 in the graph (X,R_d). Consider $R_d \circ R_2$, $R_2 \circ R_2$, $R_{d-2} \circ R_2$, $R_4 \circ R_2$, $R_{d-4} \circ R_2$, ... in turn to get the sequence of the distance relations in the graph (X,R_d) as follows: $R_0,R_d,R_2,R_{d-2},R_4,R_{d-4},\ldots,$ where the middle of the sequence is

$$\ldots\ R_{\ell-1}\ R_{\ell+2}\ R_{\ell+1}\ R_{\ell}\ R_{\ell+3}\ R_{\ell-2}\ \ldots$$

if ℓ is odd and $d = 2\ell + 1$,

$$\ldots\ R_{\ell-2}\ R_{\ell+3}\ R_{\ell}\ R_{\ell+1}\ R_{\ell+2}\ R_{\ell-1}\ \ldots$$

if ℓ is even and $d = 2\ell + 1$,

$$\ldots\ R_{\ell-3}\ R_{\ell+3}\ R_{\ell-1}\ R_{\ell+1}\ R_{\ell}\ R_{\ell+2}\ R_{\ell-2}\ R_{\ell+4}$$

if ℓ is odd and $d = 2\ell$,

$$\ldots\ R_{\ell+4}\ R_{\ell-2}\ R_{\ell+2}\ R_{\ell}\ R_{\ell+1}\ R_{\ell-1}\ R_{\ell+3}\ R_{\ell-3}$$

if ℓ is even and $d = 2\ell$.

Thus the graph (X,R_d) is distance-regular, if $d \geq 5$.

Suppose $d = 3$ and $R_3 \circ R_3 \subsetneq R_0 \cup R_2 \cup R_3$, i.e., $p^3_{13} = 0$. Then obviously the graph (X,R_3) has diameter 3 and is distance-regular. Suppose $d = 4$, $p^4_{14} = 0$ and $p^3_{14}(p^1_{11} + p^2_{12} - p^3_{13}) = p^2_{12}$. Then by (4.16), $p^4_{34} = 0$. So $R_4 \circ R_4 \subsetneq R_0 \cup R_2 \cup R_4$. Since $R_2 \circ R_4 \subsetneq R_2 \cup R_3 \cup R_4$ by the triangle inequality, the graph (X,R_4) has diameter 4 and is distance-regular.

Step 4. Suppose $\mathfrak{X}$ has a P-polynomial structure of type (IV). By Lemma 4.4(2), $k_1 \leq k_{d-1}$ and $k'_1 \leq k'_{d-1}$. Since $k'_1 = k_{d-1}$ and $k'_{d-1} = k_1$, we get $k_1 = k_{d-1}$. Since it holds that

$$k_{d-1} = k_1 \frac{b_1}{c_{d-1}} \frac{b_2}{c_{d-2}} \cdots \frac{b_{d-2}}{c_2}$$

by Proposition 1.2(ii) and $\dfrac{b_i}{c_{d-i}} \geq 1$ by Lemma 4.4(1), the equality $k_1 = k_{d-1}$ implies

$$b_i = c_{d-i} \quad \text{for} \quad 1 \leq i \leq d-2. \tag{4.17}$$

Let $d \geq 4$. Since $R_{d-1} \circ R_{d-1} \cap R_1 = \emptyset$, $p^{d-1}_{1\,d-1} = 0$ and so $a_1 = 0$ by Lemma 4.6. Since $b_1 = k_1 - 1$ and $c_{d-1} = b_1 = k_1 - 1$ by (4.17), we get $b_{d-1} = 1$. So $k_d = b_{d-1}k_{d-1}/c_d = k_1/c_d$ and k_1/c_d must be an integer. But $c_d \geq c_{d-1} = k_1 - 1$ by Proposition 1.2(iv) and hence $c_d = k_1$, $a_d = 0$. Since $(R_i \circ R_2) \cap R_{i+1} = \emptyset$ for $i \neq \ell - 1, \ell$ if $d = 2\ell$ and for $i \neq \ell, \ell \pm 1$ if $d = 2\ell + 1$,

$$p^{i+1}_{2\,i} = 0 \begin{cases} \text{for } i \neq \ell - 1, \ell \text{ if } d = 2\ell, \\ \\ \text{for } i \neq \ell, \ell \pm 1 \text{ if } d = 2\ell + 1. \end{cases} \tag{4.13)"}$$

By (4.13)" and (4.14), the equality $a_1 = a_d = 0$ implies $a_i = 0$ for $i \neq \ell$ if $d = 2\ell$ and for $i \neq \ell, \ell + 1$ if $d = 2\ell + 1$. If $d = 2\ell + 1$, then $R_{\ell-1} \circ R_2 \supseteq R_\ell$ and $R_{\ell+1} \circ R_2 \supseteq R_{\ell+2}$, i.e., $p^{\ell}_{2\,\ell-1} > 0$ and $p^{\ell+2}_{2\,\ell+1} > 0$. By (4.14), $a_\ell > 0$ and $a_{\ell+1} > 0$.

Let $d = 3$. Since $(R_2 \circ R_2) \cap R_3 = \emptyset$, we have $p^3_{22} = 0$ and so $a_1 = a_2 + a_3$ by (4.14). By (4.17), $b_1 = c_2$. So we get

$$B_1 = \begin{Bmatrix} * & 1 & k - a_1 - 1 & k - a_1 + a_2 \\ 0 & a_1 & a_2 & a_1 - a_2 \\ k & k - a_1 - 1 & a_1 - a_2 + 1 & * \end{Bmatrix}.$$

As in (4.16), $p_{33}^{3} = \frac{1}{c_3}(a_1 - a_2)(2a_1 - 2a_2 - k + 1)$. If $a_1 > a_2$, then $2a_1 - 2a_2 - k + 1 \geq 0$, which contradicts $b_1 = k - a_1 - 1 \geq b_2 = a_1 - a_2 + 1$. Therefore $a_1 = a_2$.

Conversely, suppose $b_i = c_{d-i}$ $(0 \leq i \leq d-1)$ and $a_i = 0$ $(i \neq \ell$ if $d = 2\ell$; $i \neq \ell, \ell+1$ if $d = 2\ell + 1)$. Suppose $a_\ell > 0$, $a_{\ell+1} > 0$ in case $d = 2\ell + 1$. Then $k_d = 1$ and so the graph (X, R_1) is imprimitive with blocks $\{x,y\}$, where y is the unique point at distance d from x. For a fixed point x_0, let x have distance i from x_0. Then the unique antipodal point y has distance $d - i$ from x_0 because $R_i \circ R_d \supseteq R_{d-i}$. Thus the intersection matrix with respect to R_d is

$$B_d = \begin{pmatrix} 0 & & & & 1 \\ & & & 1 & \\ & & 1 & & \\ & \cdot^{\cdot^{\cdot}} & & & \\ 1 & & & & 0 \end{pmatrix}.$$

Since $B_1 B_d = \sum_i p_{1d}^{i} B_i = B_{d-1}$,

$$B_{d-1} = \begin{pmatrix} 0 & & & & & c_1 & 0 \\ & & & & c_2 & a_1 & k \\ & & & c_{d-2} \cdots & a_2 & b_1 & \\ & & c_{d-1} & a_{d-2} \cdots & b_2 & & \\ & c_d & a_{d-1} & b_{d-2} \cdots & & & \\ a_d & b_{d-1} & & & & & 0 \end{pmatrix}.$$

The permutation

$$\begin{pmatrix} 0 & 1 & 2 & 3 & 4 & \cdots & d-3 & d-2 & d-1 & d \\ 0 & d-1 & 2 & d-3 & 4 & \cdots & 3 & d-2 & 1 & d \end{pmatrix},$$

applied to the rows and columns of B_{d-1}, makes B_{d-1} tridiagonal. Thus the graph (X, R_{d-1}) is distance-regular. This completes the proof of Theorem 4.2. □

The type (II) is the dual of type (I) and the type (IV) with $a_\ell = a_{\ell+1} = 0$ is obtained as an antipodal covering of the type (I).

QUESTIONS. (1) Is the type (III) related to (I) in any sense? Are there any examples of type (III) with large diameter? Can one get geometric insight which explains the coexistence of two P-polynomial structures of type (III)?

(2) Determine association schemes which have two Q-polynomial structures. (So far we do not know how many Q-polynomial structures can coexist in an association scheme.)

REMARK. (5) The Hamming scheme $H(n,q)$ $(n \geq 3)$ has two P-polynomial structures (i.e., type (IV)) only when $q = 2$ and $n =$ even. Since $p_{ij}^k = q_{ij}^k$ in $H(n,q)$, the same statement is true for Q-polynomial structures of $H(n,q)$.

The Johnson scheme $J(v,k)$ $(k \geq 3)$ has two P-polynomial structures only when $v = 2k+1$ (i.e., O_{k+1}) or $(v,k) = (6,3)$, and two Q-polynomial structures only when $(v,k) = (6,3)$ (see section 3.7). ($J(6,3)$ gives an example of type (III) with $d = 3$.)

For the <u>known</u> (P and Q)-polynomial schemes with double P- or Q-polynomial structures and large diameters, see section 3.6. For (P and Q)-polynomial schemes with double Q-polynomial structures, see section 3.7.

3.5 (P AND Q)-POLYNOMIAL ASSOCIATION SCHEMES AND THEOREM OF LEONARD

This section is the core of Part I. We study the parameters of (P and Q)-polynomial association schemes, and prove Leonard's Theorem (Leonard (1982)) that orthogonal polynomials which have the (P and Q)-polynomial property are the Askey-Wilson polynomials, including their certain limiting cases. This finishes Step 1 of the program for the classification of (P and Q)-polynomial association schemes explained in the Introduction.

Let $\mathfrak{X} = (X, \{R_i\}_{0 \leq i \leq d})$ be a (P and Q)-polynomial scheme with respect to the orderings $A_0, A_1, \ldots, A_d$ of the incidence matrices and $E_0, E_1, \ldots, E_d$ of the primitive idempotents. Let B_i and $\hat{B}_i$ be the ith intersection matrix and the ith dual intersection matrix respectively, where

$$B_1 = \begin{Bmatrix} * & 1 & c_2 & \cdots & c_d \\ 0 & a_1 & a_2 & \cdots & a_d \\ k & b_1 & b_2 & \cdots & * \end{Bmatrix} \qquad (c_i \neq 0,\ b_i \neq 0)$$

and

$$\hat{B}_1 = \begin{Bmatrix} * & 1 & c_2^* & \cdots & c_d^* \\ 0 & a_1^* & a_2^* & \cdots & a_d^* \\ m & b_1^* & b_2^* & \cdots & * \end{Bmatrix} \qquad (c_i^* \neq 0,\ b_i^* \neq 0).$$

Let k_i, m_i $(0 \leq i \leq d)$ be the valencies and multiplicities, respectively, and set $k = k_1$ and $m = m_1$. Let θ_i, θ_i^* $(0 \leq i \leq d)$ be the eigenvalues of B_1, $\hat{B}_1$. Then

$$B_i = v_i(B_1),$$

and

$$\hat{B}_i = v_i^*(\hat{B}_1),$$

where $v_i(x)$, $v_i^*(x)$ are polynomials of degree i satisfying

$$xv_i(x) = b_{i-1}v_{i-1}(x) + a_i v_i(x) + c_{i+1}v_{i+1}(x)$$

$$(0 \leq i \leq d-1),$$

$$xv_d(x) = b_{d-1}v_{d-1}(x) + a_d v_d(x) \quad \text{for} \quad x = \theta_j$$

$$(0 \leq j \leq d),$$

and

$$xv_i^*(x) = b_{i-1}^* v_{i-1}^*(x) + a_i^* v_i^*(x) + c_{i+1}^* v_{i+1}^*(x)$$

$$(0 \leq i \leq d-1),$$

$$xv_d^*(x) = b_{d-1}^* v_{d-1}^*(x) + a_d^* v_d^*(x) \quad \text{for} \quad x = \theta_j^*$$

$$(0 \leq j \leq d)$$

with

$$v_{-1}(x) = v_{-1}^*(x) = 0, \quad v_0(x) = v_0^*(x) = 1.$$

Notice $k = \theta_0$, $m = \theta_0^*$ and $k_i = v_i(\theta_0)$, $m_i = v_i^*(\theta_0^*)$. Let

$$u_i(x) = v_i(x)/k_i$$

and

$$u_i^*(x) = v_i^*(x)/m_i.$$

Then

$$(u_0(\theta_j), u_1(\theta_j), \ldots, u_d(\theta_j)),$$

$$(u_0^*(\theta_j^*), u_1^*(\theta_j^*), \ldots, u_d^*(\theta_j^*))$$

are the left eigenvectors of B_1, $\hat{B}_1$ belonging to the eigenvalues θ_j, θ_j^*, and $u_i(x)$, $u_i^*(x)$ satisfy

$$\left.\begin{aligned} u_i(\theta_0) &= 1 \quad (0 \le i \le d), \\ xu_i(x) &= b_i u_{i+1}(x) + a_i u_i(x) + c_i u_{i-1}(x) \\ &\quad (0 \le i \le d-1), \\ xu_d(x) &= a_d u_d(x) + c_d u_{d-1}(x) \quad \text{for} \quad x = \theta_j \\ &\quad (0 \le j \le d), \end{aligned}\right\} \tag{5.1}$$

$$\left.\begin{aligned} u_i^*(\theta_0^*) &= 1 \quad (0 \le i \le d), \\ xu_i^*(x) &= b_i^* u_{i+1}^*(x) + a_i^* u_i^*(x) + c_i^* u_{i-1}^*(x) \\ &\quad (0 \le i \le d-1), \\ xu_d^*(x) &= a_d^* u_d^*(x) + c_d^* u_{d-1}^*(x) \quad \text{for} \quad x = \theta_j^* \\ &\quad (0 \le j \le d) \end{aligned}\right\} \tag{5.2}$$

with $u_{-1}(x) = u_{-1}^*(x) = 0$. Furthermore by Theorem 3.5(i) of Chapter II, the polynomials $u_i(x)$, $u_i^*(x)$ satisfy

$$u_i(\theta_j) = u_j^*(\theta_i^*) \quad \text{for} \quad 0 \le i \le d, \quad 0 \le j \le d. \tag{5.3}$$

A <u>basic</u> <u>hypergeometric</u> <u>series</u> is, by definition,

$$
{}_{r+1}\varphi_r\left(\begin{matrix} a_1,\ldots,a_{r+1} \\ \\ b_1,\ldots,b_r \end{matrix}\ ; q,x\right) \tag{5.4}
$$

$$
= \sum_{t=0}^{\infty} \frac{(a_1\,;q)_t \cdots (a_{r+1}\,;q)_t x^t}{(b_1\,;q)_t \cdots (b_r\,;q)_t (q\,;q)_t}\ ;
$$

where

$$
(a\,;q)_t = \begin{cases} (1-a)\ \ldots\ (1-aq^{t-1}) & (t=1,2,\ldots), \\ \\ 1, & (t=0). \end{cases} \tag{5.5}
$$

THEOREM 5.1. (Leonard (1982)) Let d be an integer ≥ 3 or the infinity ∞. Let $u_i(x)$, $u_i^*(x)$ $(0 \leq i \leq d)$ be polynomials of degree i satisfying (5.1), (5.2), (5.3) for some real numbers

$$
\begin{aligned}
&a_i,\ b_i,\ c_i \quad (b_i \neq 0 \ \text{ for } \ 0 \leq i \leq d-1,\ b_d = 0, \\
&\qquad c_i \neq 0 \ \text{ for } \ 1 \leq i \leq d,\ c_0 = 0),
\end{aligned}
$$

$$
\begin{aligned}
&a_i^*,\ b_i^*,\ c_i^* \quad (b_i^* \neq 0 \ \text{ for } \ 0 \leq i \leq d-1,\ b_d^* = 0, \\
&\qquad c_i^* \neq 0 \ \text{ for } \ 1 \leq i \leq d,\ c_0^* = 0)
\end{aligned}
$$

and

$$
\theta_i,\ \theta_i^* \quad (\theta_i \neq \theta_j,\ \theta_i^* \neq \theta_j^* \ \text{ for } \ i \neq j),
$$

where i and j range from 0 to d. Then one of the following cases (I), (IA), (IB), (II), (IIA), (IIB), (IIC), (IID) and (III) holds:

(I)

$$u_i(x) = {}_4\varphi_3\left(\begin{matrix} q^{-i}, & s^*q^{i+1}, & q^{-y}, & sq^{y+1} \\ r_1q, & r_2q, & r_3q, & \end{matrix} ; q, q\right)$$

$$(0 \leq i \leq d)$$

with

$$x = \mu(y) = \theta_0 + h(1 - q^y)(1 - sq^{y+1})/q^y,$$

$$ss^* = r_1r_2r_3,$$

$$\text{one of } r_i \ (i = 1, 2, 3) = q^{-d-1} \quad \text{if } d < \infty,$$

$$\theta_i = \mu(i) \qquad (0 \leq i \leq d),$$

$$b_i = \frac{h(1 - s^*q^{i+1})(1 - r_1q^{i+1})(1 - r_2q^{i+1})(1 - r_3q^{i+1})}{(1 - s^*q^{2i+1})(1 - s^*q^{2i+2})}$$

$$(1 \leq i \leq d - 1),$$

$$b_0 = \frac{h(1 - r_1q)(1 - r_2q)(1 - r_3q)}{1 - s^*q^2},$$

$$b_d = 0,$$

$$c_i = \begin{cases} \dfrac{hq(1 - q^i)(r_1 - s^*q^i)(r_2 - s^*q^i)(r_3 - s^*q^i)}{s^*(1 - s^*q^{2i+1})(1 - s^*q^{2i})} \\ \qquad (1 \leq i \leq d - 1) \quad \text{if } s^* \neq 0, \\ \\ hq(1 - q^i)(s - (r_1r_2 + r_2r_3 + r_3r_1)q^i) \\ \qquad (1 \leq i \leq d - 1) \quad \text{if } s^* = 0, \end{cases}$$

$$c_0 = 0,$$

$$c_d = \begin{cases} \dfrac{h(1-q^d)(r_1 - s^*q^d)(r_2 - s^*q^d)}{s^*q^d(1-s^*q^{2d})} & (r_3 = q^{-d-1}) \\ \quad \text{if } s^* \neq 0 \text{ and } d < \infty, & \\ h(1-q^d)(sq - r_1 - r_2) & (r_3 = q^{-d-1}) \\ \quad \text{if } s^* = 0 \text{ and } d < \infty. & \end{cases}$$

To get $u_i^*(x)$, θ_i^*, b_i^*, c_i^*, exchange (θ_0, h, s) and (θ_0^*, h^*, s^*) and preserve r_1, r_2, r_3, q.

We allow the following limiting cases:

(A) $h \to 0$, $hs \to s'$ and $hr_2 \to r_2'$ with $s^* = r_3 = 0$,

(B) $h \to 0$, $hs \to s'$, $\sqrt{h}\, r_2 \to r_2'$ and $\sqrt{h}\, r_1 \to r_1'$ with $s^* = r_3 = 0$.

Namely,

(IA)

$$u_i(x) = {}_2\varphi_1\left(\begin{matrix} q^{-i}, & q^{-y} \\ & r_1 q \end{matrix} ; q, \frac{s'}{r_2'} q^{y+1}\right)$$

with

$$x = \mu(y) = \theta_0 - s'q(1-q^y),$$

$$r_1 = q^{-d-1} \quad \text{if } d < \infty,$$

$$\theta_i = \mu(i),$$

$$b_i = -r_2' q^{i+1}(1 - r_1 q^{i+1}),$$

$$c_i = q(1 - q^i)(s' - r_1 r_2' q^i),$$

and

$$u_i^*(x) = {}_2\varphi_1\left(\begin{matrix} q^{-i}, & q^{-y} \\ & r_1 q \end{matrix} ; q, \frac{s'}{r_2'} q^{i+1}\right)$$

with

$$x = \mu^*(y) = \theta_0^* + h^*(1 - q^y)/q^y,$$

$$\theta_i^* = \mu^*(i),$$

$$b_i^* = h^* r_2'(1 - r_1 q^{i+1})/s' q^{2i+1},$$

$$c_i^* = h^*(1 - q^i)(r_2' - s' q^i)/s' q^{2i}.$$

(IB)

$$u_i(x) = \sum_{t=0}^{i} (-1)^t \frac{(q^{-i}; q)_t (q^{-y}; q)_t}{q^{t(t+1)/2}(q; q)_t} \left(\frac{s'}{r} q^{y+1}\right)^t$$

with

$$x = \mu(y) = \theta_0 - s' q(1 - q^y),$$

$$\theta_i = \mu(i),$$

$$b_i = r q^{2i+2},$$

$$c_i = q(1 - q^i)(s' - r q^i),$$

and

$$u_i^*(x) = \sum_{t=0}^{i} (-1)^t \frac{(q^{-i};q)_t (q^{-y};q)_t}{q^{t(t+1)/2}(q;q)_t} (\frac{s'}{r} q^{i+1})^t$$

with

$$x = \mu^*(y) = \theta_0^* + h^*(1-q^y)/q^y,$$

$$\theta_i^* = \mu^*(i),$$

$$b_i^* = -h^* r/s' q^i,$$

$$c_i^* = -h^*(1-q^i)/q^i.$$

(II)

$$u_i(x) = {}_4F_3\left(\begin{matrix} -i,\ i+1+s^*,\ -y,\ y+1+s \\ r_1+1,\ r_2+1,\ r_3+1 \end{matrix} ; 1\right)$$

$$(0 \leq i \leq d)$$

with

$$x = \mu(y) = \theta_0 + hy(y+1+s),$$

$$s + s^* = r_1 + r_2 + r_3,$$

one of r_i $(i = 1, 2, 3) = -d-1$ if $d < \infty$,

$$\theta_i = \mu(i) \qquad (0 \leq i \leq d),$$

$$b_i = \frac{h(i+1+s^*)(i+1+r_1)(i+1+r_2)(i+1+r_3)}{(2i+1+s^*)(2i+2+s^*)}$$

$$(1 \leq i \leq d-1),$$

$$b_0 = \frac{h(1+r_1)(1+r_2)(1+r_3)}{2+s^*},$$

$$b_d = 0,$$

$$c_i = \frac{hi(i+s^*-r_1)(i+s^*-r_2)(i+s^*-r_3)}{(2i+1+s^*)(2i+s^*)}$$

$$(1 \leq i \leq d-1),$$

$$c_0 = 0,$$

$$c_d = \frac{hd(d+s^*-r_1)(d+s^*-r_2)}{2d+s^*} \qquad (r_3 = -d-1)$$

if $d < \infty$.

To get $u_i^*(x)$, θ_i^*, b_i^*, c_i^*, exchange (θ_0, h, s) and (θ_0^*, h^*, s^*) and preserve r_1, r_2, r_3.

We allow the following limiting cases:

(A) $h^* \to 0$, $h^*s^* \to s^{*\prime}$, $h^*r_3 \to s^{*\prime}$,

(B) $h \to 0$, $hs \to s'$, $hr_3 \to s'$,

(C) $(h,h^*) \to (0,0)$, $(hs,h^*s^*) \to (s',s^{*\prime})$, $(h^*r_2,hr_3) \to (r_2',r_3')$,

(D) $h = h^* \to 0$, $(hs,hs^*) \to (s',s^{*\prime})$, $(\sqrt{h}\, r_1,\sqrt{h}\, r_2,hr_3) \to (r_1',r_2',r_3')$.

Namely,

(IIA)

$$u_i(x) = {}_3F_2\left(\begin{matrix}-i,\ -y,\ y+1+s \\ r_1+1,\ r_2+1\end{matrix};1\right)$$

with

$$x = \mu(y) = \theta_0 + hy(y+1+s),$$

$$\text{one of } r_i \ (i = 1, 2) = -d-1 \text{ if } d < \infty,$$

$$\theta_i = \mu(i),$$

$$b_i = h(i+1+r_1)(i+1+r_2),$$

$$c_i = hi(i+r_1+r_2-s),$$

and

$$u_i^*(x) = {}_3F_2\left(\begin{matrix}-i,\ i+1+s,\ -y \\ r_1+1,\ r_2+1\end{matrix};1\right)$$

with

$$x = \mu^*(y) = \theta_0^* + s^{*\prime}y,$$

$$\theta_i^* = \mu^*(i),$$

$$b_i^* = \frac{s^{*\prime}(i+1+s)(i+1+r_1)(i+1+r_2)}{(2i+1+s)(2i+2+s)} \qquad (1 \leq i \leq d-1),$$

$$b_0^* = \frac{s^{*\prime}(1+r_1)(1+r_2)}{2+s},$$

$$b_d^* = 0,$$

$$c_i^* = \frac{-s^{*\prime} i(i+s-r_1)(i+s-r_2)}{(2i+1+s)(2i+s)} \qquad (1 \le i \le d-1),$$

$$c_0^* = 0,$$

$$c_d^* = \frac{-s^{*\prime} d(d+s-r_1)}{2d+s} \qquad (r_2 = -d-1) \quad \text{if} \quad d < \infty .$$

(IIB)

The dual of (IIA).

(IIC)

$$u_i(x) = {}_2F_1\left(\begin{matrix}-i, & -y \\ r_1+1 & \end{matrix} ; \frac{s' s^{*\prime}}{r}\right)$$

with

$$x = \mu(y) = \theta_0 + s' y,$$

$$r_1 = -d-1 \quad \text{if} \quad d < \infty,$$

$$\theta_i = \mu(i),$$

$$b_i = (i+1+r_1) r / s^{*\prime},$$

$$c_i = i(r - s' s^{*\prime}) / s^{*\prime}.$$

To get $u_i^*(x)$, θ_i^*, b_i^*, c_i^*, exchange (θ_0, s') and $(\theta_0^*, s^{*\prime})$.

(IID)

$$u_i(x) = \sum_{t=0}^{i} (-i)_t (-y)_t \frac{1}{t!} \left(\frac{s' s^{*\prime}}{r}\right)^t$$

with

$$x = \mu(y) = \theta_0 + s'y,$$

$$\theta_i = \mu(i),$$

$$b_i = r/s^{*\prime},$$

$$c_i = -s'i.$$

To get $u_i^*(x)$, θ_i^*, b_i^*, c_i^*, exchange (θ_0, s') and $(\theta_0^*, s^{*\prime})$.

(III)

For $0 \leq i \leq d$,

$$b_i = \begin{cases} \dfrac{2h(i+1+r_1)(i+1+r_2)}{(2i+2-s^*)} & (i = \text{even}), \\[2ex] \dfrac{2h(i+1-s^*)(i+1-r_3)}{(2i+2-s^*)} & (i = \text{odd}), \end{cases}$$

$$c_i = \begin{cases} \dfrac{-2hi(i-s^*+r_3)}{(2i-s^*)} & (i = \text{even}), \\[2ex] \dfrac{-2h(i-s^*-r_1)(i-s^*-r_2)}{(2i-s^*)} & (i = \text{odd}), \end{cases}$$

with

$$s + s^* = -r_1 - r_2 + r_3,$$

one of r_i $(i = 1, 2) = -d-1$ if $d = \text{even}$,

$r_3 = d+1$ if $d = \text{odd}$,

$$\theta_i = \begin{cases} \theta_0 + 2hi & (i = \text{even}), \\ \theta_0 - 2h(i+1-s) & (i = \text{odd}), \end{cases}$$

$$u_i(\theta_j) = {}_4F_3\left(\begin{matrix} -\frac{i}{2}, \frac{i+2-s^*}{2}, -\frac{j}{2}, \frac{j+2-s}{2} \\ \frac{r_1+1}{2}, \frac{r_2+1}{2}, \frac{-r_3+2}{2} \end{matrix} ; 1 \right)$$

$$- \frac{ij}{(r_1+1)(r_2+1)} \cdot$$

$${}_4F_3\left(\begin{matrix} \frac{-i+2}{2}, \frac{i+2-s^*}{2}, \frac{-j+2}{2}, \frac{j+2-s}{2} \\ \frac{r_1+3}{2}, \frac{r_2+3}{2}, \frac{-r_3+2}{2} \end{matrix} ; 1 \right)$$

$(i = \text{even}, j = \text{even})$,

$$u_i(\theta_j) = {}_4F_3\left(\begin{matrix} -\frac{i}{2}, \frac{i+2-s^*}{2}, \frac{-j+1}{2}, \frac{j+1-s}{2} \\ \frac{r_1+1}{2}, \frac{r_2+1}{2}, \frac{-r_3+2}{2} \end{matrix} ; 1 \right)$$

$$+ \frac{i(j+1-s)}{(r_1+1)(r_2+1)} \cdot$$

$${}_4F_3\left(\begin{matrix} \frac{-i+2}{2}, \frac{i+2-s^*}{2}, \frac{-j+1}{2}, \frac{j+3-s}{2} \\ \frac{r_1+3}{2}, \frac{r_2+3}{2}, \frac{-r_3+2}{2} \end{matrix} ; 1 \right)$$

$(i = \text{even}, j = \text{odd})$,

$$u_i(\theta_j) = {}_4F_3\left(\begin{matrix}\frac{-i+1}{2}, \frac{i+1-s^*}{2}, -\frac{j}{2}, \frac{j+2-s}{2} \\ \frac{r_1+1}{2}, \frac{r_2+1}{2}, \frac{-r_3+2}{2}\end{matrix} ; 1\right)$$

$$+ \frac{j(i+1-s^*)}{(r_1+1)(r_2+1)} \cdot$$

$${}_4F_3\left(\begin{matrix}\frac{-i+1}{2}, \frac{i+3-s^*}{2}, \frac{-j+2}{2}, \frac{j+2-s}{2} \\ \frac{r_1+3}{2}, \frac{r_2+3}{2}, \frac{-r_3+2}{2}\end{matrix} ; 1\right)$$

$(i = \text{odd},\ j = \text{even}),$

$$u_i(\theta_j) = {}_4F_3\left(\begin{matrix}\frac{-i+1}{2}, \frac{i+1-s^*}{2}, \frac{-j+1}{2}, \frac{j+1-s}{2} \\ \frac{r_1+1}{2}, \frac{r_2+1}{2}, \frac{-r_3+2}{2}\end{matrix} ; 1\right)$$

$$- \frac{(i+1-s^*)(j+1-s)}{(r_1+1)(r_2+1)} \cdot$$

$${}_4F_3\left(\begin{matrix}\frac{-i+1}{2}, \frac{i+3-s^*}{2}, \frac{-j+1}{2}, \frac{j+3-s}{2} \\ \frac{r_1+3}{2}, \frac{r_2+3}{2}, \frac{-r_3+2}{2}\end{matrix} ; 1\right)$$

$(i = \text{odd},\ j = \text{odd}).$

To get b_i^*, c_i^*, θ_i^*, $u_i^*(\theta_j^*)$, exchange (θ_0, h, s) and (θ_0^*, h^*, s^*) and preserve r_1, r_2, r_3.

Conversely, each of (I), (IA), (IB), (II), (IIA), (IIB), (IIC), (IID) and (III) with $b_i \neq 0$ $(i \neq d)$, $b_d = 0$, $c_i \neq 0$ $(i \neq 0)$, $c_0 = 0$, $b_i^* \neq 0$ $(i \neq d)$, $b_d^* = 0$, $c_i^* \neq 0$ $(i \neq 0)$, $c_0^* = 0$, $\theta_i \neq \theta_j$ $(i \neq j)$, $\theta_i^* \neq \theta_j^*$ $(i \neq j)$ satisfies (5.1), (5.2), (5.3).

REMARKS. (1) (II) is the limiting case of (I) as q tends to 1. Namely, let

$$(q-1)^2 h \to h', \quad (q-1)^2 h^* \to h^{*\prime},$$

$$\frac{s-1}{q-1} \to s', \quad \frac{s^*-1}{q-1} \to s^{*\prime}, \quad \frac{r_i - 1}{q-1} \to r_i' \qquad (i = 1, 2, 3),$$

as $q \to 1$ in (I). Then we get (II).

(III) is the limiting case of (I) as q tends to -1. Namely, let

$$(q+1)h \to h', \quad (q+1)h^* \to h^{*\prime},$$

$$\frac{s-1}{q+1} \to s', \quad \frac{s^*-1}{q+1} \to s^{*\prime},$$

$$\frac{r_1+1}{q+1} \to r_1', \quad \frac{r_2+1}{q+1} \to r_2', \quad \frac{r_3-1}{q+1} \to r_3',$$

as $q \to -1$ in (I). Then we get (III).

(2) In Theorem 5.1, we separate the expressions of b_i, b_i^*, c_i, c_i^* for $i = 0, d$ from those for $1 \leq i \leq d-1$. This is because the denominators of the expressions for $1 \leq i \leq d-1$ can be zero, if we simply put $i = 0, d$ in them. Notice that the expressions of b_i, b_i^*, c_i, c_i^* in Theorem 5.1 are guaranteed to have denominators all non-zero by the condition $\theta_i \neq \theta_j$, $\theta_i^* \neq \theta_j^*$ $(i \neq j,\ 0 \leq i,j \leq d)$.

It seems as if the expressions of b_i, b_i^*, c_i, c_i^* for $1 \leq i \leq d-1$ were still valid for $i = 0, d$ by regarding i as a continuous variable and taking the limits with i tending to $0, d$. But this fails in general. For example, consider the expression of type IIB with $(s', s^*; r_1, r_2) = (-4, -2d-1; -d-\frac{1}{2}, -d-1)$, which actually comes from the antipodal quotient of $H(2d,2)$. Then $c_i = i$ $(0 \leq i \leq d-1)$ and $c_d = 2d$, so the general expression $c_i = i$ is not valid for $i = d$.

(3) Suppose $u_i(x)$ and $u_i^*(x)$ come from a (P and Q)-polynomial scheme. Then $c_1 = c_1^* = 1$, $b_0 = \theta_0$ and $b_0^* = \theta_0^*$. So h, h^*, θ_0, θ_0^* are written in terms of q, r_1, r_2, r_3, s, s^*.

(4) (IB) and (IID) hold only when $d = \infty$.

(5) The polynomials $u_i(x)$ and $u_i^*(x)$ in (I) are called the <u>Askey-Wilson polynomilas</u> (Askey-Wilson (1979)), and include all "classical" orthogonal polynomials, allowing the limiting cases.

It seems that the polynomials considered in Theorem 5.1 are not exactly contained in the class of orthogonal polynomials. It is well known that polynomials $u_i(x)$ of degree i $(0 \leq i \leq d)$ are orthogonal polynomials with respect to a certain point-mass measure on the real line whose non-zero masses are defined exactly on $\{\theta_0, \theta_1, \ldots, \theta_d\}$ if and only if they staisfy a three-term recurrence

$$xu_i(x) = b_i u_{i+1}(x) + a_i u_i(x) + c_i u_{i-1}(x)$$
$$(0 \leq i \leq d-1) \quad (u_{-1}(x) = 0),$$

$$xu_d(x) = a_d u_d(x) + c_d u_{d-1}(x) \quad \text{for } x = \theta_j$$
$$(0 \leq j \leq d),$$

where θ_j $(0 \leq j \leq d)$ are the eigenvalues of the tridiagonal matrix

$$\begin{Bmatrix} * & c_1 & \cdots & c_{d-1} & c_d \\ 0 & a_1 & \cdots & a_{d-1} & a_d \\ b_0 & b_1 & \cdots & b_{d-1} & * \end{Bmatrix}$$

and

$$b_i c_{i+1} > 0 \qquad (0 \leq i \leq d-1).$$

So, if we add the assumption

$$b_i c_{i+1} > 0 \text{ and } b_i^* c_{i+1}^* > 0 \qquad (0 \leq i \leq d-1)$$

to both sides of conditions of Theorem 5.1, then all polynomials considered are orthogonal polynomials. It is also known that the point-mass on θ_i is given by the Christoffel number m_i, and in the case of Theorem 5.1, m_i is given by

$$m_i = b_0^* b_1^* \cdots b_{i-1}^* / c_1^* c_2^* \cdots c_i^* .$$

Proof of Theorem 5.1. We use the following notations:

$$(5.6)$$

$$[x]_t = \begin{cases} (x-\theta_0)(x-\theta_1) \cdots (x-\theta_{t-1}) & (t=1,2,\ldots), \\ 1 & (t=0), \end{cases}$$

$$[x]_t^* = \begin{cases} (x-\theta_0^*)(x-\theta_1^*) \cdots (x-\theta_{t-1}^*) & (t=1,2,\ldots), \\ 1 & (t=0) . \end{cases}$$

PROPOSITION 5.2. (i) The condition (5.3) holds if and only if

$$u_i(x) = \sum_{t=0}^{i} M_t[\theta^*_i]^*_t[x]_t \qquad (0 \le i \le d),$$

$$u^*_i(x) = \sum_{t=0}^{i} M_t[\theta_i]_t[x]^*_t \qquad (0 \le i \le d)$$

for some non-zero numbers $M_0, M_1, \ldots, M_d$.

(ii) The condition (5.2) holds if and only if

$$b^*_i[\theta_{i+1}]_t + a^*_i[\theta_i]_t + c^*_i[\theta_{i-1}]_t$$

$$= \theta^*_t[\theta_i]_t + \frac{M_{t-1}}{M_t}[\theta_i]_{t-1}$$

$$(0 \le i \le d-1,\ 0 \le t \le i+1),\ (i = d,\ \ 0 \le t \le d),$$

where $b^*_d = 0$, $c^*_0 = 0$, $M_{-1} = 0$, $M_0 = 1$. A similar equivalent condition holds for (5.1).

(iii) Assume (5.1), (5.2) and (5.3). Then

$$\frac{M_i}{M_{i+1}} = b^*_i \frac{[\theta_{i+1}]_{i+1}}{[\theta_i]_i} = b_i \frac{[\theta^*_{i+1}]^*_{i+1}}{[\theta^*_i]^*_i}$$

$$(0 \le i \le d-1).$$

<u>Proof</u>. (i) The "if" part is clear, so we prove the "only if" part. Let

$$u_i(x) = \sum_{t=0}^{i} M_{it}[\theta^*_i]^*_t[x]_t$$

and

$$u_j^*(x) = \sum_{t=0}^{j} M_{jt}^*[\theta_j]_t[x]_t^*.$$

We shall show $M_{it} = M_{tt} = M_{tt}^* = M_{jt}^*$ $(t \leq i \leq d,$ $t \leq j \leq d)$.

By (5.3), $0 = u_i(\theta_j) - u_j^*(\theta_i^*)$, i.e.,

$$\sum_{t=0}^{\min(i,j)} (M_{it} - M_{jt}^*)[\theta_i^*]_t^*[\theta_j]_t = 0\,. \tag{5.7}$$

Put $i = 0$ in (5.7) to get $M_{00} = M_{j0}^*$. Put $j = 0$ in (5.7) to get $M_{i0} = M_{00}^*$. Thus $M_{i0} = M_{00}^* = M_{00} = M_{j0}^*$ and the summation of (5.7) becomes over t from 1 to $\min(i,j)$.

Put $i = 1$ in (5.7) to get $(M_{11} - M_{j1}^*)[\theta_1^*]_1^*[\theta_j]_1 = 0$ for $1 \leq j \leq d$. So $M_{11} = M_{j1}^*$ for $1 \leq j \leq d$. Put $j = 1$ in (5.7) to get $M_{i1} = M_{11}^*$ for $1 \leq i \leq d$. Thus $M_{i1} = M_{11}^* = M_{11} = M_{j1}^*$ and the summation of (5.7) becomes over t from 2 to $\min(i,j)$.

Put $i = 2$ in (5.7) to get $M_{22} = M_{j2}^*$ for $2 \leq j \leq d$. Put $j = 2$ in (5.7) to get $M_{i2} = M_{22}^*$ for $2 \leq i \leq d$. Thus $M_{i2} = M_{22}^* = M_{22} = M_{j2}^*$ and the summation of (5.7) becomes over t from 3 to $\min(i,j)$. Continue this procedure to reach $M_{dd} = M_{dd}^*$. Putting $M_t = M_{it} = M_{jt}^*$ for $t \leq i \leq d$, $t \leq j \leq d$, we get the desired expressions for $u_i(x)$, $u_i^*(x)$.

(ii) Since $u_i^*(\theta_0^*) = 1$, we have $M_0 = 1$. Express $xu_i^*(x)$ as a linear combination of $[x]_t^*$. Then the coefficient of $[x]_t^*$ is

$$M_t\theta_t^*[\theta_i]_t + M_{t-1}[\theta_i]_{t-1}$$

for $0 \le t \le i+1$. Comparing the coefficients of $[x]_t^*$ in the identity (5.2), we get the desired result for $0 \le i \le d-1$, $0 \le t \le i+1$.

If $i = d$, (5.2) means that $\theta_0^*, \theta_1^*, \ldots, \theta_d^*$ are the roots of $xu_d^*(x) - a_d^*u_d^*(x) - c_d^*u_{d-1}^*(x)$, i.e.,

$$c[x]_{d+1}^* = xu_d^*(x) - a_d^*u_d^*(x) - c_d^*u_{d-1}^*(x)$$

for some $c \neq 0$. Thus when written in the linear combination of $[x]_t^*$, $xu_d^*(x)$ and $a_d^*u_d^*(x) + c_d^*u_{d-1}^*(x)$ have the same coefficients of $[x]_t^*$ for $0 \le t \le d$. So we get the desired result.

(iii) Put $t = i + 1$ in (ii). □

PROPOSITION 5.3. The condition (5.2) holds if and only if

$$b_0^* = \frac{1}{\nu_1 M_1},$$

$$b_d^* = 0,$$

$$b_i^* = \frac{[\theta_i]_{t+1}}{(\nu_{i+1} + \nu_i)\nu_{i+1}[\theta_{i-1}]_t} \cdot$$

$$\left\{\frac{M_{t+1}}{M_{t+2}} - (1 + \sigma_{it})\frac{M_t}{M_{t+1}} + \sigma_{i\ t-1}\frac{M_{t-1}}{M_t} + \tau_{it}\right\}$$

for $1 \le i \le d-1$, $0 \le t \le i-1$,

$$\frac{M_{t+1}}{M_{t+2}} - (1 + \sigma_{dt})\frac{M_t}{M_{t+1}} + \sigma_{d\ t-1}\frac{M_{t-1}}{M_t} + \tau_{dt} = 0 \quad (\#)$$

for $0 \le t \le d-2$ (only when $d < \infty$),

$$c_0^* = 0,$$

$$c_d^* = \frac{1}{-\nu_d}\left(\nu_1^*(\theta_d - \theta_0) + \frac{1}{M_1}\right),$$

$$c_i^* = \frac{\theta_i - \theta_0}{(\nu_{i+1} + \nu_i)\nu_i}\left\{\frac{M_1}{M_2} - \left(1 + \frac{\theta_{i+1} - \theta_1}{\theta_i - \theta_0}\right)\frac{1}{M_1} + \nu_2^*(\theta_i - \theta_1) - \nu_1^*(\theta_{i+1} - \theta_0)\right\}$$

for $1 \le i \le d-1$,

$$a_i^* = \theta_0^* - b_i^* - c_i^*$$

for $0 \le i \le d$, where $\nu_i = \theta_i - \theta_{i-1}$, $\nu_i^* = \theta_i^* - \theta_{i-1}^*$,

$$\sigma_{it} = \frac{\theta_{i-1} - \theta_{t+1}}{\theta_i - \theta_t} \qquad (i \ne t),$$

$$\tau_{it} = \nu_{t+2}^*(\theta_i - \theta_{t+1}) - \nu_{t+1}^*(\theta_{i-1} - \theta_t).$$

Similar equivalent conditions hold for (5.1), changing $(b_i^*, c_i^*, a_i^*, \theta_i^*, \nu_i^*, \nu_i, \sigma_{it}, \tau_{it})$ to $(b_i, c_i, a_i, \theta_i, \nu_i, \nu_i^*, \sigma_{it}^*, \tau_{it}^*)$, where

$$\sigma_{it}^* = \frac{\theta_{i-1}^* - \theta_{t+1}^*}{\theta_i^* - \theta_t^*} \qquad (i \ne t),$$

$$\tau_{it}^* = \nu_{t+2}(\theta_i^* - \theta_{t+1}^*) - \nu_{t+1}(\theta_{i-1}^* - \theta_t^*).$$

<u>Proof</u>. Let $(I_{i\,t})$ denote the identity

$$b_i^*[\theta_{i+1}]_t + a_i^*[\theta_i]_t + c_i^*[\theta_{i-1}]_t$$

$$= \theta_t^*[\theta_i]_t + \frac{M_{t-1}}{M_t}[\theta_i]_{t-1}$$

in Proposition 5.2(ii). $(I_{0\,1})$ gives $b_0^* = \frac{1}{\nu_1 M_1}$. Regard $(I_{i\,t})$, $(I_{i\,t+1})$, $(I_{i\,t+2})$ as a system of equations for b_i^*, a_i^*, c_i^*. Consider $(I_{i\,t+1})$ - $(\theta_i - \theta_t)(I_{i\,t})$ to get $(J_{i\,t+1})$:

$$b_i^*\nu_{i+1}[\theta_{i+1}]_t + c_i^*(-\nu_i)[\theta_{i-1}]_t$$

$$= \nu_{t+1}^*[\theta_i]_{t+1} + \left(\frac{M_t}{M_{t+1}} - \frac{M_{t-1}}{M_t}\,\frac{\theta_i - \theta_t}{\theta_i - \theta_{t-1}}\right)[\theta_i]_t.$$

Consider $(J_{i\,t+2}) - (\theta_{i-1} - \theta_t)(J_{i\,t+1})$ to get $(K_{i\,t+2})$:

$$b_i^*\nu_{i+1}(\nu_{i+1} + \nu_i)[\theta_{i+1}]_t$$

$$= [\theta_i]_{t+1}\{\tau_{it} + \frac{M_{t+1}}{M_{t+2}} - (1+\sigma_{it})\,\frac{M_t}{M_{t+1}}$$

$$+ \sigma_{i\,t-1}\frac{M_{t-1}}{M_t}\},$$

which is the desired identity for b_i^*. The identity (#) follows from $b_d^* = 0$.

By solving the system of equations $(I_{i\,0})$, $(I_{i\,1})$, $(I_{i\,2})$ for b_i^*, c_i^*, a_i^*, we get the desired identities for c_i^*, a_i^*. For c_d^*, solve $(J_{d\,1})$.

Conversely $(J_{i\,t+1})$ is derived from $(K_{i\,t+1})$ and $(J_{i\,t})$. $(I_{i\,t+1})$ is derived from $(J_{i\,t+1})$ and $(I_{i\,t})$. Thus $(I_{i\,t})$ are all derived from $(I_{i\,0})$, $(I_{i\,1})$ and all $(K_{i\,t})$. □

REMARK. (6) If one wants to simplify the description of Proposition 5.3, regard $\nu_{-1}, \nu^*_{-1}, \nu_{d+1}, \nu^*_{d+1}$ as indeterminates. Then the expressions of b_i, b^*_i, c_i, c^*_i for $1 \le i \le d-1$ are also valid for $i = 0, d$.

LEMMA 5.4. The following three conditions are equivalent:

(i) $\nu_t \nu^*_{t-i} - \nu^*_t \nu_{t-i} = \gamma_i$ $(0 \le i \le i_0,$ $i+1 \le t \le d)$, where γ_i is a constant which is independent of t.

(ii) $\sigma_{t\,t-i-2} = \sigma^*_{t\,t-i-2}$ $(0 \le i \le i_0,$ $i+2 \le t \le d)$.

(iii) $\tau_{t\,t-i-2} = \tau^*_{t\,t-i-2}$ $(0 \le i \le i_0,$ $i+2 \le t \le d)$.

<u>Proof</u>. Induction on i_0, using the identities

$$\sigma_{t\,t-i-2} - \sigma^*_{t\,t-i-2}$$

$$= \frac{1}{(\nu_t + \dots + \nu_{t-i-1})(\nu^*_t + \dots + \nu^*_{t-i-1})} \cdot$$

$$\sum_{\substack{1 \le r \le i \\ 0 \le s \le i+1}} (\nu_{t-r}\nu^*_{t-s} - \nu^*_{t-r}\nu_{t-s}),$$

$$\tau_{t\,t-i-2} - \tau^*_{t\,t-i-2}$$

$$= \sum_{s=0}^{i} (\nu^*_{t-i}\nu_{t-s} - \nu_{t-i}\nu^*_{t-s})$$

$$- \sum_{s=1}^{i+1} (\nu^*_{t-i-1}\nu_{t-s} - \nu_{t-i-1}\nu^*_{t-s}). \qquad \square$$

PROPOSITION 5.5. If (5.1), (5.2) and (5.3) hold, then we have the following:

(i) $\nu_t\nu^*_{t-i} - \nu^*_t\nu_{t-i} = \gamma_i$ $(0 \le i \le d-1,$ $i+1 \le t \le d)$, where γ_i is a constant which is independent of t.

(ii) $\sigma_{i+1\,i} = \sigma^*_{i+1\,i} = -1$ $(0 \le i \le d-1)$,

$\sigma_{i+1\,i-1} = \sigma^*_{i+1\,i-1} = 0$ $(1 \le i \le d-1)$,

$\sigma_{i+1\,t} = \sigma^*_{i+1\,t}$ $(0 \le t < i \le d-1)$.

(iii) $\tau_{i+1\,i} = \tau^*_{i+1\,i} = c$ $(0 \le i \le d-2)$,

$\tau_{i+1\,i-1} = \tau^*_{i+1\,i-1} = \nu_{i+1}\nu^*_{i+1} - \nu_i\nu^*_i$

$(1 \le i \le d-1)$,

$\tau_{i+1\,t} = \tau^*_{i+1\,t}$ $(0 \le t < i \le d-1)$.

Proof. Check the identities for $\sigma_{i+1\,t}$, $\sigma^*_{i+1\,t}$, $\tau_{i+1\,t}$, $\tau^*_{i+1\,t}$ $(t = i,\ i-1)$ directly.

By the expressions of b^*_i and b_i in Proposition 5.3,

$$\sigma_{i+1\,t} \tag{5.8}$$

$$= \frac{\dfrac{M_{t+1}}{M_{t+2}} - (1+\sigma_{it})\dfrac{M_t}{M_{t+1}} + \sigma_{i\,t-1}\dfrac{M_{t-1}}{M_t} + \tau_{it}}{\dfrac{M_{t+2}}{M_{t+3}} - (1+\sigma_{i\,t+1})\dfrac{M_{t+1}}{M_{t+2}} + \sigma_{it}\dfrac{M_t}{M_{t+1}} + \tau_{i\,t+1}}$$

and dually

$$\sigma^*_{i+1\,t} \tag{5.8*}$$

$$= \frac{\dfrac{M_{t+1}}{M_{t+2}} - (1+\sigma^*_{it})\dfrac{M_t}{M_{t+1}} + \sigma^*_{i\,t-1}\dfrac{M_{t-1}}{M_t} + \tau^*_{it}}{\dfrac{M_{t+2}}{M_{t+3}} - (1+\sigma^*_{i\,t+1})\dfrac{M_{t+1}}{M_{t+2}} + \sigma^*_{it}\dfrac{M_t}{M_{t+1}} + \tau^*_{i\,t+1}}$$

for $2 \leq i \leq d-1$, $0 \leq t \leq i-2$, because $\sigma_{i+1\,t} = \dfrac{[\theta_{i+1}]_t}{[\theta_i]_{t+1}} \dfrac{[\theta_i]_{t+2}}{[\theta_{i+1}]_{t+1}}$. By Lemma 5.4, it suffices to show that

$$\sigma_{i+1\,t} = \sigma^*_{i+1\,t} \quad (0 \leq t < i \leq d-1).$$

We use induction on $(i+1-t, t)$. Suppose $\sigma_{i+1,t} = \sigma^*_{i+1,t}$ holds for $i+1-t \leq i_0$. By Lemma 5.4, $\tau_{i+1\,t} = \tau^*_{i+1\,t}$ for $i+1-t \leq i_0$. Put $i = i_0$ and $t = 0$ in (5.8) and $(5.8)^*$ to get $\sigma_{i_0+1\,0} = \sigma^*_{i_0+1\,0}$. Put $i = i_0+1$ and $t = 1$ in (5.8) and $(5.8)^*$ to get $\sigma_{i_0+2\,1} = \sigma^*_{i_0+2\,1}$. Repeating this argument, we get $\sigma_{i+1\,t} = \sigma^*_{i+1\,t}$ for $i+1-t \leq i_0+1$. □

PROPOSITION 5.6. Suppose (5.1), (5.2) and (5.3) hold.

Then $(\nu_1+\nu_3)/\nu_2 = (\nu_1^*+\nu_3^*)/\nu_2^*$. Let

$$q+\frac{1}{q}=\frac{\nu_1+\nu_3}{\nu_2},$$

$$\lambda=\frac{q\nu_2-\nu_1}{q^2-1},\qquad \lambda^*=\frac{q\nu_2^*-\nu_1^*}{q^2-1},$$

$$\mu=\frac{q(q\nu_1-\nu_2)}{q^2-1},\qquad \mu^*=\frac{q(q\nu_1^*-\nu_2^*)}{q^2-1}.$$

Then we have

$$\nu_i=\lambda q^{i-1}+\mu q^{-i+1}\quad (1\le i\le d),$$

$$\nu_i^*=\lambda^* q^{i-1}+\mu^* q^{-i+1}\quad (1\le i\le d).$$

If $q=\pm 1$, the above identities mean

$$\nu_i=\lim_{q\to\pm 1}(\lambda q^{i-1}+\mu q^{-i+1}),$$

$$\nu_i^*=\lim_{q\to\pm 1}(\lambda^* q^{i-1}+\mu^* q^{-i+1}),$$

regarding $\lambda, \lambda^*, \mu, \mu^*$ as functions of q.

Proof. The identity $(\nu_1+\nu_3)/\nu_2 = (\nu_1^*+\nu_3^*)/\nu_2^*$ comes from $\sigma_{30}=\sigma_{30}^*$.

We shall prove the identities for ν_i and ν_i^* by induction on i. For $i = 1$, 2 and 3, the identities hold trivially. Put $t = 0$ in (5.8) to get

$$\sigma_{i+1\,0} = \frac{\frac{M_1}{M_2} - (1+\sigma_{i0})\frac{1}{M_1} + \tau_{i0}}{\frac{M_2}{M_3} - (1+\sigma_{i1})\frac{M_1}{M_2} + \sigma_{i0}\frac{1}{M_1} + \tau_{i1}} \tag{5.9}$$

for $2 \le i \le d-1$. Put $i = 2$ in (5.9). Then we get

$$\frac{M_2}{M_3} = \frac{q^2+q+1}{q}\left\{\frac{M_1}{M_2} - \frac{1}{M_1} + (q^2-1)(\lambda\lambda^* - \frac{\mu\mu^*}{q^2}\right\}. \tag{5.10}$$

Suppose the identities hold for ν_j and ν_j^* $(0 \le j \le i)$. Then

$$\theta_j - \theta_k = \frac{(q^{j-k}-1)(\lambda q^{j+k-1}+\mu)}{(q-1)q^{j-1}} \quad (0 \le k \le j \le i). \tag{5.11}$$

Using (5.10) and (5.11), we get that the numerator of the right hand side of (5.9) is

$$\frac{M_1}{M_2} - \frac{(q+1)(q^{i-1}-1)}{q^i-1}\frac{1}{M_1} + \frac{(q+1)(q^{i-1}-1)}{q^i} \cdot (\lambda\lambda^* q^i - \mu\mu^*)$$

and the denominator is $\frac{q^{i+1}-1}{q(q^{i-1}-1)}$ times the numerator. So we have

$$\sigma_{i+1\,0} = \frac{q(q^{i-1}-1)}{q^{i+1}-1}.$$

Since $\sigma_{i+1\,0} = \frac{\theta_i - \theta_1}{\nu_{i+1} + \theta_i - \theta_0}$, we get the desired identity for ν_{i+1}. Since $\sigma^*_{i+1\,0} = \sigma_{i+1\,0} = q(q^{i-1} - 1)/(q^{i+1} - 1)$, we get the desired identity for ν^*_{i+1}. □

To prove Proposition 5.6, we used $\sigma_{30} = \sigma^*_{30}$ but not the full information of Proposition 5.5. If we use more information of Proposition 5.5, we can prove Proposition 5.6 more quickly in case $\nu_2\nu^*_1 \neq \nu^*_2\nu_1$ as follows. Rewrite $\nu_t\nu^*_{t-1} - \nu^*_t\nu_{t-1} = \gamma_1$ as

$$\frac{\nu^*_{t-1}}{\nu_{t-1}} - \frac{\nu^*_t}{\nu_t} = \frac{\gamma_1}{\nu_{t-1}\nu_t}$$

and take the sum over t from i to $i+1$. Then

$$\frac{\nu^*_{i-1}}{\nu_{i-1}} - \frac{\nu^*_{i+1}}{\nu_{i+1}} = \gamma_1\left(\frac{1}{\nu_{i-1}\nu_i} + \frac{1}{\nu_i\nu_{i+1}}\right).$$

The left hand side is $\gamma_2/\nu_{i-1}\nu_{i+1}$. So we have

$$\frac{\gamma_2}{\gamma_1} = \frac{\nu_{i+1} + \nu_{i-1}}{\nu_i},$$

if $\gamma_1 \neq 0$. Set $q + \frac{1}{q} = \frac{\gamma_2}{\gamma_1}$. Then

$$\begin{aligned}\nu_{i+1} - q\nu_i &= \frac{1}{q}(\nu_i - q\nu_{i-1}) \\ &\cdots\cdots \\ &= \frac{1}{q^{i-1}}(\nu_2 - q\nu_1),\end{aligned}$$

and similarly

$$\nu_{i+1} - \frac{1}{q}\nu_i = q^{i+1}(\nu_2 - \frac{1}{q}\nu_1).$$

Hence $\nu_{i+1} = \lambda q^i + \mu q^{-i}$. Dually we have $\nu^*_{i+1} = \lambda^* q^i + \mu^* q^{-i}$, solving $(-\gamma_2)/(-\gamma_1) = (\nu^*_{i+1} + \nu^*_{i-1})/\nu^*_i$.

PROPOSITION 5.7. If the nine parameters θ_0, θ_1, θ_2, θ_3, θ^*_0, θ^*_1, θ^*_2, M_1 and M_2 are given, then all the θ_i, θ^*_i, M_i are uniqely determined under the condition (5.1), (5.2) and (5.3).

<u>Proof</u>. By Proposition 5.6, θ_i and θ^*_i are determined. Putting $t = 0$ in Proposition 5.3, all the b_i and b^*_i are determined and so all the M_i are determined by Proposition 5.2(iii). □

PROPOSITION 5.8. The conditions (5.1), (5.2), (5.3) hold if and only if

$$u_i(x) = \sum_{t=0}^{i} M_t[\theta^*_i]^*_t[x]_t \quad (0 \leq i \leq d),$$

$$u^*_i(x) = \sum_{t=0}^{i} M_t[\theta_i]_t[x]^*_t \quad (0 \leq i \leq d),$$

$$\theta^*_i = \theta^*_0 + \frac{(q^i - 1)(\lambda^* q^{i-1} + \mu^*)}{q^{i-1}(q-1)} \qquad (0 \leq i \leq d),$$

$$b^*_i = \frac{q^{i+1}(q^i - 1)(q^{i-1} - 1)(\lambda q^{i-1} + \mu)}{(q-1)(\lambda q^{2i-1} + \mu)(\lambda q^{2i} + \mu)} \cdot \left\{ \frac{1}{(q+1)(q^{i-1} - 1)} \frac{M_1}{M_2} - \frac{1}{q^i - 1} \frac{1}{M_1} + \lambda\lambda^* - \frac{\mu\mu^*}{q^i} \right\}$$

$$(1 \leq i \leq d-1),$$

$$b_0^* = \frac{1}{(\lambda+\mu)M_1}, \qquad b_d^* = 0,$$

$$c_i^* = \frac{(q^i-1)(\lambda q^{i-1}+\mu)(\lambda q^i+\mu)}{(q-1)(\lambda q^{2i-1}+\mu)(\lambda q^{2i-2}+\mu)} \cdot \left\{\frac{q^i}{(q+1)(\lambda q^i+\mu)}\frac{M_1}{M_2} - \frac{q^{i-1}}{\lambda q^{i-1}+\mu}\frac{1}{M_1} - \lambda^* - q^{i-1}\mu^*\right\}$$

$$(1 \le i \le d-1),$$

$$c_0^* = 0,$$

$$c_d^* = \frac{-q^{d-1}}{\lambda q^{2d-2}+\mu}\left\{\frac{1}{M_1} + (\lambda^*+\mu^*)\frac{(q^d-1)(\lambda q^{d-1}+\mu)}{(q-1)q^{d-1}}\right\},$$

$$a_i^* = \theta_0^* - b_i^* - c_i^* \quad (0 \le i \le d),$$

$$M_i = \begin{cases} \dfrac{1}{b_{i-1}^* b_{i-2}^* \cdots b_0^*[\theta_i]_i} & (1 \le i \le d), \\ 1 & (i = 0), \end{cases}$$

$$\frac{1}{(q+1)(q^{d-1}-1)}\frac{M_1}{M_2} - \frac{1}{q^d-1}\frac{1}{M_1} + \lambda\lambda^* - \frac{\mu\mu^*}{q^d} = 0 \qquad (\#_0)$$

$$(\text{only when } d < \infty),$$

and the dual identities for θ_i, b_i, c_i, a_i obtained by exchanging (θ_0, λ, μ) and $(\theta_0^*, \lambda^*, \mu^*)$. Here the nine parameters θ_0, θ_1, θ_2, θ_3, θ_0^*, θ_1^*, θ_2^*, M_1, d, of which the first eight are real numbers and d is a positive integer ≥ 3, are arbitrary as long as the numbers b_i, b_i^*, c_i, c_i^*, θ_i, θ_i^*, which are given above in terms of these nine parameters, do not violate the conditions $b_i \neq 0$, $b_i^* \neq 0$ $(0 \leq i \leq d-1)$, $b_d = b_d^* = 0$, $c_i \neq 0$, $c_i^* \neq 0$ $(1 \leq i \leq d)$, $c_1 = c_1^* = 0$ and $\theta_i \neq \theta_j$, $\theta_i^* \neq \theta_j^*$ $(i \neq j,\ 0 \leq i, j \leq d)$.

Proof. Suppose (5.1), (5.2), (5.3) hold. As mentioned in the proof of Proposition 5.7, put θ_i, θ_i^* of Proposition 5.6 in Proposition 5.3 with $t = 0$. Then b_i, b_i^*, c_i, c_i^*, a_i, a_i^* are expressed in terms of $q, \lambda, \lambda^*, \mu, \mu^*, M_1$, M_2, which are the desired results. By Proposition 5.2, $u(x)$, $u^*(x)$, M_i are determined. $(\#_0)$ follows from $(\#)$ of Proposition 5.3 with $t = 0$.

The converse is directly checked by showing the expressions of b_i, b_i^* in Proposition 5.3 are independent of t and by showing $(\#)$ of Proposition 5.3 reduces to $(\#_0)$, using (5.11) and the following:

$$\frac{M_i}{M_{i+1}} = \frac{b_i^*[\theta_{i+1}]_{i+1}}{[\theta_i]_i} \tag{5.12}$$

$$= \frac{(q^{i+1}-1)(q^i-1)(q^{i-1}-1)}{q^{i-1}(q-1)^2} \cdot \left\{ \frac{1}{(q+1)(q^{i-1}-1)} \frac{M_1}{M_2} - \frac{1}{q^i-1}\frac{1}{M_1} + \lambda\lambda^* - \frac{1}{q^i}\mu\mu^* \right\}. \qquad \square$$

REMARK. (7) In Proposition 5.3, the expressions of b_i, b_i^*, c_i, c_i^* for $1 \leq i \leq d-1$ are also valid for $i = 0, d$ by regarding ν_{-1}, ν_{-1}^*, ν_{d+1}, ν_{d+1}^* as indeterminates. In Proposition 5.8, we put the specified values of

Proposition 5.6 in ν_i, ν_i^* to get the expressions of b_i, b_i^*, c_i, c_i^*, so it is necessary to treat the case of $i = 0, d$ separately. Nevertheless, the specialization of ν_i, ν_i^* affects the validity of the general expressions for the case of $i = 0$ (resepctively $i = d$) only when the general expressions have denominators zero for $i = 0$ (respectively $i = d$). Thus in Proposition 5.8, the expressions of b_i, b_i^*, c_i, c_i^* for $1 \leq i \leq d-1$ are also valid for $i = 0$ (respectively $i = d$), as long as their denominators do not become zero for $i = 0$ (respectively $i = d$).

<u>Completion of the proof of Theorem 5.1.</u>

<u>Case 1.</u> $q \neq \pm 1$ and $\mu\mu^* \neq 0$.

Let

$$h = -\frac{q}{q-1}\mu, \qquad h^* = -\frac{q}{q-1}\mu^*,$$
$$s = -\frac{1}{q^2}\frac{\lambda}{\mu}, \qquad s^* = -\frac{1}{q^2}\frac{\lambda^*}{\mu^*}. \tag{5.13}$$

Then

$$\theta_i = \theta_0 + h(1-q^i)(1-sq^{i+1})/q^i,$$
$$\theta_i^* = \theta_0^* + h^*(1-q^i)(1-s^*q^{i+1})/q^i. \tag{5.14}$$

Let

$$f(x) = \frac{x(x-q)}{q(q+1)(q-1)}\frac{M_1}{M_2} - \frac{x(x-q^2)}{q^2(q-1)}\frac{1}{M_1} + \frac{(x-q)(x-q^2)}{q^3(q-1)}(\lambda\lambda^* x - q\mu\mu^*). \tag{5.15}$$

Then

$$b_i^* = \frac{\lambda q^{i-1} + \mu}{(\lambda q^{2i-1} + \mu)(\lambda q^{2i} + \mu)} f(q^{i+1})$$

$$(1 \le i \le d-1), \tag{5.16}$$

$$b_0^* = \frac{1}{\lambda + \mu} f(q),$$

and $(\#_0)$ of Proposition 5.8 is equivalent to

$$f(q^{d+1}) = 0 \quad (\text{if } d < \infty). \tag{5.16$'$}$$

Since $\lambda\lambda^* x - q\mu\mu^* = \frac{(q-1)^2}{q} hh^*(q^3 ss^* x - 1)$,

$$f(x) = \frac{x(x-q)}{q(q+1)(q-1)} \frac{M_1}{M_2} - \frac{x(x-q^2)}{q^2(q-1)} \frac{1}{M_1} \tag{5.15$'$}$$

$$+ \frac{q-1}{q^4} hh^*(x-q)(x-q^2)(q^3 ss^* x - 1).$$

Let

$$f(x) = \frac{1-q}{q} hh^*(1-r_1 x)(1-r_2 x)(1-r_3 x). \tag{5.15$''$}$$

Then

$$ss^* = r_1 r_2 r_3,$$

$$\text{one of } r_i \ (i = 1, 2, 3) = q^{-d-1} \tag{5.17}$$

$$(\text{if } d < \infty),$$

and b_i^* (dually b_i) is expressed in the desired form by (5.16).

By (5.15)' and Proposition 5.8,

$$c_i^* = \frac{h(1-q)(1-q^i)}{(\lambda q^{2i-1}+\mu)(\lambda q^{2i-2}+\mu)} \frac{(sq^i)^3}{(-s)} f(-\frac{1}{sq^i}) \tag{5.18}$$

$$(1 \le i \le d-1).$$

Even if $s = 0$, the polynomial $\lim_{s \to 0} s^2x^3f(\frac{1}{sx})$ is well defined and (5.18) holds in this sense. By (5.15)" and (5.18), we get the desired expression for c_i^* (dually c_i). For c_d^* (dually c_d), set $r_3 = q^{-d-1}$ in (5.17) and calculate it directly, using

$$\frac{1}{M_1} = f(q).$$

Express $u_i(x)$ (dually $u_i^*(x)$) in terms of the new parameters $h, h^*, \theta_0, \theta_0^*, s, s^*, r_1, r_2, r_3, q$ and check it is the Askey-Wilson polynomial.

<u>Case 1A</u>. $q \ne \pm 1$, $\lambda \ne 0$, $\mu = 0$, $\lambda^* = 0$, $\mu^* \ne 0$, and $\frac{1}{M_1} \ne \frac{1}{q+1}\frac{M_1}{M_2}$.

Starting from the expressions of θ_i, θ_i^* in Proposition 5.8, instead of (5.14), (5.15), we have

$$\theta_i = \theta_0 - s'q(1-q^i) \quad (s' = \frac{1}{q(q-1)}\lambda),$$

$$\theta_i^* = \theta_0^* + h^*(1-q^i)/q^i, \tag{5.14IA}$$

$$f(x) = \frac{x(x-q)}{q(q+1)(q-1)}\frac{M_1}{M_2} - \frac{x(x-q^2)}{q^2(q-1)}\frac{1}{M_1} \tag{5.15IA}$$

$$= \frac{1-q}{q} h^*(-r_2')x(1-r_1x)$$

for some r_1 and non-zero r_2'. These are realized by putting $s^* = r_3 = 0$ and letting h, hs, hr_2 tend to 0, s', r_2' in (5.14), (5.15). So $u_i(x)$, θ_i, b_i, c_i, $u_i^*(x)$, θ_i^*, b_i^*, c_i^* are obtained from (I) by putting $s^* = r_3 = 0$ and letting h, hs, hr_2 tend to 0, s', r_2'.

<u>Case IB</u>. $q \neq \pm 1$, $\lambda \neq 0$, $\mu = 0$, $\lambda^* = 0$, $\mu^* \neq 0$ and $\frac{1}{M_1} = \frac{1}{q+1}\frac{M_1}{M_2}$.

The identity (5.15) becomes

$$f(x) = \frac{1}{q^2}\frac{1}{M_1}x^2 \qquad (5.15\text{IB})$$

$$= \frac{1-q}{q}h^* r_1' r_2' x^2$$

for some $r_1' r_2'$. This is realized by putting $s^* = r_3 = 0$ and letting h, hs, $\sqrt{h}\,r_1$, $\sqrt{h}\,r_2$ tend to 0, s', r_1', r_2'. Set $r = r_1' r_2'$.

<u>Case IC</u>. $q \neq \pm 1$, $\lambda\lambda^* \neq 0$.

Change q to $\frac{1}{q}$. Then λ and μ are interchanged, and λ^* and μ^* are interchanged. So we return to Case I.

<u>Case ID</u>. $q \neq \pm 1$, $\lambda = 0$, $\mu \neq 0$, $\lambda^* \neq 0$, $\mu^* = 0$. Change q to $\frac{1}{q}$. Then we return to Case IA or IB.

Since $\lambda + \mu = \nu_1 \neq 0$ (respectively $\lambda^* + \mu^* = \nu_1^* \neq 0$), one of λ, μ (respectively λ^*, μ^*) is non-zero. So all the cases of $q \neq \pm 1$ are treated.

<u>Case II</u>. $q = 1$, $\nu_1 \neq \nu_2$, $\nu_1^* \neq \nu_2^*$.

Let q tend to 1 in (5.14). Since

$$(q-1)^2h = -q^2(q\nu_1 - \nu_2)/(q+1)$$

and

$$(s-1)/(q-1)$$
$$= -\{(q^4-1)\nu_1 - (q^3-q)\nu_2\}/q^3(q\nu_1 - \nu_2)(q-1),$$

we have

$$\lim_{q\to 1}(q-1)^2 h = -\frac{\nu_1 - \nu_2}{2}, \tag{5.19}$$
$$\lim_{q\to 1}\frac{s-1}{q-1} = -\frac{2(2\nu_1 - \nu_2)}{\nu_1 - \nu_2}.$$

So (5.14) becomes

$$\begin{aligned} \theta_i &= \theta_0 + h' i(i+1+s'), \\ \theta_i^* &= \theta_0^* + h^{*\prime} i(i+1+s^{*\prime}) \end{aligned} \tag{5.14II}$$

with

$$h' = -\frac{\nu_1 - \nu_2}{2}, \qquad h^{*\prime} = -\frac{\nu_1^* - \nu_2^*}{2},$$

$$s' = -\frac{2(2\nu_1 - \nu_2)}{\nu_1 - \nu_2}, \qquad s^{*\prime} = -\frac{2(2\nu_1^* - \nu_2^*)}{\nu_1^* - \nu_2^*}.$$

Put $x = q^{i+1}$ in (5.15)' and let q tend to 1. Then we get

$$\lim_{q \to 1} f(q^{i+1}) \tag{5.15II)'}$$

$$= \frac{i}{2} \frac{M_1}{M_2} - (i-1) \frac{1}{M_1} + h'h^{*\prime} i(i-1)(i+4+s'+s^{*\prime})$$

for all integers i. Put

$$g(i) = \lim_{q \to 1} f(q^{i+1}) .$$

Notice $h'h^{*\prime} \neq 0$ because $\nu_1 \neq \nu_2$ and $\nu_1^* \neq \nu_2^*$. Since $g(i)$ is a polynomial of degree 3 in i, set

$$g(x) = h'h^{*\prime}(x+1+r_1')(x+1+r_2')(x+1+r_3'). \tag{5.15II)''}$$

Taking the limit of $\lambda q^i + \mu = (q-1)h(sq^{i+2} - 1)/q$ as $q \to 1$, (5.16) becomes

$$b_i^* = \frac{i+1+s'}{h'(2i+1+s')(2i+2+s')} g(i) \tag{5.16II)}$$

$$(1 \leq i \leq d-1),$$

$$b_0^* = \frac{1}{h'(2+s')} g(0).$$

Also,

$$g(d) = 0 \qquad (\text{if } d < \infty). \tag{5.16II)'}$$

Comparing the coefficient of x^2 in $g(x)$ and using $g(d) = 0$,

$$s' + s^{*\prime} = r_1' + r_2' + r_3', \tag{5.17II)}$$

one of r_i' $(i = 1, 2, 3)$ $= -d-1$

(if $d < \infty$).

We want to show that as $q \to 1$, (5.18) becomes

$$c_i^* = \frac{h^{*\prime} i(i+s'-r_1')(i+s'-r_2')(i+s'-r_3')}{(2i+1+s')(2i+s')} \qquad (5.18\text{II})$$

$(1 \le i \le d-1)$.

Since $\lim_{q \to 1} f(q^{-i}) = g(-i-1)$, we have

$$\lim_{q \to 1} \prod_{\alpha=1}^{3} \frac{r_\alpha - q^i}{q-1} = \prod_{\alpha=1}^{3} (r_\alpha' - i)$$

for all i. Let σ_j (respectively σ_j') $(j = 0, 1, 2, 3)$ be the fundamental symmetric function of degree j in the variables $\frac{r_\alpha - 1}{q-1}$ (respectively r_α') $(\alpha = 1, 2, 3)$. Then

$$\lim_{q \to 1} \sum_{\alpha=0}^{3} (-1)^\alpha \left(\frac{q^i - 1}{q-1}\right)^\alpha \sigma_{3-\alpha} = \sum_{\alpha=0}^{3} (-1)^\alpha i^\alpha \sigma_{3-\alpha}'$$

for all i. Solving the above equations $(i = -1, 0, 1)$ for σ_α, we get

$$\lim_{q \to 1} \sigma_\alpha = \sigma_\alpha' \qquad (\alpha = 1, 2, 3).$$

Since $\frac{r_\alpha - 1}{q-1}$ and r_α' $(\alpha = 1, 2, 3)$ are roots of polynomials $\sum_{\alpha=0}^{3} (-1)^{3-\alpha} \sigma_{3-\alpha} x^\alpha$ and $\sum_{\alpha=1}^{3} (-1)^{3-\alpha} \sigma_{3-\alpha}' x^\alpha$ respectively, we get

$$\lim_{q \to 1} \frac{r_\alpha - 1}{q-1} = r_\alpha' \qquad (\alpha = 1, 2, 3).$$

Thus

$$\lim_{q\to 1} (sq^i)^3 f\left(\frac{1}{sq^i}\right) = h'h^{*'} \prod_{\alpha=1}^{3} (r'_\alpha - i - s'),$$

and we have (5.18II), noticing $\lim_{q\to 1} s = 1$. Similarly, c_d^* is calculated by letting q tend to 1 in Case I.

Thus everything is obtained from Case I by letting q, $(q-1)^2 h$, $(q-1)^2 h^*$, $\frac{s-1}{q-1}$, $\frac{s^*-1}{q-1}$, $\frac{r_1-1}{q-1}$, $\frac{r_2-1}{q-1}$, $\frac{r_3-1}{q-1}$ tend to 1, h', $h^{*'}$, s', $s^{*'}$, r'_1, r'_2, r'_3, respectively.

Suppose a is a function of q such that $\lim_{q\to 1} \frac{1-a}{1-q} = a'$. Then

$$\lim_{q\to 1} \frac{(a\,;q)_t}{(1-q)^t} = (a')_t\,. \tag{5.20}$$

So

$$\lim_{q\to 1} {}_4\varphi_3\left(\begin{matrix} q^{-i},\ s^* q^{i+1},\ q^{-j},\ sq^{j+1} \\ r_1 q,\ r_2 q,\ r_3 q \end{matrix}\ ;q\,,q\right)$$

$$= {}_4F_3\left(\begin{matrix} -i,\ i+1+s^{*'},\ -j,\ j+1+s' \\ r'_1+1,\ r'_2+1,\ r'_3+1 \end{matrix}\ ;1\right).$$

Case IIA. $q = 1$, $\nu_1 \neq \nu_2$, $\nu_1^* = \nu_2^*$.

Let ν_2^* tend to ν_1^* in Case II. The identities (5.14II), (5.15II)', (5.15II)", (5.16II), (5.16II)', (5.17II) and (5.18II) become those which are obtained by

$$h^{*'} \to 0,\quad h^{*'} s^{*'} \to \nu_1^* = s^{*''},\quad h^{*'} r'_3 \to s^{*''}$$

(noticing $h^{*\prime}(s' + s^{*\prime}) = h^{*\prime}(r_1' + r_2' + r_3')$). Use $(s^{*\prime} - r_3') \to (r_1' + r_2' - s')$ (because $s' + s^{*\prime} = r_1' + r_2' + r_3'$), when getting c_i.

<u>Case IIB</u>. $q = 1$, $\nu_1 = \nu_2$, $\nu_1^* \neq \nu_2^*$.

Let ν_2 tend to ν_1 in Case II. This is realized by

$$h' \to 0, \quad h's' \to \nu_1 = s'', \quad h'r_3' \to s''$$

(noticing $h'(s' + s^{*\prime}) = h'(r_1' + r_2' + r_3')$).

<u>Case IIC</u>. $q = 1$, $\nu_1 = \nu_2$, $\nu_1^* = \nu_2^*$, $\frac{1}{2}\frac{M_1}{M_2} \neq \frac{1}{M_1}$.

Let ν_2 and ν_2^* tend to ν_1 and ν_1^* in Case II, respectively. This is realized by $(h', h^{*\prime}) \to (0,0)$, $(h's', h^{*\prime}s^{*\prime}) \to (s'', s^{*\prime\prime})$, $(h^{*\prime}r_2', h'r_3') \to (r_2'', r_3'')$. Use $h'h^{*\prime}(s^{*\prime} - r_2')(s^{*\prime} - r_3') = h'h^{*\prime}\{r_2'r_3' + s^{*\prime}(r_1' - s')\} \to (r_2''r_3'' - s^{*\prime\prime}s'')$, when getting c_i.

<u>Case IID</u>. $q = 1$, $\nu_1 = \nu_2$, $\nu_1^* = \nu_2^*$, $\frac{1}{2}\frac{M_1}{M_2} = \frac{1}{M_1}$.

Let ν_2 and ν_2^* tend to ν_1 and ν_1^* in Case II, respectively. This is realized by $h' = h^{*\prime} \to 0$, $(h's', h's^{*\prime}) \to (s'', s^{*\prime\prime})$, $(\sqrt{h'}\, r_1', \sqrt{h'}\, r_2', h'r_3') \to (r_1'', r_2'', r_3'')$.

<u>Case III</u>. $q = -1$.

Let q tend to -1 in Case I. This is realized by

$$q \to -1, \quad (q+1)h \to h', \quad (q+1)h^* \to h^{*\prime},$$

$$\frac{s-1}{q+1} \to s', \quad \frac{s^*-1}{q+1} \to s^{*\prime},$$

$$\frac{r_1+1}{q+1} \to r_1', \quad \frac{r_2+1}{q+1} \to r_2', \quad \frac{r_3-1}{q+1} \to r_3'.$$

This completes the proof of Theorem 5.1. □

3.6 THE LIST OF KNOWN (P AND Q)-POLYNOMIAL ASSOCIATION SCHEMES

In this section, we give the list of known (P and Q)-polynomial association schemes with sufficiently large diameters, putting emphasis on how their intersection numbers are described by the parameters of Askey-Wilson orthogonal polynomials in Theorem 5.1. Also we discuss imprimitive C-algebras with the (P and Q)-polynomial property to study P-polynomial and Q-polynomial structures in imprimitive (P and Q)-polynomial association schemes.

The following is the core of the known (P and Q)-polynomial schemes with diameter sufficiently large. The numbers b_i, c_i, q, r_1, r_2, r_3, s, s^* are as in Theorem 5.1. (Here the s', $s^{*\prime}$, r_i' in the limiting cases are simply denoted by s, s^*, r_i.) The types of expressions in Theorem 5.1 are indicated at the end of parameters. The diameter is denoted by d in general, but for an antipodal quotient or a bipartite half, we use d for the diameter of the original association scheme, so $[\frac{d}{2}]$ is the diameter of the antipodal quotient or the bipartite half.

We notice that intersection numbers b_i, c_i (respectively b_i^*, c_i^*) depend on the P-polynomial structure (respectively Q-polynomial structure) in general, and the parameters of an expression in Theorem 5.1 depend both on the P-polynomial and Q-polynomial structures. If there exist multiple P-polynomial or Q-polynomial structures for an association scheme in the list, we include all of them in the list.

As shown later in this section, imprimitive (P and Q)-polynomial schemes are either bipartite or antipodal, and the bipartite halves and the antipodal quotients are also (P and Q)-polynomial schemes. If an association scheme in the list is imprimitive, we include the bipartite half and the antipodal quotient in the list.

(I)

(i) THE JOHNSON SCHEME $J(v,d)$:

$$r_1 = -v + d - 1, \quad r_2 = -d - 1, \tag{IIA}$$

$$s = -v - 2, \quad s^* = -v(v-1)/d(v-d).$$

If $v = 2d+1$, $J(v,d)$ has another P-polynomial structure $R_0, R_d, R_1, R_{d-1}, R_2, R_{d-2}, \ldots$ in terms of the natural one (this corresponds to the odd graph O_{d+1}):

$$r_1 = -2d - 3, \quad r_2 = -d - 1, \quad r_3 = d + 1, \tag{III}$$

$$s = 2d + 3, \quad s^* = 2d + 2 .$$

If $v = 2d$, $J(v,d)$ is antipodal.

(i)' THE ANTIPODAL QUOTIENT OF $J(2d,d)$:

$$r_1 = -\frac{d}{2} - 1, \quad r_2 = -\frac{d}{2} - \frac{1}{2}, \quad r_3 = -d - 1,$$

$$s = -d - \frac{3}{2}, \quad s^* = -d - 1.$$

(ii) THE GENERALIZED JOHNSON SCHEME (the q-analogue of the Johnson scheme):

Let V be the vector space of dimension v over $GF(p^f)$ and let X be the set of d-dimensional subspaces of V $(d \leq \frac{v}{2})$. Define the ith relation R_i on X by

$$(x,y) \in R_i \Leftrightarrow \dim(x \cap y) = d - i.$$

Then $\mathfrak{X} = (X, \{R_i\}_{0 \leq i \leq d})$ is a P-polynomial association scheme with respect to the natural ordering $R_0, R_1, \ldots, R_d$ and

$$b_i = p^{f(2i+1)}(p^{f(d-i)} - 1)(p^{f(v-d-i)} - 1)/(p^f - 1)^2,$$

$$c_i = (p^{fi} - 1)^2/(p^f - 1)^2 .$$

The full projective collineation group $P\Gamma L(v, p^f)$ acts on $\mathfrak{X}$ as a distance-transitive group. $\mathfrak{X}$ is called of type $A_{v-1}(p^f)$. The intersection numbers b_i, c_i can be written as in Theorem 5.1(I) with

$$q = p^f, \quad r_1 = q^{-d-1}, \quad r_2 = q^{-v+d-1}, \qquad \text{(I)}$$

$$s = q^{-v-2}, \quad r_3 = s^* = 0 .$$

So $\mathfrak{X}$ is a Q-polynomial association scheme as well.

(iii) THE ASSOCIATION SCHEMES OF DUAL POLAR SPACES:

Let V be one of the following vector spaces with a non-degenerate form:

name	dim	field	form	e
$B_d(p^f)$	$2d+1$	$GF(p^f)$	quadratic	0
$C_d(p^f)$	$2d$	$GF(p^f)$	symplectic	0
$D_d(p^f)$	$2d$	$GF(p^f)$	quadratic (Witt index d)	-1
${}^2D_{d+1}(p^f)$	$2d+2$	$GF(p^f)$	quadratic (Witt index d)	1
${}^2A_{2d}(p^f)$	$2d+1$	$GF(p^{2f})$	hermitian	1/2
${}^2A_{2d-1}(p^f)$	$2d$	$GF(p^{2f})$	hermitian	-1/2

and let e be as above. Let X be the set of maximal totally isotropic subspaces of V. Each element of X has dimension d. Define the ith relation R_i on X by

$$(x,y) \in R_i \Leftrightarrow \dim(x \cap y) = d-i.$$

Then $\mathfrak{X} = (X, \{R_i\}_{0 \le i \le d})$ is a P-polynomial scheme with respect to the natural ordering $R_0, R_1, \ldots, R_d$ and

$$b_i = q^{i+e+1}(q^{d-i} - 1)/(q-1),$$

$$c_i = (q^i - 1)/(q-1),$$

where $q = p^f, p^f, p^f, p^f, p^{2f}, p^{2f}$, respectively. $\mathfrak{X}$ admits $P\Gamma O(2d+1, p^f)$, $P\Gamma S_p(2d, p^f)$, $P\Gamma O^+(2d, p^f)$, $P\Gamma O^-(2d+2, p^f)$, $P\Gamma U(2d+1, p^f)$, $P\Gamma U(2d, p^f)$ as a distance-transitive group, respectively. The intersection numbers b_i, c_i can be written as in Theorem 5.1(I) with

$q = p^f, p^f, p^f, p^f, p^{2f}, p^{2f}$, respectively, and

$$r_1 = q^{-d-1}, \quad s = -q^{-d-e-2}, \qquad \text{(I)}$$

$$r_2 = r_3 = s^* = 0 .$$

So $\mathfrak{X}$ is a Q-polynomial association scheme as well.

${}^2A_{2d-1}(p^f)$ has another Q-polynomial structure $E_0, E_d, E_1, E_{d-1}, E_2, E_{d-2}, \ldots$ in terms of the natrual one:

$$q = -p^f, \quad s = p^{-2(d+1)f}, \quad r_1 = -r_2 = p^{-(d+1)f}, \qquad \text{(I)}$$

$$s^* = r_3 = 0 .$$

$D_d(p^f)$ is bipartite.

(iii)' THE BIPARTITE HALF OF $D_d(p^f)$:

$$q = p^{2f}, \quad r_1 = p^{-f(d+2)}, \quad r_2 = p^{-f(d+1)}, \qquad \text{(I)}$$

$$s = p^{-2f(d+1)}, \quad s^* = r_3 = 0 .$$

(II)

(0) ORDINARY n-GON:

$$r_1 = -1/q, \quad r_2 = 1/\sqrt{q}, \quad r_3 = -1/\sqrt{q}, \qquad \text{(I)}$$

$$s = s^* = 1/q,$$

where $d = [\frac{n}{2}]$ and q is any primitive n^{th} root of unity.

(i) THE HAMMING SCHEME $H(d,q)$:

$$r = q(q-1), \quad r_1 = -d-1, \quad s = s^* = -q . \qquad \text{(IIC)}$$

If $q = 2$ and d is even, $H(d,q)$ has another P-polynomial structure $R_0, R_{d-1}, R_2, R_{d-3}, R_4, R_{d-5}, \ldots, R_1, R_d$ and another Q-polynomial structure $E_0, E_{d-1}, E_2, E_{d-3}, E_4, E_{d-5}, \ldots, E_1, E_d$ in terms of the first ones:

$$r_1 = -d - 1, \quad r_2 = -\frac{d+1}{2}, \quad r_3 = \frac{d+1}{2}, \qquad \text{(III)}$$

$$s = s^* = d + 1 .$$

$H(d,2)$ is bipartite and antipodal.

(i') THE BIPARTITE HALF OF $H(d,2)$:

$$r_1 = -\frac{d}{2} - 1, \quad r_2 = -\frac{d}{2} - \frac{1}{2}, \quad s = -d - 1 . \qquad \text{(IIA)}$$

If d is odd $(d = 2e+1)$, the bipartite half of $H(2e+1,2)$ has another P-polynomial structure $R_0, R_d, R_1, R_{d-1}, R_2, R_{d-2}, \ldots$ and another Q-polynomial structure $E_0, E_2, E_4, \ldots, E_3, E_1$ in terms of the first natural ones. With respect to the first Q-polynomial structure and the second P-polynomial structure,

$$r_1 = -2e - 2, \quad r_2 = -e - 1, \quad r_3 = e + 1, \qquad \text{(III)}$$

$$s = s^* = 2e + 2 .$$

With respect to the first P-polynomial structure and the second Q-polynomial structure,

$$r_1 = -\frac{e}{2} - \frac{3}{4}, \quad r_2 = -\frac{e}{2} - \frac{5}{4}, \quad r_3 = -e - 1, \qquad \text{(II)}$$

$$s = s^* = -e - \frac{3}{2} .$$

With respect to the second P-polynomial structure and the second Q-polynomial structure,

$$r_1 = -\frac{d}{2} - 1, \quad r_2 = -\frac{d}{2} - \frac{1}{2}, \quad s^* = -d - 1 . \qquad \text{(IIB)}$$

(i") THE ANTIPODAL QUOTIENT OF H(d,2):

$$r_1 = -\frac{d}{2} - 1, \quad r_2 = -\frac{d}{2} - \frac{1}{2}, \quad s^* = -d - 1 . \qquad \text{(IIB)}$$

If d is odd, the antipodal quotient of H(d,2) is isomorphic to the bipartite half of H(d,2) as association schemes.

If d is even, the antipodal quotient (respectively bipartite half) of H(d,2) is bipartite (respectively antipodal), and the bipartite half of the antipodal quotient of H(d,2) is isomorphic to the antipodal quotient of the bipartite half of H(d,2) as association schemes.

(i‴) THE ANTIPODAL QUOTIENT OF (i') (d = 2e, even) ≅ THE BIPARTITE HALF OF (i") (d = 2e, even):

$$r_1 = -\frac{e}{2} - \frac{1}{2}, \quad r_2 = -\frac{e}{2} - 1, \quad r_3 = -e - \frac{1}{2}, \qquad \text{(II)}$$

$$s = s^* = -e - 1 .$$

(ii) THE ASSOCIATION SCHEME OF BILINEAR FORMS (Generalized Hamming Scheme):

Let X be the set of $d \times n$ matrices over $GF(p^f)$ $(d \leq n)$. Define the ith relation R_i on X by

$$(x,y) \in R_i \Leftrightarrow \operatorname{rank}(x - y) = i .$$

Then $\mathfrak{X} = (X, \{R_i\}_{0 \leq i \leq d})$ is a P-polynomial association scheme with respect to the natural ordering $R_0, R_1, \ldots, R_d$ and

$$b_i = p^{2fi}(p^{f(d-i)} - 1)(p^{f(n-i)} - 1)/(p^f - 1),$$

$$c_i = p^{f(i-1)}(p^{fi} - 1)/(p^f - 1) .$$

The group generated by addition and right and left multiplication of matrices acts on $\mathfrak{X}$ distance-transitively. The intersection numbers b_i, c_i can be written as in Theorem 5.1(I) with

$$q = p^f, \quad r_1 = q^{-d-1}, \quad r_2 = q^{-n-1}, \tag{I}$$

$$r_3 = s = s^* = 0 .$$

So $\mathfrak{X}$ is a Q-polynomial association scheme as well.

(iii) THE ASSOCIATION SCHEME OF ALTERNATING BILINEAR FORMS:

Let V be the vector space of dimension n over $GF(p^f)$. Let X be the set of bilinear alternating forms on V (skew symmetric matrices, if q is odd) and $d = [\frac{n}{2}]$. Define the ith relation R_i on X by

$$(x,y) \in R_i \Leftrightarrow \operatorname{rank}(x - y) = 2i.$$

Notice the rank is always even. Then $\mathfrak{X} = (X, \{R_i\}_{0 \leq i \leq d})$ is a P-polynomial association scheme with respect to the natrual ordering $R_0, R_1, \ldots, R_d$, and

$$b_i = p^{4fi}(p^{f(n-2i)} - p^f)(p^{f(n-2i)} - 1)/(p^{2f} - p^f)(p^f + 1),$$

$$c_i = p^{2f(i-1)}(p^{2fi} - 1)/(p^{2f} - 1) .$$

The group generated by $x \to {}^t a x a$ $(a \in GL(n, p^f))$ and $x \to x + y$ $(y \in X)$ acts on $\mathfrak{X}$ distance-transitively. The intersection numbers b_i, c_i can be written as in Theorem 5.1(I) with

$$q = p^{2f}, \quad r_1 = q^{-\frac{n+2}{2}}, \quad r_2 = q^{-\frac{n-1}{2}}, \tag{I}$$

$$r_3 = s = s^* = 0 .$$

So $\mathcal{X}$ is a Q-polynomial association scheme as well.

(iv) THE ASSOCIATION SCHEME OF HERMITIAN FORMS:

Let V be the vector space of dimension d over $GF(p^{2f})$. Let X be the set of hermitian forms on V (hermitian matrices). Define the ith relation R_i on X by

$$(x,y) \in R_i \Leftrightarrow \operatorname{rank}(x - y) = i .$$

Then $\mathcal{X} = (X, \{R_i\}_{0 \le i \le d})$ is a P-polynomial scheme with respect to the natural ordering $R_0, R_1, \ldots, R_d$, and

$$b_i = p^{2fi}(p^{2f(d-i)} - 1)/(p^f + 1),$$

$$c_i = p^{f(i-1)}(p^{fi} - (-1)^i)/(p^f + 1).$$

The group generated by $x \to {}^t\bar{a}xa$ $(a \in GL(d,p^{2f}))$ and $x \to x + y$ $(y \in X)$ acts on X distance-transitively. The intersection numbers b_i, c_i can be written as in Theorem 5.1(I) with

$$q = -p^f, \quad r_1 = q^{-d-1}, \quad r_2 = -q^{-d-1}, \qquad \text{(I)}$$

$$r_3 = s = s^* = 0 .$$

So $\mathcal{X}$ is a Q-polynomial association scheme as well.

(v) THE ASSOCIATION SCHEME OF QUADRATIC FORMS (see Egawa (to appear)):

Let V be the vector space of dimension $n - 1$ over $GF(p^f)$. Let X be the set of quadratic forms on V (symmetric matrices, if $p \ne 2$) and $d = [\frac{n}{2}]$. Define the ith relation R_i on X by

$$(x,y) \in R_i \Leftrightarrow \operatorname{rank}(x - y) = 2i - 1 \text{ or } 2i,$$

where "rank" means the ordinary rank if $p \neq 2$, otherwise "rank" is defined as follows. Let Q be a quadratic form, i.e., $Q(ax) = a^2Q(x)$ and let β be the bilinear form attached to Q, i.e.,

$$\beta(x,y) = Q(x + y) - Q(x) - Q(y).$$

Let

$$V(\beta) = \{v \in V \mid \beta(v,w) = 0 \text{ for } \forall w \in v\}$$

and

$$V(Q) = \{v \in V(\beta) \mid Q(v) = 0\}.$$

Then rank $Q = \dim V/V(Q)$.

$\mathfrak{X} = (X,\{R_i\}_{0 \leq i \leq d})$ is a P-polynomial association scheme with respect to the natural ordering $R_0,R_1,\ldots,R_d$ and the intersection numbers b_i, c_i are the same as those of (II)(iii), i.e., the association scheme consisting of alternating forms on an n-dimensional vector space over $GF(p^f)$. In particular, $\mathfrak{X}$ is a Q-polynomial association scheme.

$GL(n-1,p^f)$ acts naturally on X, preserving the rank. The full automorphism group of $\mathfrak{X}$ is <u>not</u> distance transitive on X.

REMARKS. (1) The reasons why we say this is a core of the list is as follows:

(i) The list contains all the known primitive (P and Q)-polynomial schemes with diameter sufficiently large. If a (P and Q)-polynomial scheme is imprimitive, the composition factors (cf. section 2.9) have primitive (P and Q)-polynomial structures and hence are contained in the list. We shall see later that the composition factors of a (P and Q)-polynomial scheme are obtained by taking bipartite halves or antipodal quotients successively. There may be some relatives which are obtained by applying a chain of antipodal and/or bipartite operations (including their inverse operations) to association schemes in the list.

(ii) There may be some association schemes with the same parameters as mentioned in the list, (so they enjoy the same algebraic property but they are not isomorphic combinatorially). See Egawa (1981) for $H(n,4)$.

We expect that this list covers all (P and Q)-polynomial schemes, allowing the possible relatives mentioned in (i) and (ii).

(2) All the association schemes in the list (II) are self-dual, i.e., the first and second eigenmatrices coincide (since $s = s^*$), while b_i^*, c_i^* are not integers in general for association schemes in the list (I) and consequently they cannot be self-dual.

Let $\mathfrak{X} = (X, \{R_i\}_{0\le i\le d})$ be an association scheme in the list (I)(ii), (iii). The main part of $\operatorname{Aut}\mathfrak{X}$ is a Chevallay group G and G is distance-transitive on the graph (X,R_1). The one-point stabilizer of G on X is a maximal parabolic subgroup P of G obtained by dropping the node of the Dynkin diagram indicated below:

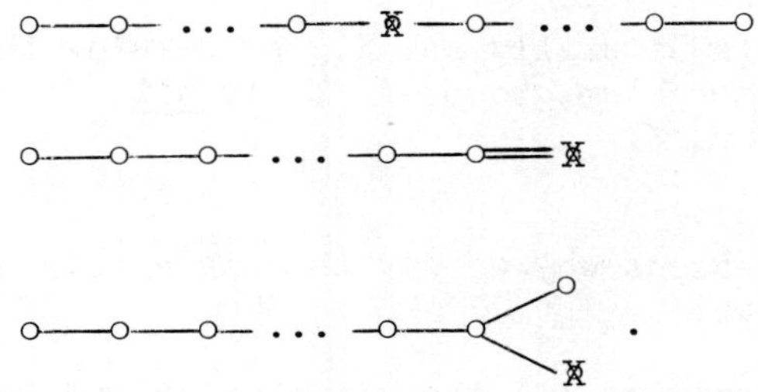

So, those in the list (I)(ii), (iii) are interpreted as obtained from a permutation representation of G acting on $X = P\backslash G$. On the other hand those in the list

(II)(ii) ~ (iv) are essentially interpreted in the following way.

Let $P = G_J \cdot U_J$ ($\triangleright U_J$) be the Levi-decomposition. Then the action of P on U_J gives the distance-regular graphs in the list (II).

(3) The imprimitive association schemes in the list are $J(2d,d)$ (antipodal), $D_d(p^f)$ (bipartite), $H(d,2)$ (bipartite and antipodal), the bipartite half of $H(2e,2)$ (antipodal) and the antipodal quotient of $H(2e,2)$ (bipartite).

$J(2d+1,d)$ is the only association scheme in the list which has double P-polynomial structures and only one Q-polynomial structure. ${}^2A_{2d-1}(p^f)$ is the only association scheme in the list which has double Q-polynomial structures and only one P-polynomial structure. The association schemes in the list which have both double P-polynomial structures and double Q-polynomial structures are $H(2e,2)$, the bipartite half of $H(2e+1,2) \cong$ the antipodal quotient of $H(2e+1,2)$. We notice that (P and Q)-polynomial schemes have at most two P-polynomial structures (cf. section 3.4) and at most two Q-polynomial structures (cf. section 3.7).

(4) A Riemannian manifold M is called 2-point homogeneous if for each $(x,y),(w,z) \in M \times M$ with $d(x,y) = d(w,z)$, there exists some $\sigma \in \mathrm{Aut}\, M$ such that $x^\sigma = w$, $y^\sigma = z$.

THEOREM. (Wang (1952))

{compact 2-point homogeneous spaces}

= {compact rank 1 symmetric spaces}.

THEOREM. (Tits (1955), cf. Helgason (1962))

{non-compact 2-point homogeneous spaces}

= {non-compact rank 1 symmetric spaces}.

Roughly speaking, the concepts of 2-point homogeneous spaces and rank 1 symmetric spaces correspond to P-polynomial and Q-polynomial structures (with group actions), respectively. Therefore, in the continuous case, one of the P-polynomial and Q-polynomial structures implies the other in the sense of the above theorems.

CONJECTURES. (1) If d is <u>sufficiently large</u>, every <u>primitive</u> P-polynomial scheme has a Q-polynomial structure and vice versa, i.e., {primitive P-polynomial schemes} = {primitive Q-polynomial schemes}.

(2) If an association scheme $\mathfrak{X}$ has a (P and Q)-polynomial structure with <u>sufficiently large</u> diameter, then $\mathfrak{X}$ is in the list given at the beginning of this section (or obtained from it as a relative in the sense of Remark (1)(i), (ii)).

To solve Conjecture (2), we need 3 steps. The first step, i.e., Theorem 5.1, has been finished by Leonard (1982). The second step is to determine the feasible parameters. We hope that most of them correspond to known association schemes. The final step is to characterize association schemes by the parameters. ($H(n,q)$ and $J(v,k)$ for $v > 4k$ have been finished. See Y. Egawa (1981), A. Moon (1983).) The second step will be discussed more in section 3.7, and the third step will be discussed more in section 3.8.

We wish this line of study could be developed further up to the point where we could classify the character tables which correspond to finite simple groups (the old theme of Brauer, cf. Brauer (1968)).

Let G be a finite group and $C_0 = \{1\}, C_1, \ldots, C_d$ be the conjugacy classes of G. Define the relation R_i on G by

$$(x,y) \in R_i \Leftrightarrow yx^{-1} \in C_i .$$

Then $\mathfrak{X} = \mathfrak{X}(G) = (G, \{R_i\}_{0 \leq i \leq d})$ becomes an association scheme and $\mathfrak{X}(G)$ is primitive if and only if G is simple. There have been a lot of works to characterize G by the character table (e.g., $G = S_n, A_n$, cf. Nagao (1957), James-Kerber (1981)).

PROBLEM. (1) Characterize the association scheme $\mathfrak{X}(G)$ by the parameters (i.e., by the character table).

As the first step of this problem, try the association scheme $\mathfrak{X}(G)$ for $G = S_n, A_n$.

REMARKS. (5) If we drop the primitivity in Conjecture (1), there are counterexamples. Let $\mathfrak{X}$ be an association scheme with incidence matrices A_i and primitive idempotents E_i. Let $\tilde{\mathfrak{X}}$ be the <u>bipartite doubling</u> of $\mathfrak{X}$, i.e., the incidence matrices of $\tilde{\mathfrak{X}}$ are A_i^+, A_i^-, where

$$A_i^+ = \begin{pmatrix} A_i & 0 \\ 0 & A_i \end{pmatrix} \quad \text{and} \quad A_i^- = \begin{pmatrix} 0 & A_i \\ A_i & 0 \end{pmatrix} . \tag{6.1}$$

Then the primitive idempotents of $\tilde{\mathfrak{X}}$ are E_i^+, E_i^-, where

$$E_i^+ = \frac{1}{2}\begin{pmatrix} E_i & E_i \\ E_i & E_i \end{pmatrix} \quad \text{and} \quad E_i^- = \frac{1}{2}\begin{pmatrix} E_i & -E_i \\ -E_i & E_i \end{pmatrix} . \tag{6.2}$$

If $\tilde{\mathfrak{X}}$ has a P-polynomial structure, the ordering must be $A_0^+, A_1^-, A_2^+, A_3^-, \ldots$ with suitable rearrangement of A_i. If $\tilde{\mathfrak{X}}$ has a Q-polynomial structure, the ordering must be $E_0^+, E_1^-, E_2^+, E_3^-, \ldots$ with suitable rearrangement of E_i. Let p_{ij}^k, q_{ij}^k be the intersection and Krein numbers of $\mathfrak{X}$.

Check the following: $\tilde{\mathfrak{X}}$ is a P-polynomial scheme with respect to $A_0^+, A_1^-, A_2^+, A_3^-, \ldots$ if and only if $\mathfrak{X}$ is a P-polynomial scheme with respect to $A_0, A_1, A_2, A_3, \ldots$ and $p_{1i}^i = 0$ $(i \neq d)$, $p_{1d}^d \neq 0$. $\tilde{\mathfrak{X}}$ is a Q-polynomial scheme with respect to $E_0^+, E_1^-, E_2^+, E_3^-, \ldots$ if and only if $\mathfrak{X}$ is a Q-polynomial scheme with respect to $E_0, E_1, E_2, E_3, \ldots$ and $q_{1i}^i = 0$ $(i \neq d)$, $q_{1d}^d \neq 0$. For example, let $\mathfrak{X} = O_k$, the odd graph, and $\tilde{\mathfrak{X}} = 2.O_k$, the double covering (cf. section 3.4, Example 4.1(4)). Then $\tilde{\mathfrak{X}}$ is a P-polynomial scheme but not a Q-polynomial scheme.

(6) The Johnson scheme $J(v,d)$ has the q analogue $A_{v-1}(q)$, but the odd graph O_{d+1} in $J(2d+1,d)$ does not have a q analogue in $A_{2d}(q)$. This is because the vector space of $\dim 2d+1$ has three mutually disjoint subspaces of dimension d, for example, $\langle u_1, u_2, \ldots, u_d\rangle$, $\langle v_1, v_2, \ldots, v_d\rangle$, $\langle u_1+v_1, u_2+v_2, \ldots, u_d+v_d\rangle$ with $u_1, u_2, \ldots, u_d, v_1, v_2, \ldots, v_d$ linearly independent. Nevertheless, the double covering $2 \cdot O_{d+1}$ of O_{d+1}, which is also the bipartite doubling of O_{d+1}, has a q analogue as shown below. Unlike $2 \cdot O_{d+1}$, the q-analogue of $2 \cdot O_{d+1}$ is not antipodal.

Let V be a vector space of dimension $2d+1$. Let X (respectively Y) be the set of d dimensional (respectively $d+1$) subspaces of V. (X and Y are q analogue of $J(2d+1,d)$.) The graph with vertex set $X \cup Y$ and edge set $\{(\alpha,\beta) \mid \alpha \in X, \beta \in Y, \alpha \subseteq \beta\}$ is bipartite and distance-transitive with diameter $2d+1$. The intersection numbers are $c_{2i-1} = c_{2i} = (q^i-1)/(q-1)$, $b_{2i-1} = b_{2i} = (q^{d+1}-q^i)/(q-1)$. This is not a Q-polynomial association scheme by Theorem 5.1.

(7) Let $\mathfrak{X}$ be a Q-polynomial association scheme with respect to the ordering $E_0, E_1, \ldots, E_d$ of the

primitive idempotents. Let a_i^*, b_i^*, c_i^* be the dual intersection numbers: $E_1 \circ E_i = b_{i-1}^* E_{i-1} + a_i^* E_i + c_{i+1}^* E_{i+1}$. Then $\mathfrak{X}$ is a Q-polynomial association scheme with respect to $E_0, E_2, E_4, \ldots, E_5, E_3, E_1$ if and only if $a_i^* = 0$ $(i = 0, 1, \ldots, d-1)$ and $a_d^* \neq 0$. The proof follows quite straight from the consideration of the products $E_1 \circ E_1 \circ \cdots \circ E_1$. The bipartite doubling of such a Q-polynomial association scheme is also a Q-polynomial association scheme as shown in Remark (5). Thus ${}^2A_{2d-1}(p^f)$ and the bipartite half of $H(2e+1,2) \cong$ the antipodal quotient of $H(2e+1,2)$ have the bipartite doubling with Q-polynomial structures. Notice the bipartite doubling of ${}^2A_{2d-1}(p^f)$ is no longer a P-polynomial scheme, because the unique P-polynomial structure of ${}^2A_{2d-1}(p^f)$ has $a_i \neq 0$ (cf. Remark (5) of section 3.6). On the other hand, the bipartite half of $H(2e+1,2)$ ($\cong$ the antipodal quotient of $H(2e+1,2)$) has the bipartite doubling ($\cong$ the antipodal double covering) which is isomorphic to $H(2e+1,2)$ and is obviously a (P and Q)-polynomial scheme.

(8) The antipodal double coverings of O_{d+1} and the q-analogues of them are the only known P-polynomial association schemes with sufficiently large diameters which have no Q-polynomial structures. The bipartite doublings of ${}^2A_{2d-1}(p^f)$ are the only known Q-polynomial association schemes with sufficiently large diameters which have no P-polynomial structures.

We have already discussed bipartite halves and antipodal quotients of imprimitive (P and Q)-polynomial association schemes without detailed explanation. Here we shall explain them systematically. The starting point is the following theorem.

THEOREM 6.1. (Biggs-Gardiner-Smith, cf. Smith (1971)) Let $\mathfrak{X} = (X, \{R_i\}_{0 \le i \le d})$ be an imprimitive P-polynomial association scheme with respect to the natural ordering, and not an ordinary n-gon. Then the distance-regular graph $\Gamma = (X,R_1)$ is either bipartite or antipodal.

Proof. Let $R_i(x) = \{y \in x \mid (x,y) \in R_i\}$. Let S be a non-trivial block. Then by the definition of blocks,

$$S = R_0(x) \cup R_{i_1}(x) \cup R_{i_2}(x) \cup \ldots \cup R_{i_s}(x)$$

for some $i_1, i_2, \ldots, i_s$ independent of $x \in S$. We may assume $1 < i_1 < i_s < \ldots$. If $i_1 = d$, then Γ is antipodal. Since $i_1 > 1$, there are no edges in $R_j(x)$ $(j = i_1, i_2, \ldots, i_s)$, i.e., the intersection number $a_j = 0$. Now $c_j + a_j + b_j = k \geq 3$ and so one of c_j, b_j is at least 2. Since $c_{j+1} \geq c_j$ and $b_{j-1} \geq b_j$ by Proposition 1.2, one of c_{j+1}, b_{j-1} is at least 2. This means $R_j(x)$ contains a pair of vertices whose distance is two. Thus $i_1 = 2$ and consequently $i_2 = 4$, $i_3 = 6$, ..., i.e., Γ is bipartite. □

Let Γ be a bipartite distance-regular graph with partition $\Gamma = \Gamma_1 \cup \Gamma_2$. Then Γ_i becomes distance-regular by defining (x,y) to be an edge if and only if $\partial(x,y) = 2$ in Γ. Γ_i is called the halved graph (or bipartite half) and denoted by Γ'. The two halved graphs of Γ have the same parameters, but they are not always isomorphic. The generalized 6-gon with 126 vertices is a counterexample (see section 2.3).

QUESTION. (1) When are the halved graphs of a bipartite distance-regular graph isomorphic?

Let Γ be an antipodal distance-regular graph. Let $x \sim y$ if and only if $\partial(x,y) = 0$ or d. Let $\overline{\Gamma} = \Gamma/\sim$. Then $\overline{\Gamma}$ becomes distance-regular by defining $(\overline{x},\overline{y})$ to be an edge if and only if there is an edge of Γ between $\overline{x}$ and $\overline{y}$. $\overline{\Gamma}$ is called the antipodal quotient graph.

Γ' and $\overline{\Gamma}$ have roles like the commutator group $[G,G]$ and the central factor group G/Z respectively, in group theory, where Z is the center of G.

Thus the classification problem of P-polynomial association schemes reduces to that of the primitive P-polynomial association schemes (Conjecture (1)) and the following:

PROBLEM. (2) Given known distance-regular graph Γ,

(a) can one determine all distance-regular graphs Π such that $\Pi' = \Gamma$,

(b) can one determine all distance-regular graphs Δ such that $\overline{\Delta} = \Gamma$?

(Π, Δ have the roles like the automorphism group of a simple group and the extension of a simple group by the Schur multiplier, respectively.)

We introduce C-algebras of P-polynomial type and of Q-polynomial type to study Q-polynomial structures of imprivitive (P and Q)-polynomial association schemes. Let $\mathfrak{A} = \langle x_i \mid 0 \leq i \leq d \rangle$ be a C-algebra. In what follows, we always assume that

$$p_{ij}^k \geq 0 \quad \text{and} \quad q_{ij}^k \geq 0.$$

$\mathfrak{A}$ is said to be <u>symmetric</u> if $x_{i'} = x_i$ for all i, or equivalently if the eigenmatrices have entries all real. A symmetric C-algebra $\mathfrak{A}$ is said to be of <u>P-polynomial type</u> with respect to the ordering $x_0, x_1, \ldots, x_d$ of the basis if $x_i = v_i(x_1)$ for polynomials $v_i(x)$ of degree i $(0 \leq i \leq d)$. As in Proposition 1.1, $\mathfrak{A}$ is of P-polynomial type if and only if

$$x_1 x_i = b_{i-1} x_{i-1} + a_i x_i + c_{i+1} x_{i+1} \qquad (6.3)$$
$$(0 \leq i \leq d)$$

with $b_{i-1} > 0$ $(1 \le i \le d)$, $c_{i+1} > 0$ $(0 \le i \le d-1)$ and $b_{-1} = c_{d+1} = 0$. Dually <u>C-algebras of Q-polynomial type</u> are defined by using the primitive idempotents $e_0, e_1, \ldots, e_d$ and the Schur product. In other words, a C-algebra is of Q-polynomial type if and only if the dual C-algebra is of P-polynomial type.

If $\mathfrak{A}$ has a C-subalgebra $\langle x_\alpha \mid \alpha \in T\rangle$ with

$$T = \{0,2,4,\ldots\},$$

then the P-polynomial structure is said to be <u>bipartite</u>. If $\mathfrak{A}$ has a C-subalgebra $\langle x_\alpha \mid \alpha \in T\rangle$ with

$$T = \{0,d\},$$

then the P-polynomial structure is said to be <u>antipodal</u>. <u>Bipartite</u> and <u>antipodal Q-polynomial structures</u> are defined similarly.

PROPOSITION 6.2. Let $\mathfrak{A} = \langle x_i \mid 0 \le i \le d\rangle$ be a C-algebra of P-polynomial type with respect to the ordering $x_0, x_1, \ldots, x_d$ of the basis. Assume $p_{ij}^k \ge 0$ and $q_{ij}^k \ge 0$ for all i, j, k. Let $\langle x_\beta \mid \beta \in T\rangle$ be a proper C-subalgebra of $\mathfrak{A}$. Then

$$T = \{0,\alpha,2\alpha,3\alpha,\ldots\}$$

for some $\alpha \in \{2, d, \frac{d}{s}, \frac{2d+1}{2s+1}, \frac{2d}{2s+1}\}$.

Let B be the tridiagonal matrix

$$\left\{\begin{matrix} * & 1 & c_2 & \cdots & c_{d-1} & c_d \\ 0 & a_1 & a_2 & \cdots & a_{d-1} & a_d \\ k & b_1 & b_2 & \cdots & b_{d-1} & * \end{matrix}\right\}$$

with a_i, b_i, c_i defined in (6.3). Then $\mathfrak{A}$ has a C-subalgebra $\langle x_\beta \mid \beta \in T\rangle$ with (i) $\alpha = 2$, (i) $\alpha = d$, (iii) $\alpha = \frac{d}{s}, \frac{2d+1}{2s+1}, \frac{2d}{2s+1}$ respectively if and only if

(i) $a_2 = a_4 = \cdots = 0$ and $a_1 = a_3 = \cdots$,

(ii)
$$B = \begin{Bmatrix} * & 1 & c_2 & \cdots & c_t & b_t & \cdots & b_2 & b_1 & k \\ 0 & a_1 & a_2 & \cdots & a_t & a_t & \cdots & a_2 & a_1 & 0 \\ k & b_1 & b_2 & \cdots & b_t & c_t & \cdots & c_2 & 1 & * \end{Bmatrix}$$

$(d = 2t+1)$,

$$B = \begin{Bmatrix} * & 1 & c_2 & \cdots & c_{t-1} & c_t & b_{t-1} & \cdots & b_2 & 1 & k \\ 0 & a_1 & a_2 & \cdots & a_{t-1} & a_t & a_{t-1} & \cdots & a_2 & a_1 & 0 \\ k & b_1 & b_2 & \cdots & b_{t-1} & b_t & c_{t-1} & \cdots & c_2 & b_1 & * \end{Bmatrix}$$

$(d = 2t)$,

(iii)
$$B = \begin{Bmatrix} * & \overbrace{1 & c_2 & \cdots & b_2 & b_1 & c_\alpha}^{D_1} & \overbrace{1 & c_2 & \cdots & b_2 & b_1 & c_{2\alpha}}^{D_2} & \cdots \\ 0 & a_1 & a_2 & \cdots & a_2 & a_1 & 0 & a_1 & a_2 & \cdots & a_2 & a_1 & 0 & \cdots \\ k & b_1 & b_2 & \cdots & c_2 & 1 & b_\alpha & b_1 & b_2 & \cdots & c_2 & 1 & b_{2\alpha} & \cdots \end{Bmatrix},$$

where each block D_i is

$$D_i = \begin{matrix} 1 & c_2 & \cdots & c_t & b_t & \cdots & b_2 & b_1 & c_{i\alpha} \\ a_1 & a_2 & \cdots & a_t & a_t & \cdots & a_2 & a_1 & 0 \\ b_1 & b_2 & \cdots & b_t & c_t & \cdots & c_2 & 1 & b_{i\alpha} \end{matrix}$$

$(\alpha = 2t+1)$,

$$D_i = \begin{matrix} 1 & c_2 & \cdots & c_{t-1} & c_{i\alpha - t} & b_{t-1} & \cdots & b_2 & b_1 & c_{i\alpha} \\ a_1 & a_2 & \cdots & a_{t-1} & a_t & a_{t-1} & \cdots & a_2 & a_1 & 0 \\ b_1 & b_2 & \cdots & b_{t-1} & b_{i\alpha - t} & c_{t-1} & \cdots & c_2 & 1 & b_{i\alpha} \end{matrix}$$

$(\alpha = 2t)$,

and the tail of B is

$$\left.\begin{matrix} \cdots & 1 & c_2 & \cdots & b_2 & b_1 & k \\ \cdots & a_1 & a_2 & \cdots & a_2 & a_1 & 0 \\ \cdots & b_1 & b_2 & \cdots & c_2 & 1 & * \end{matrix}\right\} \qquad (\alpha = \tfrac{d}{s}),$$

$$\left.\begin{matrix} \cdots & 1 & c_2 & \cdots & c_{(\alpha-3)/2} & c_{(\alpha-1)/2} \\ \cdots & a_1 & a_2 & \cdots & a_{(\alpha-3)/2} & a_{(\alpha-1)/2} + b_{(\alpha-1)/2} \\ \cdots & b_1 & b_2 & \cdots & b_{(\alpha-3)/2} & * \end{matrix}\right\}$$

$(\alpha = \frac{2d+1}{2s+1})$,

$$\left.\begin{matrix} \cdots & 1 & c_2 & \cdots & c_{(\alpha-2)/2} & c_{\alpha/2} + b_{\alpha/2} \\ \cdots & a_1 & a_2 & \cdots & a_{(\alpha-2)/2} & a_{\alpha/2} \\ \cdots & b_1 & b_2 & \cdots & b_{(\alpha-2)/2} & * \end{matrix}\right\}$$

$(\alpha = \frac{2d}{2s+1})$,

respectively. (Note that (i) and (ii) are special cases of (iii) for $\alpha = 2$ and $\alpha = d$, respectively.)

Proof. Keeping the notation of Theorem 9.9 of Chapter II, let $\langle \frac{1}{p} x_{T_i} \mid 0 \leq i \leq t \rangle$ be the quotient C-algebra $\mathfrak{A}/\langle x_\beta \mid \beta \in T \rangle$. Let $T = \{0, \alpha_1, \alpha_2, \ldots\}$ $(0 < \alpha_1 < \alpha_2 < \ldots)$. Consider the identity $x_\delta x_T = \lambda x_{T_i}$ with $\delta = \alpha_2 - \alpha_1$. By the non-negativeness condition of $p^\gamma_{\alpha\beta}$, T_i contains $\delta,\ \alpha_1 + \delta,\ \alpha_2 + \delta,\ \ldots$. So $T_i \cap T \ni \alpha_2$ and hence $T_i = T$. Since $T = T_i \ni \alpha_2 - \alpha_1$, we have $\alpha_2 - \alpha_1 = \alpha_1$.

Repeat this argument in turn for $\delta = \alpha_i - \alpha_{i-1}$ $(i = 3, 4, \ldots)$. Then we get $\alpha_i = i\alpha$ with $\alpha_1 = \alpha$.

Consider the identity $x_1 x_T = \lambda x_{T_1}$ for T_1 and some $\lambda \in \mathbb{C}$, where T_1 is the block containing 1. Express the left hand side as a linear combination of x_i, using (6.3). The right hand side implies that the coefficient of x_i is a constant λ for $i \in T_1$ and zero for $i \notin T_1$. By the 3-term recurrence (6.3) and the positiveness condition of b_i, c_i, we get

$$T_1 = \{\beta \pm 1 \mid \beta \in T\} \cap \{0, 1, 2, \ldots, d\}.$$

Furthermore

$$a_\beta = 0 \quad (\beta \in T),$$

$$\begin{cases} 1 = c_1 = c_{\beta+1} = b_{\beta-1} \quad (\beta \in T) & \text{if } \alpha > 2, \\ c_1 + b_1 = c_3 + b_3 = \cdots & \text{if } \alpha = 2, \end{cases}$$

where we only consider $\beta \pm 1$ that belong to $\{0, 1, 2, \ldots, d\}$. If $\alpha = 2$, we get $a_2 = a_4 = \cdots = 0$ and $a_1 = a_3 = \cdots$.

Consider the product $x_1 x_{T_1}$. Since $x_1 x_{T_1} = \frac{1}{\lambda} x_1^2 x_T$

and $x_1^2 = b_0x_0 + a_1x_1 + c_2x_2$, the product $x_1x_{T_1}$ is a linear combination of x_T, x_{T_1} and some x_{T_2}. By comparing the coefficients of x_i, we get

$$T_2 = \{\beta \pm 2 \mid \beta \in T\} \cap \{0,1,2,\ldots,d\},$$

and

$$\begin{cases} a_1 = a_{\beta \pm 1} \quad (\beta \in T) \quad \text{if} \quad \alpha > 3, \\ a_1 + b_1 = a_2 + c_2 = a_4 + b_4 = a_5 + c_5 = \ldots \quad \text{if} \quad \alpha = 3, \end{cases}$$

$$\begin{cases} c_2 = c_{\beta+2} = b_{\beta-2} \quad (\beta \in T) \quad \text{if} \quad \alpha > 4, \\ c_2 + b_2 = c_6 + b_6 = c_{10} + b_{10} = \ldots \quad \text{if} \quad \alpha = 4, \end{cases}$$

where we only consider $\beta \pm 1$, $\beta \pm 2$ that belong to $\{0,1,2,\ldots,d\}$.

Case 1. Assume $\alpha = \frac{d}{s}$ for some positive integer s. Consider the product $x_1x_{T_{i-1}}$ in turn for $i = 1,2,\ldots,[\frac{\alpha}{2}]$ as above, each turn defining the set

$$T_i = \{\beta \pm i \mid \beta \in T\} \cap \{0,1,2,\ldots,d\}. \tag{6.4}$$

Express $x_1x_{T_{i-1}}$ as a linear combination of $x_T, x_{T_1}, \ldots, x_{T_i}$ and compare the coefficients to get the desired form of B.

Case 2. Assume $s\alpha + \frac{\alpha - 1}{2} \leq d$, where $s = [\frac{d}{\alpha}]$. As in case 1, (6.4) holds for $i = 1, 2, \ldots, [\frac{\alpha}{2}]$.

Suppose α is odd. Consider $x_1x_{T_{t-1}}$ and $x_1x_{T_t}$

$(t = \frac{\alpha - 1}{2})$ to get $c_t = c_{s\alpha + t}$ and $a_t + b_t = a_{s\alpha + t}$. Since $k = a_i + b_i + c_i$, we get $b_{s\alpha + t} = 0$, which implies $s\alpha + t = d$, i.e., $\alpha = \frac{2d + 1}{2s + 1}$.

Suppose α is even. Consider $x_1 x_{T_{t-1}}$ and $x_1 x_{T_t}$ $(t = \frac{\alpha}{2})$ to get $c_t + b_t = c_{s\alpha + t}$ and $a_t = a_{s\alpha + t}$. So we get $b_{s\alpha + t} = 0$, which implies $s\alpha + t = d$, i.e., $\alpha = \frac{2d}{2s + 1}$.

The matrix B is determined in the same way as in case 1.

<u>Case 3</u>. Assume $s\alpha + \frac{\alpha - 1}{2} > d$ and $s\alpha \neq d$. As in case 1, (6.4) holds for $i = 1, 2, \ldots, (d - s\alpha)$. Consider $x_1 x_{T_{\delta - 1}}$ and $x_1 x_{T_\delta}$ $(\delta = d - s\alpha)$ to get $c_\delta = c_{s\alpha + \delta}$ and $a_\delta = a_{s\alpha + \delta}$. Since $c_d + a_d = k$, we get $b_\delta = 0$, which is a contradiction.

Thus we get the desired form of B in each case. Conversely, if we are given such B, set $T = \{0, \alpha, 2\alpha, \ldots\}$. Then we can trace back the above argument with T_i in (6.4) to show that $x_1 x_{T_j}$ is contained in $\langle x_{T_i} \mid 0 \leq i \leq t \rangle$ for all j. This implies that $x_k x_T$ is contained in $\langle x_{T_i} \mid 0 \leq i \leq t \rangle$ for all k. So $\langle x_\beta \mid \beta \in T \rangle$ is a C-subalgebra. □

To study C-algebras which have (P and Q)-polynomial structures, we shall briefly review section 2.9. Let $\mathfrak{A} = \langle x_i \mid 0 \leq i \leq d \rangle$ be a C-algebra with $p_{ij}^k \geq 0$ and $q_{ij}^k \geq 0$, and let $\hat{\mathfrak{A}} = \langle ne_i \mid 0 \leq i \leq d \rangle$ be the dual C-algebra of $\mathfrak{A}$ with multiplication the Schur product $\circ$, where e_i $(0 \leq i \leq d)$ are the primitive idempotents of $\mathfrak{A}$. For a subset T of $\{0, 1, \ldots, d\}$, let x_T (respectively e_T) denote the sum of x_α (respectively e_α) over $\alpha \in T$. Suppose for some subsets T, U of $\{0,1,\ldots,d\}$,

$$\frac{1}{p} x_T = e_U \tag{6.5}$$

with $p = \sum_{\alpha \in T} k_\alpha$. Then, $\langle x_\alpha \mid \alpha \in T\rangle$ and $\langle ne_\alpha \mid \alpha \in U\rangle$ are C-subalgebras of $\mathfrak{A}$ and $\hat{\mathfrak{A}}$ respectively. The existence of one of the subalgebras $\langle x_\alpha \mid \alpha \in T\rangle$, $\langle ne_\alpha \mid \alpha \in U\rangle$ implies that of the other which satisfies (6.5). Let $q = \sum_{\alpha \in U} m_\alpha$. Then

$$pq = n \tag{6.6}$$

by Lemma 9.7(ii). By Theorem 9.9 of Chapter II, the quotient C-algebra $\mathfrak{A}/\langle x_\alpha \mid \alpha \in T\rangle$ is

$$\langle \frac{1}{p} x_{T_i} \mid 0 \le i \le t\rangle \subset \mathfrak{A} \tag{6.7}$$

for some partition $T_0 = T, T_1, \ldots, T_t$ of $\{0,1,\ldots,d\}$. Moreover $x_\alpha x_T = \lambda_\alpha x_{T_i}$ for some $\lambda_\alpha \in \mathbb{C}$ $(\forall \alpha \in T_i)$. The dual C-algebra of the quotient $\mathfrak{A}/\langle x_\alpha \mid \alpha \in T\rangle$ is

$$\langle ne_\alpha \mid \alpha \in U\rangle \subset \hat{\mathfrak{A}}. \tag{6.7}'$$

The underlying vector spaces of the C-algebras in (6.7) and (6.7)' are the same. In particular, $|U| = 1 + t$. Dually, the quotient C-algebra $\hat{\mathfrak{A}}/\langle ne_\alpha \mid \alpha \in U\rangle$ is

$$\langle \frac{1}{q} ne_{U_i} \mid 0 \le i \le s\rangle \subset \hat{\mathfrak{A}} \tag{6.8}$$

for some partition $U_0 = U, U_1, \ldots, U_s$ of $\{0,1,\ldots,d\}$. Moreover $e_\alpha \circ e_U = \mu_\alpha e_{U_i}$ for some $\mu_\alpha \in \mathbb{C}$ $(\forall \alpha \in U_i)$.

The dual C-algebra of the quotient $\hat{\mathfrak{A}}/\langle ne_\alpha \mid \alpha \in U\rangle$ is

$$\langle x_\alpha \mid \alpha \in T\rangle \subset \mathfrak{A} . \tag{6.8)'$$

The underlying vector spaces of the C-algebras in (6.8) and (6.8)' are the same. In particular, $|T| = 1 + s$.

Now let us suppose $\mathfrak{A}$ is of P-polynomial type and also of Q-polynomial type. Then the following theorem holds.

THEOREM 6.3. Let $\mathfrak{A} = \langle x_i \mid 0 \leq i \leq d\rangle$ be a C-algebra of (P and Q)-polynomial type with respect to the natural orderings of x_i and e_i. Assume $p_{ij}^k \geq 0$ and $q_{ij}^k \geq 0$ for all i, j, k. Suppose d is sufficiently large, say $d \geq 45$, and also suppose $\mathfrak{A}$ is not isomorphic to the adjacency algebra of any ordinary n-gon. If $\langle x_\beta \mid \beta \in T\rangle$ is a proper C-subalgebra of $\mathfrak{A}$, then either

$$T = \{0,2,4,\ldots\}$$

or

$$T = \{0,d\}.$$

Proof. Assume neither $T = \{0,2,4,\ldots\}$ nor $T = \{0,d\}$. By Proposition 6.2, we have $\alpha = \frac{d}{s}, \frac{2d+1}{2s+1}, \frac{2d}{2s+1}$. Moreover,

$$T = \{0,\alpha,2\alpha,3\alpha,\ldots\}$$

and

$$c_{\beta+1} = b_{\beta-1} = 1 \qquad (\beta \in T)$$

$$a_\beta = 0 \qquad (\beta \in T).$$

Let B be the tridiagonal matrix defined in

Proposition 6.2 and let θ_i $(0 \leq i \leq d)$ be the eigenvalues of B. Then (5.1) defines polynomials $u_i(x)$ $(0 \leq i \leq d)$. Notice $p_j(i) = k_j u_j(\theta_i)$. Define dually θ_i^* and $u_i^*(x)$. Notice $q_j(i) = m_j u_j^*(\theta_i^*)$. Moreover (5.3) holds by Theorem 5.5(i) of Chapter II. Therefore Theorem 5.1 can be applied to describe b_i, c_i by the parameters of Askey-Wilson polynomials.

If the expression of b_i, c_i is of type (I) in Theorem 5.1, then $b_i = f(q^i)/g(q^i)$ and $c_i = d(q^i)/e(q^i)$ for certain polynomials $f(x)$, $g(x)$, $d(x)$, $e(x)$ of degree at most 4. Since $c_{\beta+1} = b_{\beta-1} = 1$ for $\beta \in T$, we get

$$f(q^{\beta-1}) = g(q^{\beta-1})$$

$$d(q^{\beta+1}) = e(q^{\beta+1})$$

for $\beta \in T$. By $\theta_i \neq \theta_0$ $(i = 1, 2, \ldots, d)$, we have

$$q^i \neq 0 \qquad (i = 1, 2, \ldots, d).$$

Thus if $[\frac{d}{\alpha}] \geq 5$, i.e., if $|T| - 1 \geq 5$, then we get

$$f(x) = g(x)$$

$$d(x) = e(x),$$

and so $b_i = c_i = 1$. Since $a_\beta = 0$ $(\beta \in T)$, we get $k = 2$, which implies $\mathfrak{A}$ comes from an ordinary n-gon.

For the rest of the expressions of b_i, c_i in Theorem 5.1, the same method can be applied to get $k = 2$, whenever $[\frac{d}{\alpha}] \geq 5$, i.e. $|T| - 1 \geq 5$. We may assume $[\frac{d}{\alpha}] < 5$.

Consider the quotient $\mathfrak{A}/\langle x_\beta \mid \beta \in T\rangle$. The dual C-algebra of this quotient is the C-subalgebra $\langle ne_\beta \mid \beta \in U\rangle \subset \hat{\mathfrak{A}}$, where U is defined by (6.5). The dimention of $\mathfrak{A}/\langle x_\beta \mid \beta \in T\rangle = \langle x_{T_i} \mid 0 \le i \le t\rangle$ is

$$1 + [\tfrac{\alpha}{2}]$$

by (6.4). So $|U| = 1 + [\frac{\alpha}{2}]$. Since $d \ge 45$ and $[\frac{d}{\alpha}] \le 4$, we have $[\frac{\alpha}{2}] \ge 5$. The previous argument for T can be applied to U. Thus we have $U = \{0,2,4,\ldots\}$ and consequently $[\frac{\alpha}{2}] = [\frac{d}{2}]$, i.e., $\alpha = d$. $\square$

The condition $d \ge 45$ in Theorem 6.3 is tentative and conventional. It is expected to be considerably weakened.

If the P-polynomial structure is bipartite, i.e., $T = \{0,2,4,\ldots\}$, then $|U| = 1 + t = 2$ and so $U = \{0,d\}$ by Theorem 6.3, i.e., the Q-polynomial structure is antipodal. The C-algebra $\langle x_\alpha \mid \alpha \in T\rangle$ is of P-polynomial type with respect to the ordering $x_0, x_2, x_4, \ldots$. Consider the dual basis $pe_{U_0}, pe_{U_1}, \ldots, pe_{U_s}$. Since $U = \{0,d\}$ and $e_\alpha \circ e_U = \mu_\alpha e_{U_i}$ $(\alpha \in U_i)$, we get, by induction on i, that

$$\begin{cases} n(e_i \circ e_U) = e_{U_i} = e_i + e_{d-i} & (i < \frac{d}{2}), \\ p(e_i \circ e_U) = e_{U_i} = e_i & (i = \frac{d}{2}). \end{cases} \tag{6.9}$$

(To calculate the coefficient p in the second identity of (6.8), apply the linear representation $ne_i \to m_i$ of $\hat{\mathfrak{A}}$.) So $\langle x_\alpha \mid \alpha \in T\rangle$ is of Q-polynomial type with respect to the ordering $pe_{U_0}, pe_{U_1}, \ldots, pe_{U_s}$ $(s = [\frac{d}{2}])$ given by (6.9).

If the P-polynomial structure is antipodal, i.e., $T = \{0,d\}$, then we have, by induction on i, that

$$\begin{cases} x_i x_T = x_{T_i} = x_i + x_{d-i} & (i < \frac{d}{2}), \\ \frac{1}{p}\, x_i x_T = x_{T_i} = x_i & (i = \frac{d}{2}). \end{cases} \tag{6.9)'$$

So the quotient C-algebra $\langle \frac{1}{p}\, x_{T_i} \mid 0 \leq i \leq t\rangle$ $(t = [\frac{d}{2}])$ is of P-polynomial type with respect to the ordering $x_{T_0}, x_{T_1}, \ldots, x_{T_t}$ given by (6.9)'. Consider the dual basis $\{ne_\alpha \mid \alpha \in U\}$. Since $|U| = 1 + t = 1 + [\frac{d}{2}]$, we get $U = \{0,2,4,\ldots\}$ by Theorem 6.3, i.e., the Q-polynomial structure is bipartite. Moreover, $\langle x_{T_i} \mid 0 \leq i \leq t\rangle$ is of Q-polynomial type with respect to the ordering ne_0, ne_2, ne_4, ...

Since an association scheme is a P-polynomial scheme (respectively a Q-polynomial scheme) if and only if its adjacency algebra is of P-polynomial type (respectively of Q-polynomial type), we have the following:

THEOREM 6.4. Let $\mathfrak{X} = (X, \{R_i\}_{0 \leq i \leq d})$ be a P-polynomial scheme with respect to the ordering $A_0, A_1, A_2, \ldots$ of the adjacency matrices. Let $\mathfrak{X}$ be also a Q-polynomial scheme with respect to the ordering $E_0, E_1, E_2, \ldots$ of the primitive idempotents. If $\mathfrak{X}$ is not an ordinary n-gon, then one of the following holds:

(i) The P-polynomial structure is bipartite and the Q-polynomial structure is antipodal. The bipartite halves are P-polynomial schemes with respect to the ordering $A_0, A_2, A_4, \ldots$ and are Q-polynomial schemes with respect to the ordering $E_0 + E_d$, $E_1 + E_{d-1}$, $E_2 + E_{d-2}$, ...

(ii) The P-polynomial structure is antipodal and the Q-polynomial structure is bipartite. The antipodal quotient is a P-polynomial scheme with respect to the ordering $A_0 + A_d, A_1 + A_{d-1}, A_2 + A_{d-2}, \ldots$ and is a Q-polynomial scheme with respect to the ordering $E_0, E_2, E_4, \ldots$.

Theorem 6.4 holds without the condition $d \geq 45$, because we have a quick proof of Theorem 6.4, using Theorem 6.1 as follows. The P-polynomial structure is known to be either bipartite or antipodal, i.e.,

$$T = \{0,2,4,\ldots\}$$

or

$$T = \{0,d\}.$$

If $T = \{0,d\}$, then $|U| = 1 + t = 1 + [\frac{d}{2}]$. This implies $U = \{0,2,4,\ldots\}$. If $T = \{0,2,4,\ldots\}$, then $|U| = 1 + t = 2$. This implies $U = \{0,\alpha\}$ for some α $(\alpha > [\frac{d}{2}])$. Since the quotient $\hat{\mathfrak{A}}/\langle e_\alpha | \alpha \in U\rangle$ has dimension $1 + s = |T|$, we get $U = \{0,d\}$ by Proposition 6.2.

The reason why we introduced Proposition 6.2 and Theorem 6.3 is to make it clear that the bipartite-antipodal property holds for imprimitive P-polynomial association schemes but not for imprimitive P-polynomial C-algebras. The Q-polynomial property is necessary for imprimitive P-polynomial C-algebras to have the bipartite-antipodal property.

QUESTION. (2) Is the proper block $U \ni 0$ necessarily either $\{0,2,4,\ldots\}$ or $\{0,d\}$ in any imprimitive Q-polynomial association scheme?

3.7 THE RATIONALITY OF EIGENVALUES OF (P AND Q)-POLYNOMIAL SCHEMES

In this section, we shall show that (P and Q)-polynomial schemes have eigenvalues θ_i all rational, and we shall also determine possible parameters of (P and Q)-polynomial schemes which have double Q-polynomial or double P-polynomial structures.

We start with two lemmas.

LEMMA 7.1. Let $f_i(x)$, $g_i(x)$ $(i = 1, 2)$ be polynomials of degree at most 4 in $\mathbb{C}[X]$, the polynomial ring over the complex number field $\mathbb{C}$, such that $f_i(x)$ and $g_i(x)$ are relatively prime for $i = 1, 2$. Suppose there exist non-zero $q_i \in \mathbb{C}$ $(i = 1, 2)$ with $q_i^\nu \neq 1$ $(\nu = 1, 2, \ldots, 32)$ such that

$$\frac{f_1(q_1^\nu)}{g_1(q_1^\nu)} = \frac{f_2(q_2^\nu)}{g_2(q_2^\nu)} \qquad (\nu = 0, 1, \ldots, 33).$$

Then unless $f_i(x)/g_i(x)$ $(i = 1, 2)$ is a constant, it holds that for some $u(x), v(x) \in \mathbb{C}[x]$, $c \in \mathbb{C}$ and $\lambda, \mu \in \mathbb{Z}$ $(1 \leq \lambda \leq 4,\ 1 \leq \mu \leq 4)$, either

(i) $q_1^\lambda = q_2^\mu$,

$$f_1(x) = u(x^\lambda), \quad g_1(x) = v(x^\lambda),$$

$$f_2(x) = cu(x^\mu), \quad g_2(x) = cv(x^\mu),$$

or

(ii) $q_1^{\lambda} = 1/q_2^{\mu}$,

$$f_1(x) = u(x^{\lambda}), \quad g_1(x) = v(x^{\lambda}),$$

$$f_2(x) = cx^r u\left(\frac{1}{x^{\mu}}\right), \quad g_2(x) = cx^r v\left(\frac{1}{x^{\mu}}\right)$$

with

$$r = \max\{\deg f_2(x), \deg g_2(x)\}.$$

Proof. Let us set

$$f_1(x)g_2(y) - g_1(x)f_2(y) = \sum_{i,j=0}^{4} c_{ij}x^i y^j .$$

Putting $x = q_1^{\nu}$, $y = q_2^{\nu}$, we get, by the assumption, that

$$\sum_{i,j=0}^{4} c_{ij} q_1^{i\nu} q_2^{j\nu} = 0 \qquad (\nu = 0, 1, \ldots, 24).$$

Regard the above as a system of equations for c_{ij}. If the determinant of the matrix

$$(q_1^{i\nu} q_2^{j\nu})_{\nu,(i,j)} \qquad (0 \leq \nu \leq 24,\ 0 \leq i \leq 4,\ 0 \leq j \leq 4)$$

is non-zero, then all c_{ij} are zero. This implies $f_1(x)/g_1(x) = f_2(y)/g_2(y)$ for all x, y, i.e., $f_i(x)/g_i(x)$ is a constant.

Assume $\det(q_1^{i\nu} q_2^{i\nu})_{\nu,(i,j)} = 0$. Since this is a Vandermonde's determinant, we have $q_1^i q_2^j = q_1^{i'} q_2^{j'}$ for some distinct column indices (i,j), (i',j') $(0 \leq i,j,i',j' \leq 4,\ (i,j) \neq (i',j'))$, i.e.,

$$q_1^{i-i'} = q_2^{j'-j} .$$

In the above identity for q_1, q_2, we have $i \neq i'$ and $j \neq j'$ because of the assumption $q_i^\nu \neq 1$ $(\nu = 1, 2, \ldots, 32)$. If $i - i'$ and $j' - j$ have opposite signs, consider $\tilde{q}_2$, $\tilde{f}_2(x)$, $\tilde{g}_2(x)$ in place of q_2, $f_2(x)$, $g_2(x)$, where

$$\tilde{q}_2 = \frac{1}{q_2},$$

$$\tilde{f}_2(x) = x^r f_2(\tfrac{1}{x}), \quad \tilde{g}_2(x) = x^r g_2(\tfrac{1}{x}),$$

with $r = \max\{\deg f_2(x), \deg g_2(x)\}$. Then $\tilde{q}_2$, $\tilde{f}_2(x)$, $\tilde{g}_2(x)$ satisfy the assumption of the lemma, i.e., $\tilde{f}_2(x)$ and $\tilde{g}_2(x)$ are relatively prime polynomials of degree at most 4 in $\mathbb{C}[x]$, $\tilde{q}_2^\nu \neq 1$ $(\nu = 1, 2, \ldots, 32)$ and

$$\begin{aligned} f_1(q_1^\nu)/g_1(q_1^\nu) &= f_2(q_2^\nu)/g_2(q_2^\nu) \\ &= \tilde{f}_2(\tilde{q}_2^\nu)/\tilde{g}_2(\tilde{q}_2^\nu) \qquad (\nu = 0, 1, \ldots, 33). \end{aligned}$$

Thus it sufficies to show (i) under the assumption that $i - i'$ and $j' - j$ have the same sign. Set

$$\lambda = |i - i'| \text{ and } \mu = |j' - j|.$$

Then $1 \le \lambda \le 4$, $1 \le \mu \le 4$ and

$$q_1^\lambda = q_2^\mu.$$

Suppose λ and μ are relatively prime. Then there exists some $q \in \mathbb{C}$ such that

$$q_1 = q^\mu$$

$$q_2 = q^\lambda.$$

(Choose $q_0 \in \mathbb{C}$ such that $q_0^\lambda = q_2$. Then $q_1 = \zeta q_0^\mu$ for some ζ, $\zeta^\lambda = 1$. Since λ and μ are relatively prime, $a\lambda + b\mu = 1$ for some integers a, b. Then $q_1 = (\zeta^b q_0)^\mu$ and $q_2 = (\zeta^b q_0)^\lambda$.) The equations $f_1(q_1^\nu)/g_1(q_1^\nu) = f_2(q_2^\nu)/g_2(q_2^\nu)$ become

$$\frac{f_1^{[\mu]}(q^\nu)}{g_1^{[\mu]}(q^\nu)} = \frac{f_2^{[\lambda]}(q^\nu)}{g_2^{[\lambda]}(q^\nu)} \qquad (\nu = 0, 1, \ldots, 28).$$

Here for $f(x) \in \mathbb{C}[x]$ and a positive integer n, $f^{[n]}(x)$ is a polynomial defined by

$$f^{[n]}(x) = f(x^n).$$

The polynomial $f_1^{[\mu]}(x)g_2^{[\lambda]}(x) - g_1^{[\mu]}(x)f_2^{[\lambda]}(x)$ is of degree at most 28 and takes zero at $x = q^\nu$ ($\nu = 0, 1, \ldots, 28$), which are distinct from each other, so it must be identically zero:

$$\frac{f_1^{[\mu]}(x)}{g_1^{[\mu]}(x)} = \frac{f_2^{[\lambda]}(x)}{g_2^{[\lambda]}(x)}$$

as rational functions. From this, it follows that $cf_1^{[\mu]}(x) = f_2^{[\lambda]}(x)$ and $cg_1^{[\mu]}(x) = g_2^{[\lambda]}(x)$ for some $c \in \mathbb{C}$, because $f^{[n]}(x)$ and $g^{[n]}(x)$ are relatively prime in $\mathbb{C}[x]$ whenever $f(x)$ and $g(x)$ are relatively prime. So we have

$$cf_1(x^\mu) = f_2(x^\lambda)$$

$$cg_1(x^\mu) = g_2(x^\lambda) .$$

Since λ and μ are relatively prime, the identity

$cf_1(x^\mu) = f_2(x^\lambda)$ implies that $f_1(x^\mu)$ and $f_2(x^\lambda)$ are polynomials in $x^{\lambda\mu}$, i.e., there exists a polynomial $u(x)$ such that $f_1(x) = u(x^\lambda)$ and $f_2(x) = cu(x^\mu)$. Similarly $g_1(x) = v(x^\lambda)$ and $g_2(x) = cv(x^\mu)$ for some $v(x) \in \mathbb{C}[x]$.

Suppose λ and μ are not relatively prime. Then $\lambda = \mu$ or $(\lambda,\mu) = (2,4), (4,2)$. First suppose $\lambda = \mu$. Then

$$q_2 = \zeta q_1 \quad \text{for some} \quad \zeta, \ \zeta^\lambda = 1.$$

We may assume ζ is a primitive λ^{th} root of unity. The identities $f_1(q_1^{\lambda\nu})/g_1(q_1^{\lambda\nu}) = f_2(q_2^{\lambda\nu})/g_2(q_2^{\lambda\nu})$ $(\nu = 0, 1, \ldots, 8)$ with $q_1^\lambda = q_2^\lambda$ imply

$$\frac{f_1(x)}{g_1(x)} = \frac{f_2(x)}{g_2(x)}$$

as rational functions, while the identities

$$f_1(q_1^{\lambda\nu+1})/g_1(q_1^{\lambda\nu+1}) = f_2(\zeta q_1^{\lambda\nu+1})/g_2(\zeta q_1^{\lambda\nu+1})$$
$$(\nu = 0, 1, \ldots, 8)$$

imply

$$\frac{f_1(x)}{g_1(x)} = \frac{f_2(\zeta x)}{g_2(\zeta x)}$$

as rational functions. So we get $cf_1(x) = f_2(x)$, $cg_1(x) = g_2(x)$ and $c'f_1(x) = f_2(\zeta x)$, $c'g_1(x) = g_2(\zeta x)$ for some non-zero c, $c' \in \mathbb{C}$. If $c \neq c'$, then the constant terms of f_i and g_i $(i = 1, 2)$ are zero, which contradicts the assumption that f_i and g_i are relatively prime. So we have $c = c'$. The lowest term of f_2 is $ax^{k\lambda}$ for some $a \in \mathbb{C}$ and non-negative integer k. From the identity $f_2(x) = f_2(\zeta x)$, it follows that if θ is a root of $f_2(x)$, then $\zeta\theta, \zeta^2\theta, \ldots, \zeta^{\lambda-1}\theta$ are also roots of $f_2(x)$. Since ζ is a primitive λ^{th} root of unity, $f_2(x)$ is a polynomial in x^λ, i.e., $f_2(x) = cu(x^\lambda)$ for some $u(x) \in \mathbb{C}[x]$. Similarly $g_2(x) = cv(x^\lambda)$ for some $v(x) \in \mathbb{C}[x]$.

Next, suppose $(\lambda,\mu) = (4,2)$. Since the case $q_2 = q_1^2$ has been already treated, we may assume

$$q_2 = -q_1^2.$$

Then the identities

$$f_1(q_1^{2\nu})/g_1(q_1^{2\nu}) = f_2(q_1^{4\nu})/g_2(q_1^{4\nu}) \qquad (\nu = 0, 1, \ldots, 12)$$

imply

$$f_1(x)/g_1(x) = f_2(x^2)/g_2(x^2),$$

while the identities

$$f_1(q_1^{2\nu+1})/g_1(q_1^{2\nu+1}) = f_2(-q_1^{4\nu+2})/g_2(-q_1^{4\nu+2})$$
$$(\nu = 0, 1, \ldots, 12)$$

imply

$$f_1(x)/g_1(x) = f_2(-x^2)/g_2(-x^2).$$

So we get $cf_1(x) = f_2(x^2)$, $cg_1(x) = g_2(x^2)$ and $cf_1(x) = f_2(-x^2)$, $cg_1(x) = g_2(-x^2)$ for some non-zero $c \in \mathbb{C}$. Therefore $f_2(x^2) = f_2(-x^2)$ and so $f_2(x) = cu(x^2)$ for some $u(x) \in \mathbb{C}[x]$. Similarly $g_2(x) = cv(x^2)$ for some

$v(x) \in \mathbb{C}[x]$. This completes the proof of the lemma. □

LEMMA 7.2. Let $f_i(x)$, $g_i(x)$ $(i = 1, 2)$ be polynomials of degree at most 4 in $\mathbb{C}[x]$. Suppose there eixsts non-zero $q \in \mathbb{C}$ such that

$$\frac{f_1(\nu)}{g_1(\nu)} = \frac{f_2(q^\nu)}{g_2(q^\nu)} \qquad (\nu = 0, 1, \ldots, 24).$$

Then either (i) $f_i(x)/g_i(x)$ $(i = 1, 2)$ is a constant or (ii) q is one of ± 1, $\omega^{\pm 1}$, $\pm\sqrt{-1}$ with $\omega^3 = 1$, $\omega \neq 1$.

Proof. Let us set

$$f_1(x)g_2(y) - g_1(x)f_2(y) = \sum_{i,j=0}^{4} c_{ij}x^i y^j .$$

Putting $x = \nu$, $y = q^\nu$, we get

$$\sum_{i,j=0}^{4} c_{ij}\nu^i q^{j\nu} = 0 \qquad (\nu = 0, 1, \ldots, 24),$$

where 0^0 is defined to be 1. Regard the above as a system of equations for c_{ij}. If the determinant of the matrix

$$(\nu^i q^{j\nu})_{\nu,(i,j)} \qquad (0 \leq \nu \leq 24,\ 0 \leq i \leq 4,\ 0 \leq j \leq 4)$$

is non-zero, then all c_{ij} are zero and so $f_1(x)/g_1(x) = f_2(y)/g_2(y)$, i.e., $f_i(x)/g_i(x)$ is a constant.

We want to show that if $\det(\nu^i q^{j\nu})_{\nu,(i,j)} = 0$, then $q = \pm 1$, $\omega^{\pm 1}$, $\pm\sqrt{-1}$. More precisely we shall show that

$$\det(\nu^i q^{j\nu})_{\nu,(i,j)} \tag{7.1}$$

$$= (2!3!4!)^5 q^{350}\{(q^4 - 1)(q^3 - 1)^2(q^2 - 1)^3 \cdot (q - 1)^4\}^{25} .$$

Let N_r be a $(r+1) \times 5$ matrix whose (ν,i) entry is ν^i $(0 \leq \nu \leq r,\ 0 \leq i \leq 4)$ and let $Q_r = \mathrm{diag}(1,q,q^2,\ldots,q^r)$. Then the matrix $(\nu^i q^{j\nu})_{\nu,(i,j)}$ is

$$(N_r,\ Q_r N_r,\ Q_r^2 N_r,\ Q_r^3 N_r,\ Q^4 N_r)$$

with $r = 24$. Subtract $(\nu - 1)^{th}$ row from ν^{th} row $(\nu = 1, 2, \ldots, r)$. Then the jth block $Q_r^j N_r$ of the matrix is transformed to have $\nu^i q^{j\nu} - (\nu - 1)^i q^{j(\nu - 1)}$ at (ν,i) entry $(\nu = 1, 2, \ldots, r)$. Since these entries are zero for $i = j = 0$, remove 0th row $(\nu = 0)$ and 0th column $(i = j = 0)$ (the expansion of the determinant with respect to 0th column). Continue the same transformations until we remove $0^{th}, 1^{st}, \ldots, 4^{th}$ rows and columns. Then the 0th block N_r disappears, leaving the contribution of $2!3!4!$ to the determinant, and the jth block $Q_r^j N_r$ $(j = 1, 2, 3, 4)$ is transformed to have

$$\nu^i q^{j\nu} - 5(\nu - 1)^i q^{j(\nu - 1)} + 10(\nu - 2)^i q^{j(\nu - 2)}$$
$$- 10(\nu - 3)^i q^{j(\nu - 3)} + 5(\nu - 4)^i q^{j(\nu - 4)}$$
$$- (\nu - 5)^i q^{j(\nu - 5)}$$

at (ν,i) entry $(\nu = 5, 6, \ldots, r)$. This (ν,i) entry is written as

$$q^{j(\nu-5)}\{(q^j-1)^5\nu^i + f_{i1}(q^j)\nu^{i-1} + f_{i2}(q^j)\nu^{i-2}$$

$$+ \dots + f_{ii}(q^j)\},$$

where $f_{ik}(x)$ is a certain polynomial determined by i and k. Subtract the $(i-1)^{th}$ column multiplied by $f_{i1}(q^j)/(q^j-1)^5$ from the ith column in the jth block. Then the (ν, i) entry becomes

$$q^{j(\nu-5)}\{(q^j-1)^5\nu^i + g_{i2}(q^j)\nu^{i-2} + g_{i3}(q^j)\nu^{i-3}$$

$$+ \dots + g_{ii}(q^j)\},$$

where $g_{ik}(x)$ is a certain polynomial determined by i and k. Continue this sort of column operations until we get

$$q^{j(\nu-5)}(q^j-1)^5\nu^i$$

at (ν, i) entry in the jth block $(\nu = 5, 6, \dots, r)$. Summarizing the above, the jth block $Q_r^j N_r$ $(j = 1, 2, 3, 4)$ is transformed to be

$$(q^j-1)^5 Q^j_{r-5} N'_{r-5},$$

where

$$N'_{r-5} = \begin{pmatrix} 1 & 5 & 5^2 & 5^3 & 5^4 \\ 1 & 6 & 6^2 & 6^3 & 6^4 \\ \cdot & \cdot & \cdot & \cdot & \cdot \\ 1 & r & r^2 & r^3 & r^4 \end{pmatrix}.$$

Since both N_{r-5} and N'_{r-5} are transformed to the same matrix

$$\begin{pmatrix} 1 & 0 & 0 & 0 & 0 \\ 1 & 1 & 0 & 0 & 0 \\ 1 & 2 & 2! & 0 & 0 \\ 1 & 3 & 6 & 3! & 0 \\ 1 & 4 & 12 & 24 & 4! \\ \cdot & \cdot & \cdot & \cdot & \cdot \\ 1 & r & \frac{r!}{(r-2)!} & \frac{r!}{(r-3)!} & \frac{r!}{(r-4)!} \end{pmatrix}$$

by column operations, N'_{r-5} is transformed to N_{r-5} by column operations. Thus the jth block $Q_r^j N_r$ $(j = 1, 2, 3, 4)$ is transformed to $(q^j - 1)^5 Q_{r-5}^j N_{r-5}$ by row and column operations and hence

$$\det(N_r, Q_r N_r, Q_r^2 N_r, Q_r^3 N_r, Q_r^4 N_r) \tag{7.2}$$

$$= 2!3!4! \prod_{j=1}^{4} (q^j - 1)^{25} \det Q_{r-5} \cdot$$

$$\det(N_{r-5}, Q_{r-5}N_{r-5}, Q_{r-5}^2 N_{r-5}, Q_{r-5}^3 N_{r-5}) .$$

Use (7.2) repeatedly to get (7.1). □

The above proof of (7.1) is due to S. Nozawa.

REMARKS. (1) Instead of the condition $f_1(q_1^\nu)/g_1(q_1^\nu) = f_2(q_2^\nu)/g_2(q_2^\nu)$ of Lemma 7.1, we only used

$$f_1(q_1^\nu)g_2(q_2^\nu) - g_1(q_1^\nu)f_2(q_2^\nu) = 0$$

$$(\nu = 0, 1, \ldots, 33)$$

for the proof. So Lemma 7.1 holds even for the case $g_i(q_i^{\nu}) = 0$. A similar remark holds for Lemma 7.2.

(2) The λ, μ in Lemma 7.1 can be chosen minimal with the property that $q_1^{\lambda} = q_2^{\mu}$ or $q_1^{\lambda} = 1/q_2^{\mu}$.

Let $\mathfrak{X}$ be a (P and Q)-polynomial association scheme with sufficiently large diameter d, say $d \geq 34$. Throughout this section we fix an ordering $A_0, A_1, \ldots, A_d$ of the adjacency matrices so that $\mathfrak{X}$ is a P-polynomial scheme with respect to this ordering. Let the intersection matrix B_1 be

$$B_1 = \begin{Bmatrix} * & 1 & c_2 & \cdots & c_d \\ 0 & a_1 & a_2 & \cdots & a_d \\ b_0 & b_1 & b_2 & \cdots & * \end{Bmatrix}.$$

The intersection numbers b_i, c_i can be expressed in the form of Theorem 5.1, but obviously the expression depends on the ordering of $\{E_i\}$, where $\{E_i\}$ is the set of primitive idempotents of the adjacency algebra. Let $\mathfrak{X}$ be a Q-polynomial scheme with respect to an ordering $E_0, E_1, \ldots, E_d$. Note that there might be more than one expression of b_i, c_i even for a fixed ordering $A_0, A_1, \ldots, A_d$ and a fixed ordering $E_0, E_1, \ldots, E_d$. As shown later, the fact is that the expression of b_i, c_i is uniquely determined by the orderings $A_0, A_1, \ldots, A_d$ and $E_0, E_1, \ldots, E_d$, if the diameter d is sufficiently large ($d \geq 34$). If $\mathfrak{X}$ with the Q-polynomial structure $E_0, E_1, \ldots, E_d$ has the intersection numbers b_i, c_i written in the form of Theorem 5.1(I), then

$$b_i = f(q^i)/g(q^i) \qquad (1 \leq i \leq d-1), \tag{7.3I}$$

$$c_i = d(q^i)/e(q^i) \qquad (1 \leq i \leq d-1),$$

where

$$f(x) = h(1 - s^*qx)(1 - r_1qx)(1 - r_2qx)(1 - r_3qx),$$

$$g(x) = (1 - s^*qx^2)(1 - s^*q^2x^2),$$

$$d(x) = \frac{hq}{s^*}(1 - x)(r_1 - s^*x)(r_2 - s^*x)(r_3 - s^*x),$$

$$e(x) = (1 - s^*x^2)(1 - s^*qx^2),$$

with $ss^* = r_1r_2r_3$. (We need to take the limit of $d(x)$ when $s^* = 0$.) The distinct eigenvalues θ_i of A_1 and the distinct dual eigenvalues θ_i^* of E_1 $(\theta_i^* = q_1(i))$ are

$$\theta_i = \theta_0 + h(1 - q^i)(1 - sq^{i+1})/q^i,$$

$$\theta_i^* = \theta_0^* + h^*(1 - q^i)(1 - s^*q^{i+1})/q^i.$$

The above expression is called <u>of type I</u> with parameters

$$(q\,; h, h^*\,; s, s^*\,; r_1, r_2, r_3).$$

Similarly, the expressions of <u>type T</u> (T = IA, II, IIA, IIB, IIC, III) are defined according to the types of b_i, c_i in Theorem 5.1. For example, the expression of type II is

$$b_i = f(i)/g(i) \qquad (1 \le i \le d - 1), \qquad (7.3\text{II})$$

$$c_i = d(i)/e(i) \qquad (1 \le i \le d - 1),$$

where

$$f(x) = h(x+1+s^*)(x+1+r_1)(x+1+r_2)(x+1+r_3),$$

$$g(x) = (2x + 1 + s^*)(2x + 2 + s^*),$$

$$d(x) = hx(x+s^* - r_1)(x+s^* - r_2)(x+s^* - r_3),$$

$$e(x) = (2x + s^*)(2x + 1 + s^*),$$

with $s + s^* = r_1 + r_2 + r_3$. The eigenvalues $\theta_i = p_1(i)$, $\theta_i^* = q_1(i)$ are

$$\theta_i = \theta_0 + hi(i + 1 + s)$$

$$\theta_i^* = \theta_0^* + h^*i(i + 1 + s^*).$$

Notice in the expressions of Type I and IA,

$$q^i \neq 1 \qquad (i = 1, 2, \ldots, d), \tag{7.4}$$

otherwise $\theta_0 = \theta_i$, which is a contradiction.

REMARK. (3) In (7.3I), it holds that if $s^* \neq 0$, then

$$d(x) = qs^{*2}x^4 f(1/qs^*x)$$

$$e(x) = qs^{*2}g(1/qs^*x).$$

The following lemma is trivial but useful.

LEMMA 7.3. There is a one-to-one correspondence between the orderings of $\{E_i\}$ and $\{\theta_i\}$ by $A_1 = \sum_{i=0}^{d} \theta_i E_i$.

LEMMA 7.4. All the parameters appearing in the expressions of b_i, c_i in Theorem 5.1 are algebraic over $\mathbb{Q}$, the field of rational numbers.

Proof. The eigenvalues of A_i are algebraic, so the first eigenmatrix P has entries all algebraic. Since $PQ = nI$, the second eigenmatrix Q has entries all algebraic. Hence $\theta_i^* = q_1(i)$ are algebraic as well as θ_i. By Proposition 5.6,

$$q + \frac{1}{q} = (\theta_3 - \theta_2 + \theta_1 - \theta_0)/(\theta_2 - \theta_1). \tag{7.5}$$

So q is algebraic (but not necessarily an algebraic integer).

Suppose the expression is of type I. By solving the equations

$$q^i(\theta_i - \theta_0)/(1 - q^i) = h - hsq^{i+1} \qquad (i = 1, 2)$$

for h and hs, we get that h and s are algebraic. Dually, h^* and s^* are algebraic. In Proposition 5.8, M_t $(t = 0, 1, \ldots, d)$ are algebraic, because $u_i(x)$ are polynomials with rational coefficients. By (5.15)' and (5.15)", we get that r_1, r_2, r_3 are algebraic.

For the expressions of other types, the proof is similar. □

We treat ordinary n-gons before investigating double Q-polynomial structures.

LEMMA 7.5. Let the expression of b_i, c_i be of type I. Let $d \geq 3$. Suppose either $f(x)/g(x)$ or $d(x)/e(x)$ is a constant function in (7.3I). Then $\mathfrak{X}$ is an ordinary n-gon and $f(x) = g(x) = (1 - x^2)(1 - qx^2)$, $d(x) = e(x) = (1 - x^2)(1 - \frac{1}{q}x^2)$, $q^n = 1$ $(n = 2d$ or $2d + 1)$.

Proof. Suppose $f(x)/g(x)$ is a constant function. Then $f(x)/g(x) = h$ by putting $x = 0$. By Remark (7) of section 3.5, it holds that $b_c = f(1)/g(1) = h$ whenever

$g(1) \neq 0$. Since $b_i = f(q^i)/g(q^i) = h$ $(1 \leq i \leq d-1)$ and $b_0 > b_j$, we have $g(1) = 0$, i.e., $s^* q = 1$ or $s^* q^2 = 1$.

Since $f(x)$ and $g(x)$ have the same roots, we have

$$\{s^* q,\ r_1 q,\ r_2 q,\ r_3 q\}$$
$$= \{\sqrt{s^* q},\ -\sqrt{s^* q},\ \sqrt{s^*}\, q,\ -\sqrt{s^*}\, q\}.$$

Suppose $s^* q^2 = 1$. Then $s^* q = \frac{1}{q}$ is one of $\pm\sqrt{s^* q}$, $\pm\sqrt{s^*}\, q$, i.e., $\pm\frac{1}{\sqrt{q}}$, ± 1, which contradicts $q \neq \pm 1$. So we have $s^* q = 1$ and

$$\{1,\ r_1 q,\ r_2 q,\ r_3 q\} = \{1,\ -1,\ \sqrt{q},\ -\sqrt{q}\}.$$

This implies

$$\frac{1}{h} f(x) = g(x) = (1 - x^2)(1 - qx^2)$$

and

$$\frac{1}{h} d(x) = e(x) = (1 - x^2)(1 - \frac{1}{q} x^2).$$

Since $d \geq 3$, $1 = c_1 = c_2 = h$. By Theorem 5.1,

$$b_0 = h(1 - r_1 q)(1 - r_2 q)(1 - r_3 q)/(1 - s^* q^2)$$

and so $b_0 = 2$. The identity $q^{2d} = 1$ or $q^{2d+1} = 1$ follows from $b_d = 0$.

Suppose $d(x)/e(x)$ is a constant function. Then $s^* \neq 0$. Putting $x = 0$, $d(x)/e(x) = hqs$. Since $d(x)$ and $e(x)$ have the same roots,

$$\{1,\ r_1/s^*,\ r_2/s^*,\ r_3/s^*\} = \{\pm 1/\sqrt{s^* q},\ \pm 1/\sqrt{s^*}\}.$$

In particular, $s^* q = 1$ or $s^* = 1$. If $s^* q = 1$, then

$$\{1, r_1 q, r_2 q, r_3 q\} = \{\pm 1, \pm \sqrt{q}\}$$

and so $f(x)/g(x)$ is a constant. If $s^* = 1$, then

$$\{1, r_1, r_2, r_3\} = \{\pm 1/\sqrt{q}, \pm 1\}$$

and so $f(x)/g(x)$ is a constant. □

Suppose $\mathcal{X}$ is also a Q-polynomial scheme with respect to an ordering $E'_0, E'_1, \ldots, E'_d$. We do not exclude the possibility of $E_i = E'_i$ because there might be more than one expression of b_i, c_i for fixed P-polynomial and Q-polynomial structures. Let $\mathcal{X}$ with the Q-polynomial structure $E_0, E_1, \ldots, E_d$ (respectively the Q-polynomial structure $E'_0, E'_1, \ldots, E'_d$) have an expression of b_i, c_i of type T (respectively T'). We use the — notation for the parameters and polynomials appearing in the expression of b_i, c_i with respect to the ordering $E'_0, E'_1, \ldots, E'_d$. (We use this notation only in this section and this should not be confused with the complex conjugation.) For example, if both the types T and T' are I, the first and second expressions have parameters

$$(q\,; h, h^*\,; s, s^*\,; r_1, r_2, r_3),$$

$$(\bar{q}\,; \bar{h}, \bar{h}^*\,; \bar{s}, \bar{s}^*\,; \bar{r}_1, \bar{r}_2, \bar{r}_3)$$

and polynomials $(f(x), g(x); d(x), e(x))$, $(\bar{f}(x), \bar{g}(x); \bar{d}(x), \bar{e}(x))$ in (7.3I) respectively.

LEMMA 7.6. Let $d \geq 34$. Let an expression of b_i, c_i with respect to $E_0, E_1, \ldots, E_d$ be of type I. Then any expression of b_i, c_i with respect to $E'_0, E'_1, \ldots, E'_d$ is also of type I. If, in addition, $\mathcal{X}$ is not an ordinary n-gon, then for some λ, μ $(\lambda, \mu = 1, 2)$ and some

$u(x), v(x), w(x), \overline{w}(x) \in \mathbb{C}[x]$, either

$$q^{\lambda} = \overline{q}^{\mu},$$

$$f(x) = w(x)u(x^{\lambda}), \quad g(x) = w(x)v(x^{\lambda}),$$

$$\overline{f}(x) = \overline{w}(x)u(x^{\mu}), \quad \overline{g}(x) = \overline{w}(x)v(x^{\mu}),$$

or

$$q^{\lambda} = 1/\overline{q}^{\mu},$$

$$f(x) = w(x)u(x^{\lambda}), \quad g(x) = w(x)v(x^{\lambda}),$$

$$\overline{f}(x) = x^{r}\,\overline{w}(\tfrac{1}{x})u(\tfrac{1}{x^{\mu}}), \quad \overline{g}(x) = x^{r}\,\overline{w}(\tfrac{1}{x})v(\tfrac{1}{x^{\mu}}),$$

with $r = \max\{\deg \overline{f}(x), \overline{g}(x)\}$, where the polynomials $u(x)$ and $v(x)$ can be chosen relatively prime.

Similar identities hold with the same λ, μ for $q, \overline{q}, d(x), e(x), \overline{d}(x), \overline{e}(x)$.

Proof. Apply Lemma 7.1 and Lemma 7.2 to

$$b_i = f(q^i)/g(q^i) = \overline{f}(\overline{q}^i)/\overline{g}(\overline{q}^i)$$

and

$$b_i = f(q^i)/g(q^i) = \overline{f}(i)/\overline{g}(i),$$

using (7.4) and Lemma 7.5. Then the conclusion of the lemma is immediately derived with $1 \leq \lambda \leq 4$, $1 \leq \mu \leq 4$. We want to show $1 \leq \lambda \leq 2$ and $1 \leq \mu \leq 2$. Suppose λ is 3 or 4. Then the identity

$$g(x) = (1 - s^{*}qx^{2})(1 - s^{*}q^{2}x^{2}) = w(x)v(x^{\lambda})$$

implies $s^{*} = 0$ and $w(x) = v(x) = 1$. Since $ss^{*} = r_1r_2r_3$, we may assume $r_3 = 0$. Then

$$f(x) = h(1 - r_1 qx)(1 - r_2 qx) = u(x^\lambda)$$

and hence $r_1 = r_2 = 0$, $u(x) = h$. This implies $f(x)/g(x)$ is a constant and hence $\mathfrak{X}$ is an ordinary n-gon. Similarly, if μ is 3 or 4, $\mathfrak{X}$ is proved to be an ordinary n-gon. □

PROPOSITION 7.7. Let $d \geq 34$. Let an expression of b_i, c_i with respect to $E_0, E_1, \ldots, E_d$ be of type I with parameters $(q; h, h^*; s, s^*; r_1, r_2, r_3)$. If $q^\nu = 1$ for some integer ν, then $\mathfrak{X}$ is an ordinary n-gon.

Proof. Suppose $\mathfrak{X}$ is not an ordinary n-gon. Let K be the normal extension field of $\mathbb{Q}$ generated by all the parameters q, h, h^*, s, s^*, r_1, r_2, r_3 (Lemma 7.4). Then all the entries of the first and second eigenmatrices P, Q belong to K and in particular each primitive idempotent E_i has entries in K. For a Galois automorphism σ of K over $\mathbb{Q}$, $\mathfrak{X}$ becomes a Q-polynomial scheme with respect to $E_0^\sigma, E_1^\sigma, \ldots, E_d^\sigma$ also, and $\mathfrak{X}$ with this Q-polynomial structure has an expression of b_i, c_i of type I with parameters

$$(q^\sigma; h^\sigma, h^{*\sigma}; s^\sigma, s^{*\sigma}; r_1^\sigma, r_2^\sigma, r_3^\sigma) .$$

By Lemma 7.6, q^σ is one of $\pm q^{\pm 1}$, $q^{\pm 2}$, $\pm q^{\pm 1/2}$. So q^{σ^2} is one of q, q^4, $q^{1/4}$. But since σ^2 is also a Galois automorphism, q^{σ^2} cannot be q^4 or $q^{1/4}$ by Lemma 7.6. Therefore q^σ is $\pm q$ or $\pm 1/q$, i.e., $\pm q$ and $\pm 1/q$ are the only possible algebraic conjugates of q in K, which is a contradiction by (7.4). □

LEMMA 7.8. Let $d \geq 34$. Let $T = T' = I$, i.e., both the expressions with respect to $E_0, E_1, \ldots, E_d$ and

$E'_0, E'_1, \ldots, E'_d$ are of type I. Suppose $\mathcal{X}$ is not an ordinary n-gon.

(i) If $E_i = E'_1$ for all i, then $\bar{q} = q$ or $\bar{q} = 1/q$. Conversely, if $\bar{q} = q$, then $E_i = E'_i$ for all i and the two expressions of b_i, c_i have the same parameters

$$(\bar{q}\,;\bar{h},\bar{h}^*\,;\bar{s},\bar{s}^*\,;\bar{r}_1,\bar{r}_2,\bar{r}_3) = (q\,;h,h^*\,;s,s^*\,;r_1,r_2,r_3).$$

If $\bar{q} = 1/q$, then $E_i = E'_i$ for all i and

$$\begin{aligned}(\bar{q}\,;\bar{h},\bar{h}^*\,;\bar{s},\bar{s}^*\,;\bar{r}_1,\bar{r}_2,\bar{r}_3)\\ = (1/q\,;hsq,h^*s^*q\,;1/s,1/s^*\,;1/r_1,1/r_2,1/r_3)\end{aligned}$$

with $ss^* \neq 0$. (This is called the <u>reciprocal parameters</u> and always exists whenever $ss^* \neq 0$.)

(ii) If $E_i \neq E'_i$ for some i, then $\bar{q} = q^2$, $\bar{q} = 1/q^2$, $\bar{q}^2 = q$ or $\bar{q}^2 = 1/q$. If $\bar{q} = q^2$, then the ordering $E'_0, E'_1, \ldots, E'_d$ is $E_0, E_2, E_4, \ldots, E_5, E_3, E_1$ and

$$\begin{aligned}(q\,;s,s^*\,;r_1,r_2,r_3)\\ = (q\,;q^{-2d-2},s^*\,;-s^*,q^{-d-1},-q^{-d-1})\end{aligned}$$

and

$$\begin{aligned}(\bar{q}\,;\bar{s},\bar{s}^*\,;\bar{r}_1,\bar{r}_2,\bar{r}_3)\\ = (q^2\,;q^{-2d-3},s^{*2}\,;q^{-2d-2},-s^*/q,-s^*).\end{aligned}$$

If $\bar{q} = 1/q^2$, then $\bar{s}\bar{s}^* \neq 0$ and so the reciprocal parameters belong to the previous case $\bar{q} = q^2$.

<u>Proof</u>. (i) Let $\theta_0, \theta_1, \ldots, \theta_d$ and $\theta'_0, \theta'_1, \ldots, \theta'_d$ be the

orderings of the eigenvalues corresponding to $E_0, E_1, \ldots, E_d$ and $E'_0, E'_1, \ldots, E'_d$ in Lemma 7.3. Then

$$\theta_i = \theta_0 + h(1-q^i)(1-sq^{i+1})/q^i,$$

$$\theta'_i = \theta'_0 + \bar{h}(1-\bar{q}^i)(1-\bar{s}\,\bar{q}^{i+1})/\bar{q}^i.$$

Suppose $E_i = E'$ for all i. Then $\theta_i = \theta'_i$ and so either

$$\bar{q} = q,\ \bar{h} = h,\ \bar{s} = s$$

or

$$\bar{q} = 1/q,\ \bar{h} = hsq,\ \bar{s} = 1/s$$

by Lemma 7.1. Similarly, from the expression of the dual eigenvalues θ^*_i (notice $E_1 = \frac{1}{n}\sum_i \theta^*_i A_i$), it follows that either

$$\bar{q} = q,\ \bar{h}^* = h^*,\ \bar{s}^* = s^*$$

or

$$\bar{q} = 1/q,\ \bar{h}^* = h^* s^* q,\ \bar{s}^* = 1/s^*.$$

In particular, if $\bar{q} = 1/q$, then $ss^* \neq 0$. Putting these identities in $b_i = f(q^i)/g(q^i) = \bar{f}(\bar{q}^i)/\bar{g}(\bar{q}^i)$, we get $\bar{r}_i = r_i$, $\bar{r}_i = 1/r_i$ according to $\bar{q} = q$, $\bar{q} = 1/q$ by Lemma 7.1.

Suppose $\bar{q} = q$. The identities $b_i = f(q^i)/g(q^i) = \bar{f}(\bar{q}^i)/\bar{g}(\bar{q}^i)$ imply $f(x)/g(x) = \bar{f}(x)/\bar{g}(x)$. Putting $x = 0$, we get $h = \bar{h}$. From the expressions of c_i, we get $d(x)/e(x) = \bar{d}(x)/\bar{e}(x)$. Putting $x = 0$, we get $hsq = \bar{h}\,\bar{s}\,\bar{q}$. So $\theta_i = \theta'_i$ for all i, i.e., $E_i = E'_i$. Similarly the assumption $\bar{q} = 1/q$ implies $f(x)/g(x) =$

$\bar{f}(\frac{1}{x}) / \bar{g}(\frac{1}{x})$. Taking the limits as x tends to 0 and ∞, we get $s^* \bar{s}^* \neq 0$, $h = \bar{h}\,\bar{s}\,\bar{q}$, $hsq = \bar{h}$ and hence $\theta_i = \theta_i'$ for all i.

(ii) By Lemma 7.6, we have $q^\lambda = \bar{q}^\mu$ or $q^\lambda = 1/\bar{q}^\mu$ $(\lambda,\mu = 1, 2)$. If $\lambda = \mu = 1$, then $E_i = E_i'$ by (i). So we may assume $\lambda = 2$. Since

$$g(x) = (1 - s^* q x^2)(1 - s^* q^2 x^2)$$

and

$$f(x)/g(x) = u(x^2)/v(x^2)$$

by Lemma 7.6, we get $f(x) = f(-x)$ and so $f(x)$ is a polynomial in x^2. From

$$f(x) = h(1 - s^* qx)(1 - r_1 qx)(1 - r_2 qx)(1 - r_3 qx),$$

we have

$$r_1 = -s^*, \quad r_2 = -r_3 = r$$

and

$$f(x) = h(1 - s^{*2} q^2 x^2)(1 - r^2 q^2 x^2), \tag{7.6}$$

$$g(x) = (1 - s^* q x^2)(1 - s^* q^2 x^2),$$

$$d(x) = hq(1 - x^2)(r^2 - s^{*2} x^2),$$

$$e(x) = (1 - s^* x^2)(1 - s^* q x^2) .$$

Here it holds that $r^2 = s$, because $d(0)/e(0) = hqs$ (even if $s^* = 0$).

The trivial solution of

$$f(q^i)/g(q^i) = \overline{f}(\overline{q}^i)/\overline{g}(\overline{q}^i),$$

$$d(q^i)/e(q^i) = \overline{d}(\overline{q}^i)/\overline{e}(\overline{q}^i)$$

is

$$(\overline{q}\,;\overline{s},\overline{s}^*;\overline{r}_1,\overline{r}_2,\overline{r}_3) = (q^2\,;r^2/q,s^{*2}\,;r^2,-s^*/q,-s^*).$$

We want to show this is the unique solution if $\overline{q} = q^2$. Suppose $\overline{q} = q^2$. Then $f(x)/g(x) = \overline{f}(x^2)/\overline{g}(x^2)$, $d(x)/e(x) = \overline{d}(x^2)/\overline{e}(x^2)$ by Lemma 7.6. Putting $x = 0$, we get $\overline{h} = h$, $\overline{s} = s/q$ and hence

$$\theta_i' = \theta_0 + h(1-q^{2i})(1-sq^{2i+1})/q^{2i}.$$

So the ordering $\theta_0',\theta_1',\ldots,\theta_d'$ is $\theta_0,\theta_2,\theta_4,\ldots,\theta_5,\theta_3,\theta_1$, i.e., $E_1' = E_2$. Let

$$E_1' = \frac{1}{n}\sum_i \theta_i^{*\prime} A_i.$$

Then $\theta_i^{*\prime} = q_2(i) = v_2^*(\theta_i^*)$, where

$$v_2^*(x) = \frac{1}{c_2^*}(x^2 - a_1^* x - b_0^*).$$

Comparing

$$\theta_i^{*\prime} = \theta_0^* + \overline{h}^*(1-q^{2i})(1-\overline{s}^* q^{2i+2})/q^{2i}$$

with

$$v_2^*(\theta_i^*) = \frac{1}{c_2^*}h^{*2}q^{-2i} + \frac{1}{c_2^*}h^{*2}s^{*2}q^{2i+2} + \ldots,$$

we get $\overline{h}^* = h^{*2}/c_2^*$, $\overline{s}^* = s^{*2}$. From the identity $f(x)/g(x) = \overline{f}(x^2)/\overline{g}(x^2)$ with $\overline{h} = h$, $\overline{s}^* = s^{*2}$, it follows that

$$(\bar{r}_1, \bar{r}_2, \bar{r}_3) = (r^2, -s^*/q, -s^*),$$

i.e., the trivial solution. Since one of r_1, r_2, r_3 is q^{-d-1} by (I) of Theorem 5.1, either $s^* = -q^{-d-1}$ or $r = q^{-d-1}$. If $s^* = -q^{-d-1}$, then $\bar{s}^* = q^{-2d-2}$ and so $\theta_d^{*\prime} = \theta_0^{*\prime}$, which is a contradiction. So $r = q^{-d-1}$.

Suppose $\bar{q} = 1/q^2$. Then

$$f(x)/g(x) = \bar{f}\left(\frac{1}{x^2}\right)/\bar{g}\left(\frac{1}{x^2}\right)$$

by Lemma 7.6. Taking the limits as x tends to 0 and ∞, we get $s^*\bar{s}^* \neq 0$, $h = \bar{h}\,\bar{s}\,\bar{q}$, $hsq = \bar{h}$ and so $s\bar{s} = q$. Consider the reciprocal parameters with respect to the second ordering. Then we return to the previous case $\bar{q} = q^2$.

Suppose $\bar{q} = -q$. Then $\lambda = \mu = 2$ in Lemma 7.6. So $f(x)/g(x) = \bar{f}(x)/\bar{g}(x)$ and $d(x)/e(x) = \bar{d}(x)/\bar{e}(x)$. Putting $x = 0$, we get $h = \bar{h}$, $-s = \bar{s}$. Since (7.6) holds for $\bar{f}(x)$, $\bar{g}(x)$, $\bar{d}(x)$, $\bar{e}(x)$, the identity $f(x)/g(x) = \bar{f}(x)/\bar{g}(x)$ becomes

$$\frac{(1 - s^{*2}q^2y)(1 - r^2q^2y)}{(1 - s^*qy)(1 - s^*q^2y)} = \frac{(1 - \bar{s}^*q^2y)(1 + r^2q^2y)}{(1 + \bar{s}^*qy)(1 - \bar{s}^*q^2y)}$$

with $y = x^2$, because $\bar{r}^2 = \bar{s} = -s = -r^2$. First verify that neither $s^* = \bar{s}^*$ nor $-s^* = \bar{s}^*$ is possible, otherwise $f(x)/g(x)$ is a constant. Since one of r_i is q^{-d-1} by (I) of Theorem 5.1, we have $s^* = -q^{-d-1}$ or $r^2 = q^{-2d-2}$. Similarly $\bar{s}^* = -q^{-d-1}$ or $r^2 = -q^{-2d-2}$. So we may assume $s^* = -q^{-d-1}$ and $r^2 = -q^{-2d-2}$. Then the identity $f(x)/g(x) = \bar{f}(x)/\bar{g}(x)$ becomes

$$(1+q^{-2d}y)/(1+q^{-d}y)(1+q^{-d+1}y)$$

$$= (1-\bar{s}^* q^2 y)/(1+\bar{s}^* qy)(1-\bar{s}^* q^2 y),$$

which is a contradiction.

Suppose $\bar{q} = -1/q$. Then $f(x)/g(x) = \bar{f}(\frac{1}{x}) / \bar{g}(\frac{1}{x})$. Considering the limits as x tends to 0 and ∞, we get $\bar{s}^* \neq 0$ and $\bar{s} \neq 0$. Since the reciprocal parameters have been proved impossible, this is not the case. □

LEMMA 7.9. Let $d \geq 34$. Let the expressions of b_i, c_i with respect to $E_0, E_1, \ldots, E_d$ and $E'_0, E'_1, \ldots, E'_d$ be of type T and T' respectively. Then either

$$T = T'$$

or

$$(T, T') \quad (\text{or } (T', T))$$

$$= (II, IIA), (IIB, IIC), (IIB, III), (IIC, III).$$

If $T = T' \neq I$, then $E_i = E'_i$ for all i and both the expressions have the same parameters.

Proof. The former half of the assertion immediately follows from Lemma 7.1 and Lemma 7.2.

Suppose $T = T' = II$. Then by (7.3II), $b_i = f(i)/g(i) = \bar{f}(i)/\bar{g}(i)$ and $c_i = d(i)/e(i) = \bar{d}(i)/\bar{e}(i)$, so

$$f(x)/g(x) = \bar{f}(x)/\bar{g}(x) \tag{7.7}$$

$$d(x)/e(x) = \bar{d}(x)/\bar{e}(x).$$

Let $S(f)$ be the sum of the roots of $f(x)$ and consider

$$S(f) - S(g) = S(\overline{f}) - S(\overline{g}) \tag{7.8}$$

$$S(d) - S(e) = S(\overline{d}) - S(\overline{e}).$$

Then we get $s^* = \overline{s}^*$. Back to (7.7), we get that both the expressions have the same parameters and hence $E_i = E_i'$.

For the rest of the types of $T = T'$, the proof is similar. □

THEOREM 7.10. Let $\mathfrak{X}$ be a (P and Q)-polynomial association scheme with diameter $d \geq 34$. Let $E_0, E_1, \ldots, E_d$ be the original ordering of the primitive idempotents for the Q-polynomial structure. Suppose $\mathfrak{X}$ is also a Q-polynomial scheme with respect to another ordering $E_0', E_1', \ldots, E_d'$ of the primitive idempotents. Then unless $\mathfrak{X}$ is an ordinary n-gon, we have the following assertions (1), (2) and (3):

(1) One of the following holds for the intersection matrix

$$B_1 = \begin{Bmatrix} * & 1 & c_2 & \cdots & c_d \\ 0 & a_1 & a_2 & \cdots & a_d \\ b_0 & b_1 & b_2 & \cdots & * \end{Bmatrix} :$$

(i) $$b_i = \frac{h(1 - s^{*2}q^{2i+2})(1 - r^2q^{2i+2})}{(1 - s^*q^{2i+1})(1 - s^*q^{2i+2})},$$

$$c_i = \frac{hq(1 - q^{2i})(r^2 - s^{*2}q^{2i})}{(1 - s^*q^{2i+1})(1 - s^*q^{2i})}$$

with $r = q^{-d-1}$. This is realized by putting

$$(q\,; r^2, s^*\,; -s^*, r, -r)$$

and

$$(q^2\,;\,r^2/q, s^{*2}\,;\,r^2, -s^*/q, -s^*)$$

in $(q\,;\,s, s^*\,;\,r_1, r_2, r_3)$ of Thecrem 5.1(I).

(ii) $b_i = d - i,$

$c_i = i,$

with d even. This is realized by putting

$$(-2,\ -1,\ -d-1)$$

in $(s',\ r/s^{*\prime},\ r_1)$ of Theorem 5.1(IIC), and

$$(d+1, d+1;\ -d-1,\ -\frac{d+1}{2},\ \frac{d+1}{2})$$

in $(s, s^*;\ r_1, r_2, r_3)$ of Theorem 5.1(III).

(iii) $b_i = h(i-d)(i+1+s^*),$

$c_i = hi(i+s^*+d+1).$

This is realized by putting

$$(-2d-2;\ s^*,\ -d-1)$$

in $(s;\ r_1, r_2)$ of Theorem 5.1(IIA), and

$$(-d-\frac{3}{2},\ s^*;\ \frac{s^*}{2},\ \frac{s^*-1}{2},\ -d-1)$$

in $(s, s^*;\ r_1, r_2, r_3)$ of Theorem 5.1(II).

(iv) $b_i = h(d-i)(i+1+s^*)/(2i+2+s^*),$

$c_i = hi(i+s^*+d+1)/(2i+s^*).$

This is realized by putting

$$(2d+2, -s^*;\ s^*, -d-1,\ d+1)$$

in $(s, s^*; r_1, r_2, r_3)$ of Theorem 5.1(III), and

$$(-2h, s^*; \frac{s^* - 1}{2}, -d - 1)$$

in $(s', s^*; r_1, r_2)$ of Theorem 5.1(IIB).

(2) Each case of (i)-(iv) has the ordering $E'_0, E'_1, \ldots, E'_d$ as follows:

(i) $E_0, E_2, E_4, \ldots, E_5, E_3, E_1$

or its dual

$$E_0, E_d, E_1, E_{d-1}, E_2, E_{d-2}, \ldots .$$

(ii) $E_0, E_{d-1}, E_2, E_{d-3}, E_4, E_{d-5}, \ldots, E_1, E_d$ (d is even).

(iii) The same as (i).

(iv) The same as (i).

(3) $\mathfrak{X}$ has at most two Q-polynomial structures.

Proof. (1) and (2). We fix the ordering $A_0, A_1, \ldots, A_d$ of the adjacency matrices so that $\mathfrak{X}$ is a P-polynomial scheme with respect to this ordering. Let the expressions of b_i, c_i with respect to $E_0, E_1, \ldots, E_d$ and $E'_0, E'_1, \ldots, E'_d$ be of type T and T' respectively. If T = T', then T = T' = I and (i) holds by Lemma 7.9 and Lemma 7.8. We may assume

$$(T, T') = (II, IIA), (IIB, IIC), (IIB, III) \text{ or } (IIC, III).$$

Suppose (T, T') = (II, IIA). Then (7.7) holds. Considering the roots of $f(x)\bar{g}(x) = \bar{f}(x)g(x)$ and $d(x)\bar{e}(x) = \bar{d}(x)e(x)$, we have

$$\{1+s^*,\ 1+r_1,\ 1+r_2,\ 1+r_3\} = \{\frac{1+s^*}{2},\ \frac{2+s^*}{2},\ 1+\bar{r}_1,\ 1+\bar{r}_2\}, \tag{7.9}$$

$$\{s^*-r_1,\ s^*-r_2,\ s^*-r_3\} = \{\frac{s^*}{2},\ \frac{1+s^*}{2},\ \bar{r}_1+\bar{r}_2-\bar{s}\}.$$

By the second identity of (7.9), we have

$$\{r_1,r_2,r_3\} = \{\frac{s^*}{2},\ \frac{s^*-1}{2},\ s^*+\bar{s}-\bar{r}_1-\bar{r}_2\}.$$

Putting this in the first identity of (7.9), we get

$$\{1+s^*,\ 1+s^*+\bar{s}-\bar{r}_1-\bar{r}_2\} = \{1+\bar{r}_1,\ 1+\bar{r}_2\}.$$

We may assume $\bar{r}_1 = s^*$ and $\bar{r}_2 = s^*+\bar{s}-\bar{r}_1-\bar{r}_2$, i.e., $\bar{r}_2 = \frac{\bar{s}}{2}$. Then $r_1 = \frac{s^*}{2}$, $r_2 = \frac{s^*-1}{2}$ and $r_3 = \frac{\bar{s}}{2}$ by symmetry of r_1, r_2, r_3. By (IIA) of Theorem 5.1, one of $\bar{r}_i$ $(i = 1, 2)$ is $-d-1$. If $\bar{r}_2 = -d-1$, then we get (iii). If $\bar{r}_1 = -d-1$, then

$$r_1 = -\frac{d+1}{2},\quad r_2 = -\frac{d+2}{2},\quad r_3 = \frac{\bar{s}}{2}.$$

Since one of r_i $(i = 1, 2, 3)$ is $-d-1$, we get $r_3 = \frac{\bar{s}}{2} = -d-1$ and return to the previous case $\bar{r}_2 = -d-1$. Since

$$\theta_i = \theta_0 + hi(i-d-\tfrac{1}{2})$$

and

$$\theta'_i = \theta_0 + \frac{h}{4}\, i(i-2d-1),$$

the ordering $\theta_0,\theta_1,\theta_2,\ldots,\theta_d$ is $\theta'_0,\theta'_2,\theta'_4,\ldots,\theta'_5,\theta'_3,\theta'_1$.

Suppose $(T,T') = (\text{IIB},\text{IIC})$. Then

$$\{1+s^*, 1+r_1, 1+r_2\} = \{\frac{1+s^*}{2}, \frac{2+s^*}{2}, 1+\bar{r}_1\},$$

$$\{s^*-r_1, s^*-r_2\} = \{\frac{1+s^*}{2}, \frac{s^*}{2}\}.$$

So $\{r_1, r_2\} = \{\frac{s^*-1}{2}, \frac{s^*}{2}\}$ by the second identity. Putting this in the first identity, we get $\bar{r}_1 = s^*$. By (IIC) of Theorem 5.1, we get $\bar{r}_1 = -d-1$, which leads to a contradiction $\theta^*_d = \theta^*_0 + h^*d(d+1+s^*) = \theta^*_0$. Notice a formal expression of type IIB with parameters

$$(s', s^*; r_1, r_2) = (-4, -d-1; -\frac{d+2}{2}, -\frac{d+1}{2})$$

gives $b_i = d-i$ and $c_i = i$, but the condition $\theta^*_i \neq \theta^*_j$ $(i \neq j)$ eliminates this case.

Suppose $(T, T') = (\text{IIB}, \text{III})$. By (7.7), the parameters of type III must satisfy $\bar{r}_1 = -\bar{s}^*$, $\bar{r}_2 = -\bar{r}_3$ and

$$\{1+s^*, 1+r_1, 1+r_2, \frac{2-\bar{s}^*}{2}\}$$

$$= \{\frac{1+s^*}{2}, \frac{2+s^*}{2}, 1-\bar{s}^*, 1-\bar{r}_3\},$$

$$\{s^*-r_1, s^*-r_2, -\frac{\bar{s}^*}{2}\} = \{\frac{1+s^*}{2}, \frac{s^*}{2}, -\bar{s}^*+\bar{r}_3\}.$$

By (7.8), we get $\bar{s}^* = -s^*$ and $\bar{r}_3 = \frac{s^*}{2} - r_1 - r_2 - \frac{1}{2}$. This leads to $r_1 = \frac{s^*-1}{2}$. By (III) of Theorem 5.1, either $1-\bar{s}^* = -d$ or $1-\bar{r}_3 = -d$, i.e., $1+s^* = -d$ or $1+r_2 = -d$, but if $1+s^* = -d$, we get a contradiction of $\theta^*_d = \theta^*_0$ in IIB. Thus (iv) holds. By

$$\theta_i = \theta_0 + s'i$$

and

$$\theta_i' = \theta_0 + \frac{s'}{2} i \qquad (i = \text{even}),$$

$$\theta_i' = \theta_0 + \frac{s'}{2}(2d+1-i) \qquad (i = \text{odd}),$$

the ordering $\theta_0, \theta_1, \ldots, \theta_d$ is $\theta_0', \theta_2', \theta_4', \ldots, \theta_5', \theta_3', \theta_1'$.

Suppose $(T, T') = (\text{IIC}, \text{III})$. Then $\bar{r}_1 = -\bar{s}^*$, $\bar{r}_2 = -\bar{r}_3$ and $\{1 + r_1, (2 - \bar{s}^*)/2\} = \{1 - \bar{s}^*, 1 - \bar{r}_3\}$, $-\bar{s}^* + \bar{r}_3 = -\bar{s}^*/2$, i.e., $\bar{r}_3 = \bar{s}^*/2$, $\bar{s}^* = -r_1$. In type IIC, $1 + r_1 = -d$ and by $c_1 = 1$, $\frac{r}{s^{*\prime}} - s' = 1$. Taking the limits of $f(x)/xg(x) = \bar{f}(x)/x\bar{g}(x)$ and $d(x)/xe(x) = \bar{d}(x)/x\bar{e}(x)$ as x tends to ∞, we get $r/s^{*\prime} = \bar{h}$ and $1 = -\bar{h}$. Thus we have $b_i = d - i$, $c_i = i$. Since $s' = \frac{r}{s^{*\prime}} - 1 = -2$, $\bar{h} = -1$ and $\bar{s} = -\bar{r}_1 - \bar{r}_2 + \bar{r}_3 - \bar{s}^* = d + 1$, we have $\theta_i = \theta_0 - 2i$, $\theta_i' = \theta_0 - 2i$ $(i = \text{even})$, $\theta_i' = \theta_0 - 2(d - i)$ $(i = \text{odd})$. If d is odd, then $\theta_d' = \theta_0$ which is a contradiction. Thus d is even and (ii) holds with the ordering $\theta_0', \theta_1', \ldots, \theta_d'$ being $\theta_0, \theta_{d-1}, \theta_2, \theta_{d-3}, \theta_4, \theta_{d-5}, \ldots, \theta_1, \theta_d$. This completes the proof of (1) and (2).

(3) Let $E_0'', E_1'', \ldots, E_d''$ be the ordering for the third Q-polynomial structure and let T'' be the type of the third expression of b_i, c_i. If two of T, T', T'' coincide, then $T = T' = T''$ by Lemma 7.9 and we get a contradiction by Lemma 7.8. By Lemma 7.9, it is impossible that T, T', T'' are all different. □

THEOREM 7.11. Let $\mathfrak{X}$ be a (P and Q)-polynomial association scheme of diameter $d \geq 34$ and let A_i $(i = 0, 1, \ldots, d)$ be the adjacency matrices of $\mathfrak{X}$. Suppose $\mathfrak{X}$ is not an ordinary n-gon. Then all the eigenvalues of A_i are rational integers.

<u>Proof</u>. We fix an ordering $A_0, A_1, \ldots, A_d$ so that $\mathcal{X}$ is a P-polynomial scheme with respect to this ordering. Notice that the eigenvalues of A_i are algebraic integers, and that if an algebraic integer is rational, then it is a rational integer. Since A_i is a polynomial in A_1 with rational coefficients, it suffices to show that every eigenvalue of A_1 is rational.

We fix an ordering $E_0, E_1, \ldots, E_d$ of the primitive idempotents so that $\mathcal{X}$ is a Q-polynomial scheme with respect to this ordering. Let the expression of the intersection numbers b_i, c_i be of type T in Theorem 5.1. Let K be the normal extension field generated by the parameters of Theorem 5.1 over $\mathbb{Q}$ (Lemma 7.4). For any Galois automorphism of K over $\mathbb{Q}$, $\mathcal{X}$ is also a Q-polynomial scheme with respect to $E_0^\sigma, E_1^\sigma, \ldots, E_d^\sigma$, where E_i^σ is the matrix obtained by applying σ to E_i entry-wise. Since b_i, c_i are fixed by σ, the σ-images of the original parameters give an expression of b_i, c_i of the same type T with respect to $E_0^\sigma, E_1^\sigma, \ldots, E_d^\sigma$.

If $T \neq I$, Lemma 7.9 tells us that $E_i^\sigma = E_i$. By the correspondence of Lemma 7.3, $\theta_i^\sigma = \theta_i$. Thus, the eigenvalues θ_i of A_1 are rational.

If $T = I$, we have $q^\sigma = \pm q$ or $\pm \frac{1}{q}$ by the same argument as in the proof of Proposition 7.7. By Lemma 7.8, we get $E_i^\sigma = E_i$ and hence $\theta_i^\sigma = \theta_i$. □

REMARKS. (4) Let $\mathcal{X} = (X, \{R_i\}_{0 \leq i \leq d})$ be a (P and Q)-polynomial scheme which has double P-polynomial structures. For a fixed Q-polynomial structure, let the dual intersection matrix be

$$B_1 = \begin{Bmatrix} * & c_1^* & \cdots & c_d^* \\ 0 & a_1^* & \cdots & a_d^* \\ b_0^* & b_1^* & \cdots & * \end{Bmatrix}.$$

Then Theorem 7.10 holds for b_i^* and c_i^*, exchanging (h,s) and (h^*,s^*) and exchanging the types IIA and IIB.

(5) Each of (i), (ii), (iii), (iv) in Theorem 7.10 has the following examples.

(i) ${}^2A_{2d-1}(p^f)$ $\quad (q = -p^f,\ s^* = 0)$:

$$b_i = -p^{(2d+1)f}(1 - p^{(-2d+2i)f})/(1 - p^{2f}),$$

$$c_i = (1 - p^{2if})/(1 - p^{2f}).$$

(ii) $H(d,2)$ $\quad (d = \text{even})$.

(iii) The bipartite half of $H(2e+1,2)$
$(s^* = -e - \frac{3}{2})$:

$$b_i = (e - i)(2e + 1 - 2i),$$

$$c_i = i(2i - 1).$$

(iv) The bipartite half of $H(2e+1,2)$
$(s^* = -2e - 2)$:

$$b_i = 2e + 1 - i,$$

$$c_i = i.$$

The dual version of Theorem 7.10(i) has no examples. $H(d,2)$ $(d = \text{even})$ and the bipartite half of $H(2e+1,2)$ are also examples for the dual version of Theorem 7.10(ii) and (iii), (iv), because they are self-dual, i.e., the first and second eigenmatrices coincide or equivalently they have certain expressions in Theorem 5.1 with $s = s^*$. In addition to them, $J(2d+1,d)$ is an example of the dual version of Theorem 7.10(iv).

(6) Let $\mathfrak{X} = (X,\{R_i\}_{0 \le i \le d})$ be a (P and Q)-polynomial scheme and let b_i, c_i $(0 \le i \le d)$ be the intersection numbers for a fixed P-polynomial structure. Assume the P-polynomial structure is antipodal. Let b_i', c_i' $(0 \le i \le [\frac{d}{2}])$ be the intersection numbers of the antipodal quotient. Then

$$b_i' = b_i \quad \text{and} \quad c_i' = c_i \qquad (0 \le i \le [\tfrac{d}{2}] - 1).$$

The expression of b_i', c_i' in Theorem 5.1 is different from that of b_i, c_i $(0 \le i \le [\frac{d}{2}] - 1)$, because the numerator of the expression of b_i' becomes zero for $i = [\frac{d}{2}]$. So we have two different expressions of b_i, c_i for $0 \le i \le [\frac{d}{2}] - 1$. If d is sufficiently large, say $d \ge 68$, the argument of this section is valid up to Theorem 7.10 with slight modifications and we get the following theorem.

THEOREM 7.12. Let $\mathfrak{X} = (X,\{R_i\}_{0 \le i \le d})$ be a (P and Q)-polynomial scheme with respect to an ordering $A_0,A_1,A_2,\ldots$ of adjacency matrices and an ordering $E_0,E_1,E_2,\ldots$ of primitive idempotents. Assume the P-polynomial structure is antipodal with diameter sufficiently large, say $d \ge 68$. Then the parameters in Theorem 5.1 are one of the following (i)-(vi), where (a) is the parameters for $\mathfrak{X}$ with respect to the orderings $A_0,A_1,A_2,\ldots$ and $E_0,E_1,E_2,\ldots$ and (b) is the parameters for the antipodal quotient of $\mathfrak{X}$ with respect to the orderings A_0+A_d, A_1+A_{d-1}, A_2+A_{d-2}, $\ldots$ and $E_0,E_2,E_4,\ldots$:

(i) (a) $(q;\ r^2,s^*;\ -s^*,r,-r)$ for

$$(q;\ s,s^*;\ r_1,r_2,r_3) \quad \text{(I)},$$

(b) $(q^2;\ \frac{r^2}{q}, s^{*2};\ r^2, -\frac{s^*}{q}, -s^*)$ for

$(q;\ s, s^*;\ r_1, r_2, r_3)$ (I).

(ii) (a) $(-2d-2;\ s^*, -d-1)$ for $(s;\ r_1, r_2)$ (IIA),

(b) $(-d-\frac{3}{2}, s^*;\ \frac{s^*}{2}, \frac{s^*-1}{2}, -d-1)$ for

$(s, s^*;\ r_1, r_2, r_3)$ (II).

(iii) (a) $(s;\ -d-1, \frac{s}{2})$ for $(s;\ r_1, r_2)$ (IIA),

(b) $(\frac{s-1}{2}, -d-1;\ -\frac{d+1}{2}, -\frac{d+2}{2}, \frac{s}{2})$ for

$(s, s^*;\ r_1, r_2, r_3)$ (II).

(iv) (a) $(-2, -1, -d-1)$ for $(s, \frac{r}{s^*}, r_1)$ (IIC),

(b) $(-d-1;\ -\frac{d+2}{2}, -\frac{d+1}{2})$ for

$(s^*;\ r_1, r_2)$ (IIB).

(v) (a) $(2(d+1), -s^*;\ s^*, -d-1, d+1)$ for

$(s, s^*;\ r_1, r_2, r_3)$ (III),

(b) $(s^*;\ \frac{s^*-1}{2}, -d-1)$ for $(s^*;\ r_1, r_2)$ (IIB).

(vi) (a) $(-2r, d+1;\ -d-1, r, -r)$ for

$(s, s^*;\ r_1, r_2, r_3)$ (III),

(b) $(-d-1;\ -\frac{d+2}{2}, r)$ for $(s^*;\ r_1, r_2)$ (IIB).

Since the proof is quite similar to that of Theorem 7.10, we omit it. In this proof, the following argument is useful. Let θ_i^* and $\theta_i^{*\prime}$ be the dual eigenvalues of $\mathfrak{X}$ and the antipodal quotient respectively. Since

$$E_2 = \frac{1}{n} \sum_{i=0}^{[\frac{d}{2}]} \theta_i^{*\prime} \frac{1}{p}(A_i + A_{d-i})$$

with $n = |X|$, $p = 1 + k_d$, we get $\frac{1}{p}\theta_i^{*\prime} = q_2(i)$ $(0 \leq i \leq [\frac{d}{2}] - 1)$. Since $q_2(i) = v_2^*(\theta_i^*) = \frac{1}{c_2^*}(\theta_i^{*2} - a_1^*\theta_i^* - b_0^*)$ and $a_1^* = 0$ by the bipartite property of the Q-polynomial structure, the following useful identity holds:

$$\theta_i^{*\prime} = \frac{p}{c_2^*}(\theta_i^{*2} - b_0^*).$$

For (P and Q)-polynomial schemes with Q-polynomial structures antipodal, Theorem 7.12 holds, exchanging s and s^* and exchanging the types IIA and IIB. Since imprimitive (P and Q)-polynomial schemes have either antipodal P-polynomial structure or antipodal Q-polynomial structure, they are contained in Theorem 7.12 or in its dual version.

The parameters in (b) of Theorem 7.12 are real ones for the antipodal quotients but fake ones for the original association schemes. For example, it seems as if $J(2k,k)$ had double Q-polynomial structures with the second expression of b_i, c_i given by $(s, s^*; r_1, r_2, r_3) = (-k - \frac{3}{2}, -k - 1; -\frac{k}{2} - 1, -\frac{k}{2} - \frac{1}{2}, -k - 1)$ of type II. But these are fake parameters which violate the condition $\theta_d^* \neq \theta_0^*$. Similarly $H(d,2)$ has fake parameters $(s, s^*; r_1, r_2) = (-4, -d - 1; -\frac{d}{2} - 1, -\frac{d}{2} - \frac{1}{2})$ of type IIB. If we take the antipodal quotients, the intersection matrices become the left top quarters of the original ones. So the original parameters become fake ones which violate the condition $b_{[\frac{d}{2}]} = 0$ for the antipodal quotients, and the fake parameters are promoted to the real ones.

The parameters of the Askey-Wilson polynomials in Theorem 5.1 satisfy the following property, if they come from (P and Q)-polynomial schemes:

(i) b_i, c_i, k_i, m_i are positive integers, where

$$k_i = b_0 b_1 \dots b_{i-1}/c_1 c_2 \dots c_i$$

and

$$m_i = b_0^* b_1^* \dots b_{i-1}^*/c_1^* c_2^* \dots c_i^* .$$

(ii) b_i^*, c_i^* are positive rational numbers.

(iii) $b_0 \geq b_1 \geq b_2 \geq \dots$ and $c_0 \leq c_1 \leq c_2 \leq \dots$.

These are called <u>feasibility</u> conditions. Besides the feasibility, there is a general condition: $\theta_i \neq \theta_j$ $(i \neq j)$, $\theta_0 > \theta_i \geq -\theta_0$ $(i \neq 0)$, $\theta_i^* \neq \theta_j^*$ $(i \neq j)$ and $\theta_0^* > \theta_i^* \geq -\theta_0^*$ $(i \neq 0)$, coming from Perron-Frobenius' theorem.

PROBLEMS. (1) Are ${}^2A_{2d-1}(p^{\bar{2}})$, $H(d,2)$ $(d = \text{even})$, and the bipartite half of $H(2e+1,2)$ the only (P and Q)-polynomial schemes with double Q-polynomial structures and with large diameters? Similarly classify (P and Q)-polynomial schemes with double P-polynomial structures.

(2) If a (P and Q)-polynomial scheme has an expression of b_i, c_i of type III in Theorem 5.1, then is it $J(v,d)$ or $H(d,2)$? Many problems of this type can be asked.

CONJECTURE. Let $\mathfrak{X}$ be a (P and Q)-polynomial scheme with large diameter. Suppose $\mathcal{X}$ has an expression of b_i, c_i of type I in Theorem 5.1. Then

(1) $s^* = 0$.

(2) q is a rational integer. Moreover $q = \pm p^f$ with p a prime number for some $f \in \mathbb{Z}$.

(3) r_i is a rational integer. Moreover $r_i = 0$ or $r_i = \pm p^\ell$ with p the same prime number as in (2) for some $\ell \in \mathbb{Z}$.

PROBLEM. (4) Determine all the feasible parameters in Theorem 5.1.

3.8 A REVIEW OF THE CHARACTERIZATION PROBLEM OF KNOWN (P AND Q)-POLYNOMIAL SCHEMES BY THE PARAMETERS

One of the main goals of Part I is to start the classification of (P and Q)-polynomial association schemes. At the end of section 3.6, we proposed the following 3 steps for this problem:

(i) Determination of the orthogonal polynomials which have the (P and Q)-polynomial property.

(ii) Determination of the feasible parameters of (P and Q)-polynomial association schemes.

(iii) The characterization of the known (P and Q)-polynomial association schemes by the parameters.

Step (i) was completed by D. Leonard in the form of Theorem 5.1. A partial solution of Step (ii) is given in section 3.7. In this section, we discuss the last step, assuming that Step (ii) will be successfully completed in the near future. (Even if Step (ii) should not be completed, Step (iii) is of independent interest.)

The characterization problems of known important classes of association schemes (or distance-regular graphs, or their attached finite geometric structures) have a long history in combinatorics. There are many different philosophies and motivations of why and how to characterize these objects. Consequently, there are many characterizations other than that by the parameters, which we are interested in. But we believe that the characterization by the parameters is one of the most important. Here when we say the characterization by the parameters, we assume that all the parameters p^k_{ij} are given. In many cases, however, what we actually need is only some of them.

The characterization problem of class 2 association schemes (i.e., strongly regular graphs) has been an important topic in combinatorics. The readers are referred to Seidel (1979), Brower and van Lint (to appear), and others for a survey on this vast area.

The class 2 Hamming schemes $H(2,q)$ (the lattice graphs L_2) and Johnson schemes $J(v,2)$ (the triangular graphs T_2) were characterized by the parameters. (See Shrikhande (1959), Connor (1958), Chang (1959), Hoffman (1960), etc.) The results are as follows. $H(2,q)$ is unique except for $q = 4$ and there are exactly 2 non-isomorphic ones for $q = 4$. $J(v,2)$ is unique except for $v = 8$ and there are exactly 4 non-isomorphic ones for $v = 8$.

Since then, there have been many interesting researches, too many to be reviewed here, which also deal with other classes of association schemes, including those with large class numbers d.

The complete result for $H(n,q)$ was obtained by Egawa (1981) recently. The result is as follows. The association scheme $H(n,q)$ is uniquely characterized by the parameters if $q \neq 4$. For $q = 4$, there are exactly $[\frac{n}{2}] + 1$ non-isomorphic ones with the same parameters.

The result for $J(v,k)$ is not yet completed at the time of this writing, although there are many recent developments on this problem. The current best result available is as follows: The association scheme $J(v,k)$ $(k \leq \frac{v}{2})$ is unique if one of the following conditions holds:

(i) $v > 4k$ by Moon (1983b) (cf. Dowling (1969)),

(ii) $v = 2k + 1$ by Moon (1982),

(iii) $v = 2k + 2$ by Moon (unpublished),

(iv) $v = 3k$ and $3k + 1$ by Moon (1983a),

(v) $k = 3$ by Moon (1983a) and by P. Rolland (unpublished) (cf. Liebler (1977)).

It seems that these results are by no means best possible and should be improved for other values of (v,k). One test case is $J(8,4)$. This is the smallest open case and there seems to be a chance that $J(8,4)$ is not uniquely characterized by the parameters. (We believe that it will heavily depend on the case $J(8,4)$ whether all the $J(v,k)$ with $k \geq 4$ are uniquely characterized by the parameters or not.) Note that $J(8,4)$ is imprimitive and the antipodal factor graph is a strongly regular graph of 35 vertices. It is known that there are at least several thousands of non-isomorphic strongly regular graphs with the parameters of the antipodal factor graph of $J(8,4)$ (cf. Seidel (1979, page 178)). So it would be interesting to know which of these strongly regular graphs allow the antipodal coverings with the parameters of $J(8,4)$.

Bose and Lasker (1967) is one of the original papers dealing with the characterization of $H(2,q)$ and $J(v,2)$. As in this paper, one of the crucial steps to characterize association schemes (and graphs) is to show the existence of maximal cliques of the right size. This technique is called the Bose-Laskar argument. In order to use this argument, we usually need some restrictions on the parameters. For example, q must be sufficiently large compared with n in $H(n,q)$ and v must be sufficiently large compared with k in $J(v,k)$. Dowling (1969) used this argument to show that $J(v,k)$ is uniquely characterized if $v > 2k^2 - 2k + 4$. As is shown in Dowling (1969), once the maximal cliques of the right size are found, then they sometimes lead to the determination of the exact neighborhood structure of a point in the graph, and often this in turn leads to the characterization of the graph by geometric argument. It seems that there are several different stages in the characterization of association schemes. We propose to distinguish the following:

(a) Geometric characterization. (The meaning of "geometric" is not unique, and several geometric characterizations are possible.)

(b) Characterization by the parameters, assuming the structure of the neighborhood of a point. (Here the neighborhood usually means the induced graph $x \cup \Gamma_1(x)$, i.e., the neighborhood of distance ≤ 1. Sometimes we consider the neighborhood of distance ≤ 2.)

(c) Characterization by the parameters, assuming the existence of maximal cliques of the right size.

(d) Characterization only by the parameters with some arithmetic restrictions on the parameters.

(e) Complete characterization only by the parameters.

Note that the characterization gets more difficult (i.e., stronger) as the alphabetical order proceeds. That is, characterization (b) is usually more difficult than and is usually completed through characterization (a), and characterization (c) is usually more difficult than and is usually completed through characterization (b), and so on. In this sense, characterization (e) is the most difficult and it is what we need when classifying (P and Q)-polynomial schemes. Geometric characterization (a) is usually far easier than that by the parameters (e), but it is extremely important as a starting point of the whole characterization.

Let us discuss how much was done in the characterizations of each family of (P and Q)-polynomial schemes listed in section 3.6.

Except for the trivial case (II)(0) (ordinary n-gon), H(n,q) (i.e., (II)(i)) is the only family listed in section 3.6 for which the characterization of the last Stage (e) was completed. J(v,k) is in Stage (d) as was just discussed. If one reads Dowling (1969), we will see that (c) ⇒ (b) and (b) ⇒ (a) are the important parts of the proof of his result, and that Stage (a) essentially means the argument in pages 261-262 in Dowling (1969).

Let us consider the generalized Johnson schemes $A_{v-1}(q)$, i.e., (I)(ii) in section 3.6. Since the underlying geometric structures of these association schemes are nothing but the finite projective spaces over GF(q), many different types of geometric characterizations are already available. For our purpose of the characterization of $A_{v-1}(q)$, the most relevant ones are those which use the set of k-dimensional subspaces and the set of (k - 1)-dimensional subspaces, because the graph (association scheme) consists of the k-dimensional subspaces of a v-dimensional vector space over GF(q) with two k-dimensional subspaces being adjacent if and only if their intersection is (k - 1)-dimensional. In fact a characterization of this type has been carried out by Ray-Chaudhuri and Sprague (1976). Using this geometric characterization, Sprague (1978) gave the characterization of Stage (d) for $A_{v-1}(q)$ under the assumption $v > 3k$ which was necessary to use Bose-Lasker argument. No improvement of this result seems to be available at the time of this writing. We note that Cohen (1981b) gave other geometric characterizations of these structures. Actually, to find geometric characterizations of these and related structures is one of the main problems in finite geometries, and there is a long history of this line of research.

Recently Cohen (1981a) shed new light on distance-regular graphs, from a similar motivation to ours. The paper (Cohen (1981a)) is an excellent review on this topic, and we should like to recommend it to the reader. It contains detailed proofs that the examples listed in section 3.6 are distance-regular graphs, together with detailed geometric analysis of each example. Since this excellent review is already available, we have omitted the details of geometric and combinatorial analysis of each example in the list in section 3.6. (An extended version of Cohen (1981a) is planned by A. Brower, A. Cohen and A. Neumaier (to appear).)

As is explained in Cohen (1981a), some geometric characterizations are already obtained for all the families in (I) in section 3.6. For further reference on this topic, the reader is referred to Cohen (1981a), Camerom (1982).

Cohen and Cooperstein (to appear), Cooperstein (1977), Shult (1974, to appear), etc. It seems to be a very interesting question how much these geometric characterizations can be extended to those of Stages (b), (c), (d) and (e). To answer this is not an easy task, because the geometric conditions are in many cases much stronger than those on the parameters. For example, neat and convenient as they are, Pasch's axiom and the dual of Pasch's axiom (which are often used in geometric characterizations) are very strong conditions, and it is not easy at all to recover these axioms just from the conditions on the parameters, even if we assume some additional conditions such as (b), (c), (d).

We remark that the maximal cliques and the structure of the neighborhood of a point in the family of examples in (I)(iii) in section 3.6 (i.e., the dual polar spaces) are easily determined. In fact there is only one kind of maximal cliques, while there are two kinds of maximal cliques in (I)(i) and (ii). The sizes of the maximal cliques are:

I(i) (Johnson scheme) $J(v,k)$:

2 types of maximal clique sizes: $k+1$ and $v-k+1$

I(ii) (Generalized Johnson schemes):

2 types of maximal clique sizes: $\dfrac{q^{k+1}-1}{q-1}$

$$\frac{q^{v}-q^{k-1}}{q^{k}-q^{k-1}}$$

I(iii) (Dual polar spaces):

unique maximal clique size: $q^{e+1}+1$.

(These remarks are due to Joe Hemmeter.) This difference between the number of the types of maximal cliques seems to be an interesting fact, and perhaps it is because of this fact that the Bose-Laskar type argument works for (I)(i) and (ii) but not directly for (I)(iii).

For the families of association schemes in (II) in section 3.6, it seems that not much work has been done about geometric characterizations, except for (II)(0) (ordinary n-gons) and (II)(i) (Hamming schemes). Two of the few exceptions are Ray-Chaudhuri and Sprague (1979) and Sprague (1979) which treat (II)(ii) in the case $d = 3$. Sprague (1979b, 1981b) gave some geometric characterizations for general d (but with different axioms). (We think that these interesting and pioneering results are still preliminary and experimental, and that some other definite geometric characterizations are yet to come, even for (II)(ii), not to mention the other open cases (II)(iii), (iv) and (v).) Consequently, the characterizations of stages (b), (c), (d) and (e) are all open at the time of this writing. Since a lot of work is going to start on this problem, for example by some graduate students at OSU, some progress may be expected in the near future. In connection with this problem, Joe Hemmeter and Yiming Hong have determined the maximal cliques and the structure of the neighborhood of a point for all the families in (II) except for II(v). Results are summarized as follows:

II(i) (Hamming scheme)

unique maximal clique size: q

the number of maximal cliques: nq^{n-1}

II(ii) ($d \times n$ $(d < n)$ bilinear forms)

2 types of maximal clique sizes: q^n, q^d

the numbers: $\dfrac{q^{dn}(q^d - 1)}{q^n(q - 1)}$, $\dfrac{q^{dn}(q^n - 1)}{q^d(q - 1)}$

II(iii) ($n \times n$ alternating bilinear forms)

2 types of maximal clique sizes: q^3, q^{n-1}

the numbers: $q^{\frac{(n-3)(n+2)}{2}} \left[{n \atop 3}\right]_q$, $q^{\frac{(n-2)(n-1)}{2}} \left[{n \atop 1}\right]_q$

II(iv) ($n \times n$ hermitian forms)

unique maximal clique size: q

the number: $q^{\frac{n^2-2}{2}} \left[{n \atop 1}\right]_q$

II(v) (quadratic forms):

Not yet completely finished at this time. (There are at least several different types of maximal cliques.)

(In the above, $\left[{n \atop r}\right]_q = \prod_{i=0}^{r-1} (q^n - q^i)/(q^r - q^i)$.)

It seems that the Bose-Laskar type argument works for the family (II)(ii), if n is sufficiently large compared with d. However, the size of maximal cliques which is obtained by simply applying the Bose-Laskar argument is not the desired size (slightly smaller), and so some further argument will be necessary to overcome this difficulty. Also, even if we assume the condition on the size of maximal cliques, it is still not easy to translate it into reasonable geometric conditions. It seems that unlike the cases (II)(i) and (ii), the Bose-Laskar type argument does not work directly for families (II)(iii), (iv) and (v).

We believe that the characterization problems for families in (II) are very important and interesting.

3.9 NOTES TO CHAPTER III

The concept of distance-regular graphs is a generalization of that of strongly regular graphs, which is a distance-regular graph with diameter 2. It also gives a combinatorial setting for distance-transitive graphs, which were called permutation groups with maximal diameter in Higman (1967). The terminology of P-polynomial association schemes is due to Delsarte (1973b). (We sometimes abbreviate P-polynomial association schemes to P-polynomial schemes. There is no difference between these two terms.)

Distance-regular and transitive digraphs (directed graphs) were defined and their fundamental properties were studied by Damerell (1981) (see also Lam (1980)). Not many such examples are known. Bannai-Cameron-Kahn (1981) proved the nonexistence of distance-transitive digraphs with odd girth > 3. (The even girth case is still open.) The existence problem of nontrivial distance-regular digraphs is open. If we define the P-polynomial structure for nonsymmetric association schemes in the natural way, i.e., $A_i = v_i(A_1)$ for some polynomial $v_i(x)$ of degree i, then the concept coincides with that of Damerell (1981).

As orthogonal polynomials are, in one definition, a set of polynomials satisfying certain 3-term recurrence relations, the theory of distance-regular graphs (and of Q-polynomial schemes) is closely related to orthogonal polynomials. For example, the multiplicities m_i are nothing but the Christoffel numbers. (Readers are referred to Szegö (1939), Erdélyi, et al. (1953), etc. for the definition of Christoffel numbers and numerical quadretures.)

QUESTION. Can orthogonal polynomials be related to distance-regular digraphs in any sense?

Yes. But the orthogonal polynomials relevant are not those usual orthogonal polynomials on a real line but other kinds of (orthogonal) polynomials which are more closely related to the orthogonal polynomials on the unit circle in the complex plane. We hope to discuss this furthermore on some other occasion.

The unimodal property of the valencies k_i of distance-regular graphs seems interesting, though the proof is very easy, and this type of unimodality is expected in many places of combinatorics (see, for example, Stanley (1982)). The unimodality of the multiplicities m_i of distance-regular graphs is expected to hold in general but not always true. (For example, the 3-fold covering of $K_{3,3}$ with 18 vertices has multiplicities $m_0 = 1$, $m_1 = 6$, $m_2 = 4$, $m_3 = 6$, and $m_4 = 1$.)

The unimodality of Christoffel numbers is well known to hold for Hermite, Laguerre and Jacobi (including Gegenbauer) polynomials (see Szegö (1939)). Bannai and Damerell (unpublished manuscript) proved the following.

THEOREM. Let $f_i(x)$ $(i = 0, 1, \ldots, n)$ be the orthogonal polynomials defined by

$$f_i(x) = H_i(x) \qquad (i = 0, 1, \ldots, n-1)$$

and

$$f_n(x) = H_n(x) + aH_{n-1}(x) + bH_{n-2}(x) \qquad (b \le 0),$$

where $H_i(x)$ are the Hermite polynomials. Then the Christoffel numbers m_i of $\{f_i\}_{0 \le i \le n}$ are unimodal. (See below for the definition of m_i.)

The polynomials $\{f_i\}_{0 \le i \le n}$ are orthogonal with respect to the integral defined by a point-mass measure which has positive masses exactly on $\{\theta_0, \theta_1, \ldots, \theta_n\}$ $(\theta_0 > \theta_1 > \ldots > \theta_n)$. Then m_i is the mass at θ_i and calculated by

$$m_i = \frac{n}{\sum_{\nu=0}^{n} \frac{1}{k_\nu} f_\nu(\theta_i)^2},$$

where

$$n = \sum_{\nu=0}^{n} k_\nu$$

and

$$k_\nu = \frac{b_0 b_1 \cdots b_{\nu-1}}{c_1 c_2 \cdots c_\nu}$$

with b_j, c_i the coefficients of the 3-term recurrence for $\{f_i\}_{0 \le i \le n}$.

CONJECTURE. For Laguerre or Jacobi polynomials (including Gegenbauer), define the orthogonal polynomials $\{f_i\}_{0 \le i \le n}$ as above (confining to the case $b = 0$ if necessary). Then the Christoffel numbers $m_0, m_1, \ldots, m_n$ are unimodal in the following sense: either

(1) $m_0 \le m_1 \le \cdots \le m_i$ and $m_{i+1} \ge m_{i+2} \ge \cdots \ge m_n$ for some i

or

(2) $m_0 \ge m_1 \ge \cdots \ge m_i$ and $m_{i+1} \le m_{i+2} \le \cdots \le m_n$ for some i.

For some cases, e.g., Tchebycheff polynomials, the conjecture is proved to be true.

Higman (1967) (based on Feit-Higman (1964)) gives another method to calculate the multiplicities $m(\theta_i)$ of distance-regular graphs. (Cf. Higman (1967), (1975).)

Proposition 1.2(iii), (iv) is due to Biggs (1976). See Gardiner (1978) and Terwilliger (1983) for further restrictions on the parameters a_i, b_i, c_i.

The polynomials $v_i(x)$ for $J(v,k)$ seem to have been first calculated by Ogasawara (1965), S. Yamamoto-Fujii-Hamada (1965) and K. Yamamoto (1965). The treatment of $H(n,q)$ and $J(v,k)$ we chose in section 3.2 is different from that of Delsarte (1973b). The calculation of $u_i(x)$ for $J(v,k)$ is based on Dunkl (1979).

The present proof of Theorem 3.1 (for $d \ge 3$) is a combination of Damerell (1973) and Bannai-Ito (1973). This kind of argument works for wider classes of graphs. For other types of extremal graphs which are not distance-regular but algebraically close to them, see Bannai-Ito (1981) and Biggs-Ito (1980).

Moore geometries are interesting objects of finite geometries in connection with Buekenhout's description of sporadic simple groups in terms of Buekenhout's diagrams (see Buekenhout (1979, 1982)).

There are many different proofs of Feit-Higman's theorem on generalized n-gons (see Higman (1967, 1974), Kilmoyer-Solomon (1973), Stanton (1982), etc.). Let us discuss the difference between the method of Feit-Higman (1964) and Kilmoyer-Solomon (1973). Roughly speaking, Feit-Higman (1964) (and Higman (1967)) decomposed the character of the permutation representation on the set of points (i.e., on the quotient space by the maximal parabolic subgroup), whereas Kilmoyer-Solomon (1973) did it on the set of flags (incident point-line pairs, i.e., the quotient space by the Borel subgroup). We feel that Feit-Higman's method is philosophically simpler but involves more complicated calculation than Kilmoyer-Solomon's. Although the decomposition of the representation on points contains less information in general than that on flags, it is enough to prove the result.

Ott (1981b) defines "Hecke algebra" for geometries. His definition is a generalization of the idea of Iwahori (1964) and Kilmoyer-Solomon (1973) (see Curtis-Iwahori-Kilmoyer (1971)). Ott's Hecke algebra is a certain subalgebra of the Bose-Mesner algebra of the association scheme attached to the permutation representation on flags. In general, the representations of Ott's Hecke algebra are decomposed more roughly than the Bose-Mesner algebra on flags and it is sometimes possible to calculate the decompositions explicitly. (Ott (1981a) announced the complete determination of Moore geometries of diameter $d \geq 3$. As the details of his proof are not available at the present time, we are not yet completely convinced that his method works, in particular for the difficult cases of $d = 3, 4$ in Fuglister-Damerell-Geogiacodis' approach. As is sometimes the case, the decomposition of Ott's Hecke algebra might give no more useful information than the permutation representation on points (i.e., the quotient by the maximal parabolic subgroup), which is what Fuglister-Damerell-Georgiacodis used and where they found trouble for small d; for $d = 3$ and 4.) Fuglister (1977) exploited very deep number theoretical result by Mordell on diophantine equations for $d = 3$. In the case $d = 4$ by Fuglister (1982), the argument is also number theoretical and involved. For large d, Damerell-Georgiacodis used the Newton-polygon method to study the factorization (over 2-adic integers) of the minimal polynomial of the adjacency matrix A_1. Ott (1981a) contains many interesting ideas and we recommend it to the reader.

Our main concern in Chapter III has been not distance-transitive but distance-regular graphs. If we have recourse to the classification of finite simple groups, we believe that the classification of distance-transitive graphs is essentially possible and perhaps a matter of time, though actual finishing is not easy at all.

For rank 3 group, see Kantor-Liebler (1982), and for multiplicity-free subgroups of certain groups (in particular, symmetric and alternating groups), see Saxl (1981).

The concept of distance-transitive and distance-regular graphs are naturally defined for locally finite connected graphs of infinite diameter $(d = \infty)$. MacPherson (1982), by using a result of Dunwoody (1982) concerning cuts of infinite graphs, determined all distance-transitive graphs of infinite diameter:

THEOREM. (MacPherson (1982)) If Γ is an infinite distance-transitive graph of finite valency, then

(i) there are no non-degenerate cycles in Γ of length greater than 3, and

(ii) there exist integers s, t $(s \geq 2, t \geq 1)$ such that the vertices adjacent to any vertex carry a subgraph consisting of s disjoint copies of the t-vertex complete graph.

The graph with the properties (i), (ii) is clearly unique for each pair of s, t. (If $t = 1$, then Γ is the regular tree of valency s.) The intersection matrix is

$$B = \begin{Bmatrix} * & 1 & 1 & \cdots \\ 0 & t-1 & t-1 & \cdots \\ st & (s-1)t & (s-1)t & \cdots \end{Bmatrix}$$

(Note that if we replace s, t by $t+1, s,$ respectively, the matrix B becomes that of an "infinite Moore geometry" (cf. section 3.3).)

PROBLEM. Can one determine all infinite distance-regular graphs of finite valency? Are they all distance-transitive?

It is easy to see that the distance-regular graph with intersection matrix B given above is unique and hence distance-transitive. The classification of infinite distance-regular graphs seems possible and much easier than that of finite ones. Notice that the intersection numbers b_i, c_i of infinite distance-regular graphs cannot be expressed by the parameters of Askey-Wilson polynomials as in Theorem 5.1. This means the non-existence of "infinite (P and Q)-polynomial association schemes".

It is interesting to study when all p_{ij}^k are non-negative for a C-algebra of P-polynomial type with a given $B = (p_{ij}^k)$:

$$B = \begin{Bmatrix} * & 1 & c_2 & \cdots & c_{d-1} & c_d \\ 0 & a_1 & a_2 & \cdots & a_{d-1} & a_d \\ k & b_1 & b_2 & \cdots & b_{d-1} & b_d \end{Bmatrix}$$

with $b_i > 0$, $c_i > 0$, $a_i \geq 0$ (d may be ∞). In the field of orthogonal polynomials, this problem has been also interested in and extensively studied. It is known that many classical orthogonal polynomials have p_{ij}^k all non-negative. The reader is referred to Szegö (new editions), Askey (1975), Gasper (1975, and other papers), Lasser (to appear) and the papers referred to in Lasser (to appear). (Lasser considers the hypergroups on $\mathbb{N}$, which is essentially C-algebras with d infinite.) Sometimes, the existence of combinatorial structures such as P-polynomial association schemes implies the non-negativeness of all the p_{ij}^k.

The following is a partial answer to this non-negativeness problem:

THEOREM. (Askey (1975)) Let $\mathfrak{A}$ be a C-algebra of P-polynomial type with B given above and $d = \infty$. Then p_{ij}^k are all non-negative, if $c_{i+1}b_i \geq c_i b_{i-1}$ and $a_{i+1} \geq a_i$ $(i = 1, 2, \ldots)$.

Theorem 4.2 is based on Bannai-Bannai (1980) (and partly on subsequent unpublished work). This result was also obtained by A. Gardiner independently (unpublished). We hope that this kind of technique might be useful to study association schemes with multiple Q-polynomial structures. The difficulty of the corresponding problem for Q-polynomial structures lies in the fact that so far, we do not have the unimodality theorem for the m_i's of Q-polynomial schemes. For example, if we assume $b_i \leq b_{i-1}$ and $c_i \leq c_{i+1}$, it is easily proved by imitating the argument in Theorem 6.1 and Theorem 6.2 that imprimitive C-algebras of P-polynomial type are antipodal or bipartite, which is a stronger version of Theorem 6.2 and a generalization of Theorem 6.1. Ordinary n-gons are the only known examples of triple Q-polynomial structures.

The first version of Theorem 5.1 was that all the parameters of (P and Q)-polynomial schemes are described by using only 7 parameters (including the diameters) (Leonard (to appear)). R. Askey noticed that a class of orthogonal polynomials, which we now call the Askey-Wilson polynomials, has 5 parameters and satisfies the duality property of Leonard (cf. Theorem 5.1), and he also noticed that Leonard's polynomials are described essentially by 5 parameters as well. With this notice, Leonard succeeded in proving that the orthogonal polynomials with the dual property are Askey-Wilson polynomials, allowing the limiting cases (Leonard (1982)).

Theorem 5.1 is based on Leonard (1982). There are some improvements in the treatment of Theorem 5.1:

(i) There is no condition on the diameter d for the theorem to hold. (In the original paper, it was assumed that d is not too small.)

(ii) Each limiting case of Askey-Wilson polynomials is precisely described so that we can choose free parameters for (P and Q)-polynomial association schemes.

It is interesting to remark that we cannot easily eliminate the possibility that the number q is not a real number. (If q is not real, then we have $|q| = 1$.) Except for ordinary n-gons, we have no examples of (P and Q)-polynomial schemes in which q is not real.

CONJECTURE. If $q \notin \mathbb{R}$ in a (P and Q)-polynomial association scheme $\mathfrak{X}$, then $\mathfrak{X}$ is an ordinary n-gon.

We remark that if a (P and Q)-polynomial scheme $\mathfrak{X}$ with large diameter is not an ordinary n-gon, then all the $p_j(i)$ and $q_j(i)$ are in the rational field $\mathbb{Q}$, which is a main consequence of section 3.7. This implies that q is at most in a quadratic field over $\mathbb{Q}$. Except for ordinary n-gons, we do not know any example of (P and Q)-polynomial schemes with diameter ≥ 3 in which q is not a rational number. (In known examples, q (or $1/q$) is always an integer, furthermore a prime power.)

A hypergeometric series is $\Sigma\, a_n$ with a_{n+1}/a_n a rational function of n, and a basic hypergeometric series is $\Sigma\, a_n$ with a_{n+1}/a_n a rational function of q^n. It seems that the functions which are expressed by using hypergeometric or basic hypergeometric series form a fairly wide class but are important and interesting. They include all classical orthogonal polynomials and many other important orthogonal polynomials. There has been a lot of work to find new orthogonal polynomials which are expressed by hypergeometric or basic hypergeometric series. One important family of such orthogonal polynomials were recently found by J. Wilson (1978) in his thesis (see also Askey and Wilson (1979)). His polynomials are

$$p_n(\lambda(x)) = {}_4F_3\left(\begin{matrix} -n,\ n+\alpha+\beta+1,\ -x,\ x+\gamma+\delta+1 \\ \alpha+1,\ \beta+\delta+1,\ \gamma+1 \end{matrix}\ ;1\right),$$

where $\lambda(x) = x(x+\gamma+\delta+1)$ and one of $\alpha+1$, $\beta+\delta+1$, $\gamma+1$ is $-N$. The $p_n(\lambda(x))$ $(0 \leq n \leq N)$ are polynomials of degree n in $\lambda(x)$ and orthogonal on $x = 0,1,\ldots,N$. The orthogonality relation is equivalent to Racah's orthogonality for functions that are usually called Racah coefficients or (Wigner's) 6-j symbols.

Askey-Wilson polynomials (or q-Racah polynomials) are q-analogues of them, defined as

$$p_n(\mu(x);\, a,b,c,d;\, q) = p_n(\mu(x))$$
$$= {}_4\varphi_3\left(\begin{matrix} q^{-n},\ q^{n+1}ab,\ q^{-x},\ q^{x+1}cd \\ aq,\ bdq,\ cq \end{matrix};\, q,\, q\right),$$

where $\mu(x) = q^{-x} + q^{x+1}cd$. The $p_n(\mu(x))$ are orthogonal polynomials of degree n in the variable $\mu(x)$. Many important properties of them are discussed in Askey and Wilson (1979, to appear), etc., as well as how they include other important orthogonal polynomials as their special cases or limiting cases. It is a remarkable coincidence that Askey-Wilson polynomials already existed, just found in the field of orthogonal polynomials, when Leonard grasped them through the study of (P and Q)-polynomial association schemes.

It seems that the class of Askey-Wilson polynomials is the widest known of orthogonal polynomials which are expressed by hypergeometric series or basic hypergeometric series. So far it seems difficult to find such examples furthermore.

The following interesting observation by Koornwinder (1981, 1982) is useful to understand Askey-Wilson polynomials in relation to their special cases:

(i) Krawtchouk polynomials ${}_2F_1$ have orthogonality equivalent to the unitaryness of unitary representations of SU(2).

(ii) Hahn polynomials ${}_3F_2$ have orthogonality equivalent to that of the Clebsch-Gordan coefficients for SU(2) or 3-j symbols.

(iii) Racah polynomials (balanced $_4F_3$) have orthogonality equivalent to that of the Racah coefficients for SU(2) or 6 - j symbols.

For details, the reader is referred to Koornwinder (1981, 1982). Perhaps it is difficult to find further orthogonal polynomials by considering "higher" Racah coefficients for SU(2) or "further" Racah coefficients for other groups. (Racah coefficients are defined not only for SU(2) but also for other groups, cf. Derome and Sharp (1964).) Dunkl (1980) gives another interpretation of Askey-Wilson polynomials (balanced $_4\varphi_3$).

Stanton (1980) calculated those polynomials which are attached to finite Chavalley groups acting on the quotients by certain parabolic subgroups and those which are related to type (II) of section 3.6. All the examples were of type at most $_3F_2$ or $_3\varphi_2$, and $_4F_3$, $_4\varphi_3$ did not appear. This may suggest the nonexistence of (P and Q)-polynomial schemes which actually have $_4F_3$ or $_4\varphi_3$.

The list presented here was first compiled by Bannai in 1979, and announced at several conferences: Oberwolfach (May, 1980), AMS Ann Arbor meeting (August, 1980), etc. The compilation of the list was inspired by the work of Stanton (1980), in which the (P and Q)-structures are proved to exist for most of the members in the present list. The quadratic form examples of Egawa (to appear) were found while this list was being compiled. Actually, it was conjectured from the list under compilation that there were (P and Q)-polynomial schemes related to quadratic forms in the way of gluing some relations, and Egawa succeeded in proving it.

Cohen (1981a), partially inspired with the talk of Bannai at Oberwolfach, also obtained the same list (of distance-regular graphs with large diameters). His treatment is very thorough, and his geometric explanation is very precise and clear. We recommend the reader to be referred to Cohen (1981a). Since this paper is already available, we have discussed these examples in relation to the Askey-Wilson polynomials rather than the detailed description of each. An extended version of Cohen (1981a) is being prepared by A. Brower, A. Cohen and A. Neumaier (to appear).

The present list is just the core of (P and Q)-polynomial schemes. We believe that all (P and Q)-polynomial schemes are obtained in the sense of Remark (1)(i), (ii) in section 3.6. Recently, J. Hemmeter (to appear) determined distance-regular graphs Γ whose bipartite half is one of the known distance-regular graphs in the list of section 3.6 except for (I)(iii)' and II(v) (cf. Problem (2)(a) in section 3.6). Given a distance-regular graph Σ, to determine the antipodal covering Γ is essentially a very special case of the characterization problem by the parameters (see section 3.8). The second problem (cf. Problem (2)(b) in section 3.6) is not yet completely solved at the time of this writing. However, we have that the neighborhood of distance at most $d - 1$ of a point in Π is the same as that of a point in Γ. Since this condition is really strong, we expect that the second part of the question should be solved without difficulty once a geometric characterization is obtained for the association scheme Γ. Actually, we think that all the relatives which can be obtained by this method may be already listed in the core examples in section 3.6. (Note that the q-analogue of $2.O_k$ is missing in the P-polynomial scheme list of Cohen (1981a).)

We should like to finish Chapter III with our dream. The scope of the algebraic approach described in these lecture notes should not be limited to (P and Q)-polynomial schemes (or P-polynomial or Q-polynomial schemes). These classes are, very roughly speaking, nothing but "finite symmetric spaces of rank 1". Clearly, we should have theorems for higher ranks, and we should have theories for more general association schemes, in particular for commutative primitive association schemes. So, the classification of (P and Q)-polynomial schemes is by no means the final goal but the start of many really interesting problems. As we mentioned in Chapter II, any finite simple group gives a commutative primitive association scheme (of very special type), and our dream is to classify the latter structure algebraically and combinatorially. Then, it might make it possible to give an alternative approach to the classification of finite simple groups along the line of algebraic combinatorics.

REFERENCES

1. K. Asano and H. Nagao (1965): 群論 (Group Theory), Iwanani, Tokyo.

2. M. Aschbacher (1971): The non-existence of rank three permutation groups of degree 3250 and subdegree 57, J. Algebra 19, 538-540.

3. R. Askey (1975): Orthogonal polynomials and special functions, Regional Conference Series in Applied Mathematics, vol. 21, SIAM, Philadelphia.

4. R. Askey (to appear): Orthogonal polynomials old and new and some combinatorial connections, (preprint).

5. R. Askey and M. E. H. Ismail (to appear): Recurrence relations, continued fractions and orthogonal polynomials, (preprint).

6. R. Askey and J. Wilson (1979): A set of orthogonal polynomials that generalize the Racah coefficients or 6-j symbols, SIMA J. Math Anal. 10, 1008-1016.

7. R. Askey and J. Wilson (to appear): Some basic hypergeometric orthogonal polynomials that generalize Jacobi polynomials, Memoirs Amer. Math. Soc.

8. L. Babai (1981): On the order of uniprimitive permutation groups, Annals of Math. 113, 553-568.

9. L. Babai, P. J. Cameron, M. Deza and N. M. Singhi (1981): On sharply edge-transitive permutation groups, J. Algebra 73, 573-585.

10. W. N. Bailey (1964): Generalized hypergeometric series, Stechert-Hafner Serivce Agency N.Y. (original edition 1935, Cambridge Univ. Press).

11. E. Bannai (1977a): On perfect codes in the Hamming schemes H(n,q) with q arbitrary, J. of Combinatorial Theory (A), 23, 52-67.

12. E. Bannai (1977b): Codes in bipartite distance regular graphs, J. London Math. Soc. 16, 197-202.

13. E. Bannai (1977c): On tight designs, Quart. J. Math. (Oxford), 28, 433-448.

14. E. Bannai (1979a): On tight spherical designs, J. of Combinatorial Theory (A), 26, 38-47.

15. E. Bannai (1979b): On some spherical t-designs, J. of Combinatorial Theory (A) 26, 157-161.

16. E. Bannai (1979c): 代数的組合せ論 (Algebraic Combinatorics), Sugaku (Mathematics, a publication of Math. Soc. of Japan), 31, 126-143.

17. E. Bannai (1980a): On the weight distribution of spherical t-designs, Europ. J. of Combinatorics 1, 19-26.

18. E. Bannai (1980b): Orthogonal polynomials, algebraic combinatorics and spherical t-designs, Proc. Symp. in Pure Math. 37, Amer. Math. Soc., 465-468.

19. E. Bannai (1982a): On s-distance subsets in real hyperbolic space, Hokkaido Math. J. 11, 201-204.

20. E. Bannai (1982b): Tannaka-Krein duality for association schemes, Linear Algebra and its Applications, 46, 83-90.

21. E. Bannai (preprint): Spherical t-designs which are orbits of finite groups.

22. E. Bannai and E. Bannai (1980): How many P-polynomial structures can an association scheme have? Europ. J. of Combinatorics, 1, 289-298.

23. E. Bannai and E. Bannai (1981): An upper bound for the cardinality of an s-distance subset in real Euclidean space, Combinatorica (Hungary), 1, 99-102.

24. E. Bannai, E. Bannai and D. Stanton (to appear): An upper bound for the cardinality of an s-distance subset in real Euclidean space, Combinatorica (Hungary).

25. E. Bannai, A. Blokhuis, P. Delsarte and J. J. Seidel (to appear): An addition formula for hyperbolic space, J. of Combinatorial Theory (A).

26. E. Bannai, P. J. Cameron and J. Kahn (1981): Non-existence of certain distance-transitive digraphs, J. of Combinatorial Theory (B), 31, 105-110.

27. E. Bannai and R. M. Damerell (1979): Tight spherical designs, I, J. Math. Soc. Japan, 31, 199-207.

28. E. Bannai and R. M. Damerell (1980): Tight spherical designs, II, J. London Math. Soc. 21, 13-30.

29. E. Bannai and T. Ito (1973): On finite Moore graphs, J. Fac. Sci. Univ. Tokyo, Sect IA 20, 191-208.

30. E. Bannai and T. Ito (1979): On the spectra of certain distance-regular graphs, J. of Combinatorial Theory (B), 27, 274-293.

31. E. Bannai and T. Ito (1981a): On the spectra of certain distance-regular graphs, II, Quart. J. Math. (Oxford) 32, 389-411.

32. E. Bannai and T. Ito (1981b): Regular graphs with excess one, Discrete Math. 37, 147-158.

33. E. Bannai and N. J. A. Sloane (1981): Uniqueness of certain spherical codes, Canad. J. Math. 33, 437-449.

34. M. R. Best (1978): On the existence of perfect codes, Math. Centre Report ZN82/78, Amsterdam.

35. M. R. Best (1982): A contribution to the non-existence of perfect codes, Ph.D. Thesis, Amsterdam.

36. T. Bier (to appear): A distribution-invariant for association schemes and strongly regular graphs, Linear Alg. and its Appls.

37. N. L. Biggs (1973): Perfect codes in graphs, J. of Combinatorial Theory (B) 15, 289-296.

38. N. L. Biggs (1974): Algebraic Graph Theory, Cambridge Univ. Press, Cambridge, 1974.

39. N. L. Biggs (1975): Designs, factors and codes in graphs, Quart. J. Math. (Oxford), 26, 113-119.

40. N. L. Biggs (1976): Automorphic graphs and the Krein condition, Geom. Dedicata 5, 117-127.

41. N. L. Biggs and T. Ito (1980): Graphs with even girth and small excess, Math. Proc. Cambridge Phil. Soc., 88, 1-10.

42. N. L. Biggs and D. H. Smith (1971): On trivalent graphs, Bull. London Math. Soc. 3, 155-158.

43. N. L. Biggs and A. White (1979): Permutation groups and combinatorial structures, London Math. Soc. Lecture Note Series 33, Cambridge Univ. Press.

44. A. Blokhuis (1981a): An upper bound for the cardinality of a set of equiangular lines in $\mathbb{R}^{d,1}$, T.H.E. Memorandum 1981, No. 2.

45. A. Blokhuis (1981b): A new upper bound for the cardinality of 2-distance sets in Euclidean space, T.H.E. Memorandum 1981, No. 4.

46. A. Blokhuis (1982): An upper bound for the cardinality of s-distance sets in E^d and H^d, T.H.E. Memorandum 1982, No. 8.

47. A. Blokhuis (1983): Few distance sets, Ph.D. Thesis, Eindhohen.

48. A. Blokhuis and N. M. Singhi (1981): Bounds on sets with few distances modulo a prime in metric spaces of strength t, T.H.E. Memorandum 1981, No. 16.

49. R. C. Bose (1963): Strongly regular graphs, partial geometries and partially balanced designs, Pacific J. Math. 13, 389-419.

50. R. C. Bose and T. A. Dowling (1971): A characterization of Moore graphs of diameter two, J. of Combinatorial Theory (B), 11, 213-226.

51. R. C. Bose and R. Laskar (1967): A characterization of tetrahedral graphs, J. of Combinatorial Theory, 3, 366-385.

52. R. C. Bose and D. M. Mesner (1959): On linear associative algebras corresponding to association schemes of partially balanced designs, Ann. Math. Statist 30, 21-38.

53. R. C. Bose and K. R. Nair (1939): Partially balanced incomplete block designs, Sankhya, 4, 337-372.

54. R. C. Bose and T. Shimamoto (1952): Classification and analysis of partially balanced incomplete block designs with two associate classes, J. Amer. Statist. Assoc., 47, 151-184.

55. N. Bourbaki (1968): Groupes et algebres de Lie, Chap. 4.5.6. Hermann.

56. R. Brauer (1945): On the representation of a group of order g in the field of the g-th root of unity, Amer. J. Math 67, 461-471.

57. R. Brauer (1947): Applications of induced characters, Amer. J. Math, 69, 709-716.

58. R. Brauer (1963): Representations of Finite groups, Lectures on Modern Mathematics, vol. I, 133-175, Wiley, N.Y.

59. R. Brauer (1968): On pseudo groups, J. Math. Soc. Japan, 20, 13-22.

60. R. Brauer (1980): Collected papers, vols. I, II, III, MIT Press.

61. R. Brauer and H. S. M. Coxeter (1940): A generalization of theorems of Schönhardt and Mehmke polytopes, Trans. R. S. Canad. 34, 29-34.

62. A. Bremner (1979): A diophantine equation arising from tight 4-designs, Osaka J. Math. 16, 353-356.

63. A. Brower, A. Cohen and A. Neumaier (to appear): The known distance-regular graphs, (in preparation).

64. A. Brower and J. H. van Lint (to appear): Strongly regular graphs and partial geometries, (preprint).

65. A. Brower and H. A. Wilbrink (1983): The structure of near polygons with quads, Geometriae Dedicata 14, 145-176.

66. F. Buekenhout (1979): Diagrams for geometries and groups, J. of Combinatorial Theory (A) 27, 121-151.

67. F. Buekenhout (1982): (g,d^*,d)-gons, in Proceedings of a conference in honor of T. G. Ostrom, M. Dekker.

68. F. Buekenhout (to appear): Diagram geometries for sporadic groups, (preprint).

69. F. Buekenhout and E. Shult (1974): On the foundations of polar geometry, Geom. Dedicata 3, 155-170.

70. A. R. Calderbank and W. M. Kantor (to appear): The geometry of two-weight codes, (preprint).

71. P. J. Cameron (1972): Permutation groups with multiply transitive suborbits, Proc. London Math. Soc. (3) 25, 427-440.

72. P. J. Cameron (1973a): Near-regularity conditions for designs, Geom. Dedicata 2, 213-223.

73. P. J. Cameron (1973b): Extending symmetric designs, J. of Combinatorial Theory (A) 14, 215-220.

74. P. J. Cameron (1975): Two remarks on Steiner systems, Geom. Dedicata 4, 403-418.

75. P. J. Cameron (1982a): Dual polar spaces, Geometriae Dedicata 12, 75-85.

76. P. J. Cameron (1982b): There are only finitely many distance-transitive graphs of given valency greater than two, Combinatorical (Hungary), 2, 9-13.

77. P. J. Cameron and M. Deza (1979): On permutation geometries, J. London Math. Soc. (2) 20, 373-386.

78. P. J. Cameron, J. M. Goethals and J. J. Seidel (1978a), Strongly regular graphs having strongly regular subconstituents, J. Algebra 55, 257-280.

79. P. J. Cameron, J. M. Goethals and J. J. Seidel (1978b), The Krein condition, spherical designs, Norton algebras and permutation groups, Indag. Math. 40, 196-206.

80. P. J. Cameron, J. M. Goethals, J. J. Seidel and E. E. Shults (1976): Line graphs, root systems, and elliptic geometry, J. Algebra 43, 305-327.

81. P. J. Cameron and W. M. Kantor (1978): Rank 3 groups and biplanes, J. of Combinatorial Theory (A) 24, 1-23.

82. P. J. Cameron and J. H. van Lint (1975): Graphs, codes and designs, London Math. Soc. Lecture Notes 43, Cambridge Univ. Press, Cambridge 1980. (Older version, London Math. Soc. Lecture Notes 19, 1975).

83. P. J. Cameron, C. E. Praeger, J. Saxl and G. M. Seitz (to appear): On a conjecture of Sims, (in preparation).

84. P. J. Cameron and J. J. Seidel (1973): Quadratic forms over GF(2), Indag. Math., 35, 1-8.

85. L. Carlitz (1954a): Representations by quadratic forms in a finite field, Duke Math. J. 21, 123-137.

86. L. Carlitz (1954b): Representations by skew forms in a finite field, Arch. Math. 5, 19-31.

87. L. Carlitz and J. Hodges (1955): Representations by Hermitian forms in a finite field, Duke Math. J. 22, 393-405.

88. E. Cartan (1929): Sur la determination d'un systeme orthogonel complet dans un espace de Riemann symmetrique clos, Rend. Circ. Mat. Palermo, 53, 217-252.

89. P. Cartier (1973): Harmonic analysis on trees, Harmonic Analysis on homogeneous spaces, ed. C. C. Moore: Proc. Sym. Pure Math. 26, Amer. Math. Soc. Providence, R.I., 419-424.

90. L. C. Chang (1959): The uniqueness and nonuniqueness of triangular association schemes, Sci. Record 3, 604-613.

91. T. S. Chihara (1978): An introduction to orthogonal polynomials, Gordon and Breach, N.Y.

92. A. M. Cohen (1976): Finite complex reflection groups, Ann. Sci. Ecole Norm. Sup 9, 379-436.

93. A. M. Cohen (1980): Finite Quaternionic reflection groups, J. Algebra, 64, 293-324.

94. A. M. Cohen (1981a): A synopsis of known distance-regular graphs with large diameters, Mathematisch Centrum Report ZW 168/81, Amsterdam, 1981.

95. A. M. Cohen (1981b): A characterization of subspaces of given rank in a projective space, Mathematisch Centrum Report ZW 165/81, Amsterdam, 1981.

96. W. S. Connor (1958): The uniqueness of the triangular association scheme, Ann. Math. Stat. 29, 262-266.

97. J. H. Conway (1969a): A group of order 8, 315, 553, 613, 086, 720, 000, Bull. London Math. Soc. 1, 79-88.

98. J. H. Conway (1969b): A characterization of Leech's lattice, Invent. Math. 7, 137-142.

99. J. H. Conway (1971): Three lectures on exceptional groups, in Finite Simple Groups, edited by Powell and Higman, Academic Press, 215-247.

100. J. H. Conway, R. A. Parker and N. J. A. Sloane (to appear): The covering radius of the Leech lattice, Proc. Royal Soc., (preprint).

101. J. H. Conway and N. J. A. Sloane (1982): Lorentzian forms for the Leech lattice, Bull. Amer. Math. Soc. 6, 215-127. (See many subsequent papers by them on related topics.)

102. J. H. Conway and N. J. A. Sloane (to appear): Twenty-three constructions for the Leech lattice, Proc. Royal Soc., (preprint).

103. B. N. Cooperstein (1977): A characterization of some Lie incidence structures, Geom. Dedicata 6, 205-258.

104. H. S. M. Coxeter (1934): Discrete groups generated by reflections, Annals Math. 35, 588-621.

105. H. S. M. Coxeter and W. O. J. Moser (1957): Generators and relations for discrete groups, Springer, N.Y. 3rd edition, 1972 (1st edition in 1957).

106. C. W. Curtis and T. V. Fossum (1968): On centralizer rings and characters of representations of finite groups, Math. Z. 107, 402-406.

107. C. W. Curtis, N. Iwahori and R. Kilmoyer (1971): Hecke algebras and characters of parabolic type of finite groups with (B,N)-pairs, Publ. Math. I.H.E.S. 40, 81-116.

108. C. W. Curtis and I. Reiner (1962): Representation Theory of finite groups and associative algebras, Interscience, N.Y.

109. D. M. Cvetkovic, M. Doob and H. Sachs (1980): Spectra of Graphs, Academic Press, N.Y.

110. D. M. Cvetkovic and J. H. van Lint (1977): An elementary proof of Lloyd's Theorem, Indag. Math. 39, 6-10.

111. R. M. Damerell (1973): On Moore graphs, Proc. Cambridge Philos. Soc. 74, 227-236.

112. R. M. Damerell (1981a): On Moore geometries, II, Math. Proc. Cambridge Phil. Soc. 90, 33-40.

113. R. M. Damerell (1981b): Distance-transitive and distance-regular digraphs, J. of Combinatorial Theory (B) 31, 46-53.

114. R. M. Damerell and M. A. Georgiacodis (1981a): On the maximum diameter of a class of distance-regular graphs, Bull. London Math. Soc. 13, 316-322.

115. R. M. Damerell and M. A. Georgiacodis (1981b): Moore geometries, I, J. London Math. Soc. (2) 23, 1-9.

116. P. Delsarte (1972): Bounds for unrestricted codes by linear programming, Philips Res. Rep. 27, 272-289.

117. P. Delsarte (1973a): Four fundamental parameters of a code and their combinatorial significance, Inform. Control 23, 407-438.

118. P. Delsarte (1973b): An algebraic approach to the association schemes of coding theory, Philips Research Reports Supplements, No. 10.

119. P. Delsarte (1974): The association schemes of coding theory, in Combinatorics (M. Hall, Jr. and J. H. van Lint, eds), Mathematical Center Tracts 55, Amsterdam, 139-157.

120. P. Delsarte (1976a): Association schemes and t-designs in regular semilattices, J. of Combinatorial Theory (A), 20, 230-243.

121. P. Delsarte (1976b): Properties and applications of the recurrence $F(i+1,k+1,n+1) = q^{k+1}F(i,k+1,n) - q^kF(i,k,n)$, SIMA J. Appl. Math. 31, 262-270.

122. P. Delsarte (1977): Pairs of vectors in the space of an association scheme, Philips Res. Rep. 32, 373-411.

123. P. Delsarte (1978a): Hahn polynomials, discrete harmonics, and t-designs, SIAM J. Appl. Math. 34, 157-166.

124. P. Delsarte (1978b): Bilinear forms over a finite field with applications to coding theory, J. of Combinatorial Theory (A), 25, 226-241.

125. P. Delsarte and J. M. Goethals (1975): Alternating bilinear forms over GF(q), J. of Combinatorial Theory (A) 19, 26-50.

126. P. Delsarte, J. M. Goethals and J. J. Seidel (1975): Bounds for systems of lines, and Jacobi polynomials, Philips Res. Reports 30, 91*-105*, Bouwkamp volume.

127. P. Delsarte, J. M. Goethals and J. J. Seidel (1977): Spherical codes and designs, Geom. Dedicata 6, 363-388.

128. P. Delsarte and R. J. McEliece (1976): Zeros of functions in finite abelian group algebras, Amer. J. Math. 98, 197-224.

129. P. Dembowski (1968): Finite geometries, Springer-Verlag, N.Y.

130. J. R. Derome and W. T. Sharp (1965): Racah algebra for arbitrary groups, J. Math. Phys. 6, 1584-1590.

131. M. Deza, P. Erdős and P. Frankl (1978): Intersection properties of systems of finite sets, Proc. London Math. Soc. 36, 369-384.

132. M. Deza and P. Frankl (1977): On the maximum number of permutations with given maximal or minimal distance, J. of Combinatorial Theory (A), 22, 352-360.

133. J. Dieudonné (1955): Le geometrie des groupes classiques, Springer.

134. L. Dornhoff (1971, 1972): Group representation theory (A), (B), Marcel Dekker, Inc., N.Y.

135. T. A. Dowling (1969): A characterization of the T_m-graph, J. of Combinatorial Theory 6, 251-263.

136. D. Drucker and D. Frohardt (1982): Irreducible root systems and finite linear groups of degree two, Bull. London Math. Soc., 142-148.

137. C. F. Dunkl (1976): A Krawtchouk polynomial addition theorem and wreath products of symmetric groups, Indiana Univ. Math. J. 25, 335-358.

138. C. F. Dunkl (1977): An addition theorem for some q-Hahn polynomials, Monats. Math. 85, 5-37.

139. C. F. Dunkl (1978): An addition theorem for Hahn polynomials: the spherical functions, SIAM J. Math. Anal. 9, 627-637.

140. C. F. Dunkl (1979a): Discrete quadrature and bounds on t-design, Mich. Math. J. 26, 81-102.

141. C. F. Dunkl (1979b): Orthogonal functions on some permutation groups, Proc. Symp. Pure Math. 34, Amer. Math. Soc., R.I., 129-147.

142. C. F. Dunkl (1980): Orthogonal polynomials in two variables of q-Hahn and q-Jacobi type, SIAM J. Alg. Disc. Math. 1, 137-151.

143. Y. Egawa (1981): Characterization of H(n,q) by the parameters, J. of Combinatorial Theory (A), 31, 108-125.

144. Y. Egawa (to appear): Association schemes of quadratic forms, J. of Combinatorial Theory (A), (preprint 1980).

145. H. Enomoto (1973a): Characterization of families of finite permutation groups by the subdegrees, II, J. Fac. Sci. Uni. Tokyo, Sect IA, 1-11.

146. H. Enomoto (1973b): Primitive extensions of maximal diameter of multiply transitive permutation groups, Proc. second int. Conf. Camberra, 288-297 (Springer Lecture Notes Series, 372.

147. H. Enomoto, N. Ito and R. Noda (1979): Tight 4-designs, Osaka J. Math. 16, 39-43.

148. A. Erdelyi, W. Magnus, F. Oberhettinger and F. G. Tricomi (1953): Higher Transcendental Functions (Bateman Manuscript Project), McGraw-Hill.

149. P. Erdős, Chao Ko and R. Rado (1961): Intersection theorems for systems of finite sets, Quart. J. Math. (Oxford), 12, 313-320.

150. W. Feit (1967): Characters of finite groups, Benjamin, N.Y. - Amsterdam.

151. W. Feit (1982): The Representation Theory of Finite Groups, North Holland, Amsterdam.

152. W. Feit and G. Higman (1964): The non-existence of certain generalized polygons, J. Algebra 1, 114-131.

153. J. S. Frame (1941): The double cosets of a finite group, Bull. Amer. Math. Soc., 47, 458-467.

154. P. Frankl (1976): The Erdős-Ko-Rado theorem is true for n = ckt, in Combinatorics, Proc. 5th Hungarian Coll. Combinatorics, Keszthely, 1976, Amsterdam, 365-375.

155. P. Frankl and R. M. Wilson (1981): Intersection theorems with geometric consequences, Combinatorica 1, 357-368.

156. G. Freud (1971): Orthogonal polynomials, Pergamon Press, Oxford.

157. F. G. Frobenius (1968): Gesammelte Abhandlungen I, II, III, Springer.

158. F. J. Fuglister (1977): On finite Moore geometries, J. of Combinatorial Theory (A), 23, 187-197.

159. F. J. Fuglister (1983): The nonexistence of Moore geometries of diameter 4, Discrete Math. 45, 229-238.

160. R. Fuji-Hara and S. A. Vanstone (1981): Equidistant permutation arrays from finite geometries, Congressus Numerantium 32, 333-345.

161. P. X. Gallagher (1979): Invariants for finite groups, Advances in Math. 34, 46-57.

162. P. X. Gallager and R. J. Proulx (1977): Orthogonal and unitary invariants of families of subspaces, in Contribution to Algebra: A collection of papers dedicated to Ellis Kolchin, Academic Press, N.Y.

163. R. Gangolli (1967): Positive definite kernels on homogeneous spaces and certain stochastic processes related to Levy's Brownian motion of several parameters, Ann. Inst. Henri Poincare 3, 121-225.

164. A. Gardiner (1973): Arc transitivity in graphs, Quart. J. Math. (Oxford) (2) 24, 399-407.

165. A. Gardiner (1974a): Arc transitivity in graphs, II, Quart. J. Math. (Oxford) 25, 163-167.

166. A. Gardiner (1974b): Antipodel covering graphs, J. of Combinatorial Theory (B) 16, 255-273.

167. A. Gardiner (1974c): Imprimitive distance-regular graphs and projective planes, J. of Combinatorial Theory (B), 16, 274-281.

168. A. Gardiner (1976): Arc transitivity in graphs, III, Quart. J. Math. (Oxford) 27, 313-323.

169. A. Gardiner (1978): When is an array realized by a distance-regular graph?, in Colloquia Math. Soc. Janos Bolyai, 25 Algebraic Methods in Graph Theory, Szeged (Hungary).

170. G. Gasper (1974): Projective formulas for orthogonal polynomials of a discrete variable, J. Math. Anal. Appl. 45, 176-198.

171. G. Gasper (1975): Positivity and special functions, Theory and Applications of Special Functions, R. Askey, ed., Academic Press, N.Y., 375-433.

172. Ja. L. Geronimus (1960): Polynomials orthogonal on a circle and interval, English translation, Pergamon Press, Oxford, 1960.

173. G. Glauberman (1969): Normalizers of p-subgroups in finite groups, Pacific J. Math. 29, 137-144.

174. R. Godement (1952): A theory of spherical functions I, Trans. Amer. Math. Soc. 73, 496-556.

175. J. M. Goethals (1969): A polynomial approach to linear codes, Philips Res. Rep. 24, 145-159.

176. J. M. Goethals and J. J. Seidel (1979a): Spherical designs, Proc. Symp. Pure Math. 34, Amer. Math. Soc., Providence, R.I., 255-272.

177. J. M. Goethals and J. J. Seidel (1979b): Cubature formulae, polytopes and spherical designs, in Proc. Coxeter Symposium, Toronto, May 1979, to appear.

178. J. M. Goethals and H. C. A. van Tilborg (1975): Uniformly packed codes, Philips Res. Repts 30, 9-36.

179. D. Gorenstein (1968): Finite Groups, Harper and Row, N.Y.

180. R. Griess (1981): A construction of F_1 as automorphisms of 196883 dimensional algebra, Proc. Nat. Acad. Sci. USA 78, 686-691.

181. W. Haemers (1981): Eigenvalue techniques in design and graph theory, Math. Cent. Tracts, 21, Amsterdam (thesis).

182. J. I. Hall and E. E. Shult (1981): Locally cotriangular graphs in Finite Geometries and designs, London Math. Soc. Lecture Note Series 49, 128-133.

183. M. Hall, Jr. (1959): The theory of groups, Macmillan, N.Y.

184. M. Hall, Jr. (1967): Combinatorial Theory, Blaisdell.

185. P. Hammond and D. H. Smith (1975a): Perfect codes in graphs O_k, J. of Combinatorial Theory (B), 19, 239-255.

186. P. Hammond and D. H. Smith (1975b): An analogue of Lloyd's theorem for completely regular codes, Proc. 5th British Combinatorial Conference, Aberdeen, 261-267.

187. D. Happel, U. Preiser and C. M. Ringel (1980): Binary polyhedral groups and Euclidean diagrams, Manuscripta Math. 31, 317-329.

188. D. K. Harrison (1979): Double coset and orbit spaces, Pacific J. Math. 80, 451-491.

189. H. Hasse (1949): Verallgemeinerung des Dualitätsatzes für die charactere endliche Gruppen, Math. Nachr. 3, 1-3.

190. S. Helgason (1978): Differential geometry, Lie groups, and symmetric spaces, Academic Press, N.Y., (Old version in 1962).

191. J. Hemmeter (to appear): Halved graphs, Johnson and Hamming graphs, Utilitas Math.

192. J. Hemmeter (in preparation).

193. E. Hewitt and K. A. Ross (1963, 1970): Abstract Harmonic Analysis, Springer vol. I (1963, 2nd edition 1979), vol. II (1970).

194. D. G. Higman (1964): Finite permutation groups of rank 3, Math. Z. 86, 145-156.

195. D. G. Higman (1967): Intersection matrices for finite permutation groups, J. Algebra 6, 22-42.

196. D. G. Higman (1974): Invariant relations, coherent configurations, and generalized polygons, Math. Cent. Tract, No. 57, 27-43.

197. D. G. Higman (1975): Coherent configurations, Part I, Ordinary representation theory, Geom. Dedicata 4, 1-32.

198. D. G. Higman (1976): Coherent configurations, Part II, Weights, Geom. Dedicata 5, 413-424.

199. G. Higman (1971): Construction of simple groups from character table, in Finite Simple Groups, edited by Powell and Higman, Academic Press, 205-214.

200. J. H. Hodges (1955): Exponential sums for symmetric matrices in a finite field, Math. Nachr. 14, 331-339.

201. J. H. Hodges (1956): Exponential sums for skey matrices in a finite field, Arch. Math. 7, 116-121.

202. A. J. Hoffman (1960): On the uniqueness of the triangular association schemes, Ann. Math. Stat. 31, 492-497.

203. A. J. Hoffman (1963): On the polynomial of a graph, Amer. Math. Monthly 70, 30-36.

204. A. J. Hoffman and R. R. Singleton (1960): On Moore graphs with diameters 2 and 3, IBM J. Res. Develop. 4, 497-504.

205. A. J. Hoffman and D. K. Ray-Chaudhuri (1965): On the line graph of a symmetric balanced incomplete block design, Trans. Amer. Math. Soc. 116, 238-252.

206. S. G. Hoggar (1978): Bounds for quaternionic line systems and reflection groups, Math. Scand. 43, 241-249.

207. S. G. Hoggar (1982): t-designs in projective spaces, Europ. J. Combinatorics 3, 233-254.

208. S. G. Hoggar (to appear): Parameters of t-designs, Europ. J. Combinatorics.

209. S. G. Hoggar (to appear): Parameters of association schemes, (preprint).

210. G. Hoheisel (1939): Über Charaktere, Monatsch f. Math. Phys. 48, 448-456.

211. Y. Hong (1982): On spherical t-designs in $\mathbb{R}^2$, Europ. J. of Combinatorics 3, 255-258.

212. Y. Hong (to appear): On the nonexistence of unknown perfect 6- and 8-codes in Hamming schemes $H(n,q)$ with q arbitrary, (preprint).

213. X. L. Hubaut (1975): Strongly regular graphs, Discrete Math. 13, 357-381.

214. D. R. Hughes (1965): On t-designs and groups, Amer. J. Math. 87, 761-778.

215. B. Huppert (1967): Endlichen Gruppen, I, Springer-Verlag.

216. M. E. H. Ismail and J. A. Wilson (1982): Asymptotic and generating relations for the q-Jacobi and ${}_4\varphi_3$ polynomials, J. of Approximation Theory 36, 43-54.

217. I. M. Issacs (1976): Character theory of finite groups, Academic Press, N.Y.

218. N. Ito (1975a): On tight 4-designs, Osaka J. Math. 12, 493-522. (Corrections and supplements, Osaka J. Math. 15 (1978), 693-697.)

219. N. Ito (1975b): ある種の4-デザインについて (On certain 4-designs), Sugaku (Mathematics, a publication of Math. Soc. of Japan) 27, 23-30 (in Japanese).

220. T. Ito (1974): Primitive rank 5 permutation groups with two doubly transitive constituents of different sizes, J. Fac. Sci. Univ. Tokyo, Sect IA 21, 271-277.

221. T. Ito (1982): Bipartite distance-regular graphs of valency three, Linear algebra and its application, 46, 195-213.

222. T. Ito and M. Kiyota (1981): Sharp permutation groups, J. of Math. Soc. Japan, 33, 435-444.

223. N. Iwahori (1964): On the structure of a Hecke ring of a Chevalley group over a finite field, J. Fac. Sci. Univ. Tokyo, Sect. I, 10, 215-236.

224. N. Iwahori and T. Yokonuma (1982): On self-dual, completely reducible finite subgroups of GL(2,k), J. Fac. Sci. Univ. Tokyo, Sect IA, 28, 829-842.

225. G. James and A. Kerber (1981): The representation theory of the symmetric group, Addison-Wesley Publ. Company, in Encyclopedia of Mathematics and its applications.

226. K. Johnson (1981): S-rings over loops, right mapping groups and transversals in permutation group, Math. Proc. Camb. Phil. Soc., 433-442.

227. W. M. Kantor (1972a): On incidence matrices of finite projective and affine spaces, Math. Z. 124, 315-318.

228. W. M. Kantor (1972b): k-homogeneous groups, Math. Z. 124, 261-265.

229. W. M. Kantor (1975): Rank 3 characterizations of classical geometries, J. Algebra 36, 309-313.

230. W. M. Kantor (1977): Moore geometries and rank 3 groups having $\mu = 1$, Quart. J. Math. (Oxford), 28, 309-328.

231. W. M. Kantor and R. A. Liebler (1982): The rank 3 permutation representations of the finite classical groups, Trans. Amer. Math. Soc. 271, 1-71.

232. S. Karlin and J. L. McGregor (1961): The Hahn polynomials, Formulas and an Application, Scripta Math. 26, 33-46.

233. Y. Kawada (1942): Über den Dualitätssatz der Charaktere nichtcommutativer Gruppen, Proc. Phys. Math. Soc. Japan (3), 24, 97-109.

234. R. Kilmoyer and L. Solomon (1973): On the theorem of Feit-Higman, J. of Combinatorial Theory (A), 15, 310-322.

235. M. Kiyota (1979): An inequality for finite permutation groups, J. of Combinatorial Theory (A), 27, 119.

236. T. H. Koornwinder (1973): The addition formula for Jacobi polynomials and spherical harmonics, SIAM J. Appl. Math. 25, 236-246.

237. T. H. Koornwinder (1976): A note on the absolute bound for systems of lines, Indag. Math. 38, 152-153.

238. T. H. Koornwinder (1981): Clebsch-Gordan coefficients for SU(2) and Hahn polynomials, Nieuw Archief Voor Wiskunde (3), 29, 140-155.

239. T. H. Koornwinder (1982): Krawtchouk polynomials, a unification of two different group theoretical interpretations, SIAM J. Math. Anal. 13, 1011-1023.

240. M. G. Krein (1949): A principle of duality for a bicompact group and square block-algebra, Dokl. Akad. Nauk SSSR, N.5, 69, 725-728 (in Russian).

241. M. G. Krein (1949, 1950): Hermitian-positive kernels on homogeneous spaces, I and II (in Russian), Ukranian Math. J. (No. 4, 1949), 64-94 and (No. 1, 1950), 10-59, English translation in Amer. Math. Soc. Translations (2) 34 (1963), 69-108 and 109-164.

242. C. W. H. Lam (1980): Distance transitive digraphs, Discrete Math. 29, 265-274.

243. D. G. Larman, C. A. Rogers and J. J. Seidel (1977): On two distance sets in Euclidean spaces, Bull. London Math. Soc. 9, 261-267.

244. R. Lasser (to appear): Orthogonal polynomials and hypergroups: contribution to analysis and probability theory, Technische Universität München, (preprint)

245. J. Leech (1967): Notes on sphere packings, Canad. J. Math. 19, 251-267.

246. P. W. H. Lemmens and J. J. Seidel (1973a): Equi-angular lines, J. Algebra 24, 494-512.

247. P. W. H. Lemmens and J. J. Seidel (1973b): Equi-isoclinic subspaces of Euclidean spaces, Indag. Math. 35, 98-107.

248. H. W. Lenstra, Jr. (1972): Two theorems on perfect codes, Discrete Math. 3, 125-132.

249. D. A. Leonard (to appear): Parameters of association schemes that are both P- and Q-polynomial, J. of Combinatorial Theory (A), (preprint, 1980).

250. D. A. Leonard (1982): Orthogonal polynomials, duality, and association schemes, SIAM J. Math. Anal. 13, 656-663.

251. R. A. Liebler (1977): On the uniqueness of the tetrahedral association schemes, J. of Combinatorial Theory (B) 22, 246-262.

252. J. H. van Lint (1971): Coding Theory, Lecture Notes in Math. No. 201, Springer, N.Y./Berlin.

253. J. H. van Lint (1974): Recent results on perfect codes and related topics, Math. Center Tracts 55, Mathematisch Centrum, Amsterdam, 158-178.

254. J. H. van Lint and J. J. Seidel (1966): Equilateral point sets in elliptic geometry, Indag. Math. 28, 335-348.

255. D. Livingstone and A. Wagner (1965): Transitivity of finite permutation groups on unordered sets, Math. Z. 90, 393-403.

256. S. P. Lloyd (1957): Binary block coding, Bell System Tech. J. 36, 517-535.

257. P. Lorimer (1973): Finite projective planes and sharply 2-transitive subsets of finite groups, Proc. Second Intern. Conf. Theory of Groups, Camberra, 1973, Springer Lecture Notes Series of Math., 372, 432-436.

258. L. Lovász (1979): On the Shannon capacity of a graph, IEEE Trans. Inform. Theory IT-25, 1-7.

259. H. D. MacPherson (1982): Infinite distance-transitive graphs of finite valency, Combinatorica (Hungary) 2, 63-69. (See also, ibid. 15-23.)

260. F. J. MacWilliams (1963): A theorem on the distribution of weights in a systematic code, Bell Syst. Tech. J. 42, 79-94.

261. F. J. MacWilliams and N. J. A. Sloane (1977): The theory of error-correcting codes, North-Holland, Amsterdam and Elsevier/North-Holland, N.Y. (See many references in the bibliography of this book.)

262. F. J. MacWilliams, N. J. A. Sloane and J. M. Goethals (1972): The MacWilliams identities for nonlinear codes, Bell System Tech. J. 51, 803-819.

263. R. Mathon (1975): Three-class association schemes, in Proc. Conf. on Algebraic Aspect of Combinatorics, Toronto, 123-155.

264. G. Maxwell (1978): Hyperbolic trees, J. Algebra 54, 46-49.

265. R. J. McEliece (1977): The theory of information and coding, A mathematical framework for communication, in Encyclopedia of Math. and its Applications, Addison-Wesley, Mass.

266. R. J. McEliece (1979): The bounds of Delsarte and Lovasz and their applications to coding theory, CISM Courses and Lectures No. 258, Springer-Verlag, Wien and N.Y.

267. R. J. McEliece, E. R. Rodemich, H. C. Rumsey, Jr. and L. R. Welch (1977): New upper bound on the rate of a code via the Delsarte-MacWilliams inequalities, IEEE Trans. Inf. Theory It-23, 157-166.

268. J. McKay (1980): Graphs, singularities, and finite groups, Proc. of Symp. in Pure Math. 37, 183-186.

269. J. McKay (1981): Cartan matrices, finite groups of quaternions, and Kleinian Singularities, Proc. Amer. Math. Soc. 81, 153-154.

270. A. Moon (1982): Characterization of the odd graphs O_k by parameters, Discrete Math. 42, 91-97.

271. A. Moon (1983a): Characterization of the graphs of the Johnson schemes $G(3k,k)$ and $G(3k+1,k)$, J. of Combinatorial Theory (B), 213-222.

272. A. Moon (1983b): On the uniqueness of the graphs $G(n,k)$ of the Johnson schemes, J. of Combinatorial Theory (B), 256-264.

273. H. Nagao (1957): On the groups with the same table of characters as symmetric groups. J. Inst. Polytech. Osaka City Univ. Ser. A 8, 1-8.

274. H. Nagao (1974): 群とデザイン (Groups and Designs), Iwanani, Tokyo (in Japanese).

275. A. Neumaier (1980): Distances, graphs and designs, Europ. J. Combinatorics 1, 163-174.

276. A. Neumaier (1981a): Distance matrices and n-dimensional designs, Europ. J. Combinatorics 2, 165-172.

277. A. Neumaier (1981b): New inequalities for the parameters of an association scheme, in Combinatorics and Graph Theory, Springer Lecture Notes in Math. 885, 365-367.

278. A. Neumaier (1981c): Combinatorial configurations in terms of distances, T.H.E. Memorandum 81, No. 09, p. 98.

279. A. Neumaier: Pseudo classical distance-regular graphs, (Manuscript).

280. A. Neumaier (1982): The second largest eigenvalue of a tree, Lin. Alg. and its Appl. 46, 9-25.

281. A. Neumaier (to appear): Regular sets and quasi-symmetric 2-designs, (preprint).

282. A. Neumaier and J. J. Seidel (to appear): Discrete hyperbolic geometry, Combinatorica (Hungary).

283. H. V. Niemeier (1973): Definite quadratische Formen der Dimension 24 und Diskriminante 1, J. Number Theory 5, 142-178.

284. P. M. Neumann (1975): Generosity and characters of multiply transitive permutation groups, Proc. London Math. Soc. 31, 457-481.

285. R. Noda (1972): Steiner systems which admit block transitive automorphism groups of small rank, Math. Z. 125, 113-121.

286. R. Noda (1976): Some inequalities for t-designs, Osaka J. Math. 13, 361-366.

287. R. Noda (1979): On orthogonal arrays of strength 4 achieving Rao's bound, J. London Math. Soc., 19, 385-390.

288. A. M. Odlyzko and N. J. A. Sloane (1979): New bounds on the number of unit spheres that can touch a unit sphere in n dimensions, J. of Combinatorial Theory (A), 26, 210-214.

289. M. Ogasawara (1965): A necessary condition for the existence of regular and symmetrical PBIB designs of T_m type, Univ. North Carolian Inst. Statistics, Mimeo Series No. 418.

290. U. Ott (1981a): Some remarks on representation theory in finite geometry, in Geometries and Groups, Springer Lecture Notes 893, 68-110.

291. U. Ott (1981b): Bericht über Hecke Algebren und Coxeter Algebren endlicher Geometrien, in Finite Geometries and Designs, London Math. Soc. Lecture Note Series 49, 260-271.

292. U. Ott (1981c): Eine Bemerkung über Polaritäten eines verallgemeinerten Hexagons, Geom. Dedicata, 11, 341-345.

293. T. Oyama (1964): On the groups with the same table of characters as alternating groups, Osaka J. Math. 1, 91-101.

294. C. L. Peterson (1977): On tight 6-designs, Osaka J. Math. 14, 417-435.

295. A. Ja. Petrenjuk (1968): Fisher's inequality for tactical configurations, Math. Zametki, 4, 417-424 (in Russian).

296. W. L. Quirin (1971): Primitive permutation groups with small orbitals, Math. Z. 122, 267-274.

297. S. B. Rao, D. K. Ray-Chaudhuri and N. M. Singhi (to appear): On imprimitive association schemes, (preprint).

298. D. K. Ray-Chaudhuri and A. P. Sprague (1976): Characterization of projective incidence structures, Geom. Dedicata 5, 361-376.

299. D. K. Ray-Chaudhuri and A. P. Sprague (1979): A combinatorial characterization of attenuated spaces, Utilitas Math. 15, 3-29.

300. D. K. Ray-Chaudhuri and R. M. Wilson (1975): On t-designs, Osaka J. Math. 12, 737-744 (original version: mimeographed notes, 1971).

301. H. F. H. Reuvers (1977): Some non-existence theorems for perfect codes over arbitrary alphabets. Ph.D. Thesis, Tech. Univ. of Eindhoven, January.

302. C. Roos and A. J. van Zanten (to appear): On the existence of certain generalized Moore geometries, I, II, (preprint).

303. K. A. Ross (1977): Hypergroups and centers of measure algebras, in Symposia Mathematics, Instituto Nazionale di Alta Matematica, vol. XXII, 189-203, Academic Press, London, (see also many papers on hypergroups in the references of this paper).

304. J. Saxl (1975): Characters of multiply transitive permutation groups, J. Algebra 34, 528-539.

305. J. Saxl (1981): On multiplicity-free permutation representations, in Finite Geometries and Designs, London Math. Soc. Lecture Notes Series 49, Cambridge Univ. Press, 337-353.

306. A. Schrijver (1979a): Packing and covering in combinatorics, Math. Center Tracts 106, Mathematisch Centrum, Amsterdam.

307. A. Schrijver (1979b): A comparison of the Delsarte and Lovasz bounds, IEEE Trans. Inform. Theory IT-25, 425-429.

308. A. Schrijver (1981): Association schemes and the Shannon capacity: Eberlein polynomials and the Erdős-Ko-Rado theorem, in Algebraic Method in Graph Theory, L. Lovasz and V. T. Sos, eds., North-Holland, Amsterdam-Oxford-N.Y., 671-688.

309. I. Schur (1933): Zur Theorie der einfach transitiven Permutationsgruppen, S. B. Preuss. Akad. Wiss., Phys.-Math. Kl, 598-623.

310. I. Schur (1973): Gesammelte Abhandlungen I, II, III, Springer.

311. L. L. Scott (1971): The modular theory of permutation representations, Proceedings of Symp. in Pure Math. 21, 137-144.

312. L. L. Scott (1973): Modular permutation representations, Trans. Amer. Math. Soc. 175, 101-121.

313. L. L. Scott (1977): Some properties of character products, J. Algebra 45, 259-265 (original version: A condition on Higman's parameters, Amer. Math. Soc. Notices, January 1973, 701-20-45).

314. W. S. Scott (1964): Group Theory, Prentice-Hall, New Jersey.

315. J. J. Seidel (1955): Angles and distances in n-dimensional Euclidean and non-Euclidean geometry, I, II, III, Indag. Math. 17, 329-335, 336-340, 535-541.

316. J. J. Seidel (1976a): A survey of two graphs, Proc. Intern. Coll. Theorie Combinatorie (Roma 1973), Accad. Naz. Lincei, 481-511.

317. J. J. Seidel (1976b): Eutactic stars (1976): Proc. 5th Hungarian Colloquium on Combinatorics, Keszthely.

318. J. J. Seidel (1979): Strongly regular graphs, 157-180 in B. Bollobas, Surveys in Combinatorics, London Math. Soc. Lecture Note Series 38.

319. J. J. Seidel and D. E. Taylor (1981): Two graphs, a second survey, in Algebraic methods in graph theory, vol. 2, eds. Lovasz and Sos, North-Holland, 689-711.

320. A. Selberg (1956): Harmonic analysis and discontinuous group in weakly symmetric Riemannian space with applications to Dirichlet series, J. Ind. Math. Soc. 20, 47-87.

321. J. P. Serre (1970): Cours d'Arithmetique, Presses Univ. de France.

322. J. P. Serre (1971): Representations lineaires des groupes finis, Hermann, Paris.

323. P. Seymour and T. Zaslavsky (to appear): Averaging sets: A generalization of mean values and spherical designs, Advances in Math.

324. G. C. Shephard and J. A. Todd (1954): Finite unitary reflection groups, Canad. J. Math. 6, 274-304.

325. S. S. Shrikhande (1959): The uniqueness of the L_2 association schemes, Ann. Math. Stat. 30, 781-798.

326. E. E. Shult (1974): Groups, polar spaces and related structures, 130-161, Math. Centre Tracts 57, Math. Centrum, Ansterdam.

327. E. E. Shult and A. Yanushka (1980): Near n-gons and line systems, Geom. Dedicata, 9, 1-72.

328. V. M. Sidel'nikov (1974): New bounds for the density of sphere packings in an n-dimensional Euclidean space, Mat. Sbornik 95 (1974), English translation Math. USSR Sbornik 24, 147-157.

329. C. C. Sims (1967): Graphs and finite permutation groups, I, Math. Z. 95, 76-86.

330. C. C. Sims (1968): Graphs and finite permutation groups, II, Math. Z. 103, 276-281.

331. R. R. Singleton (1966): On minimal graphs of maximal even girth, J. of Combinatorial Theory 1, 306-332.

332. N. J. A. Sloane (1975): An introduction to association schemes and coding theory, in Theory and Application of special functions, edited by R. A. Askey, Academic Press, N.Y., 225-260.

333. N. J. A. Sloane (1977): Error-correcting codes and invariant theory: New applications of a nineteenth-century technique, Amer. Math. Monthly, 84, 82-107.

334. N. J. A. Sloane (1982): Recent bounds for codes, sphere packings and related problems obtained by linear programming and other methods, Amer. Math. Soc., (Contemporary Math 9), 153-185.

335. D. H. Smith (1971): Primitive and imprimitive graphs, Quart. J. Math. (Oxford) 22, 551-557.

336. D. H. Smith (1973): On tetravalent graphs, J. London Math. Soc. 6, 659-662.

337. D. H. Smith (1974a): Distance-transitive graphs of valency four, J. London Math. Soc. 8, 377-384.

338. D. H. Smith (1974b): On bipartite tetravalent graph, Discrete Math. 10, 167-172.

339. D. H. Smith (1976): An improved version of Lloyd's theorem, Discrete Math. 15, 175-184.

340. J. D. H. Smith (1976): Centralizer rings of multiplicative groups on quasigroups, Math. Proc. Comb. Phil. Soc. 79, 427-431.

341. S. D. Smith (1977): Non-associative commutative algebras for triple covers of 3-transposition groups, Michigan Math. J. 24, 273-287.

342. A. P. Sprague (1978): Characterization of Projective graphs, J. of Combinatorial Theory (B), 24, 294-300.

343. A. P. Sprague (1979a): A characterization of 3-nets, J. of Combinatorial Theory (A), 27, 223-254.

344. A. P. Sprague (1979b): Incidence structures with specific planes, Proc. Sympos. Pure Math. 34, Amer. Math. Soc., Providence, R.I.

345. A. P. Sprague (1981a): Pasch's axiom and projective spaces, Discrete Math. 33, 79-87.

346. A. P. Sprague (1981b): Incidence structures whose planes are nets, Europ. J. Combinatorics 2, 193-204.

347. R. P. Stanley (1979): Invariants of finite groups and their applications to combinatorics, Bull. Amer. Math. Soc. 1, 475-511.

348. R. P. Stanley (1982): Some aspects of groups acting on finite posets, J. of Combinatorial Theory (A), 32, 132-161.

349. D. Stanton (1980a): Some q-Krawtchouk polynomials on Chevalley groups, Amer. J. Math. 102, 625-662.

350. D. Stanton (1980b): Some Erdős-Ko-Rado theorems for Chevalley groups, SIAM J. Alg. Disc. Math. 1, 160-163.

351. D. Stanton (1981a): Three addition theorems for some q-Krawtchouk polynomials, Geom. Dedicata 10, 403-425.

352. D. Stanton (1981b): A partially ordered set and q-Krawtchouk polynomials, J. of Combinatorial Theory (A), 30, 276-284.

353. D. Stanton (1983): Generalized n-gons and Chebychev polynomials, J. of Combinatorial Theory (A), 34, 15-27.

354. D. Stanton (to appear): Orthogonal polynomials and Chevalley groups, (preprint).

355. D. Stanton (to appear): Harmonics on posets, (preprint).

356. R. Steinberg (1951): A geometric approach to the representations of the full linear group over a Galois field, Trans. Amer. Math. Soc. 41, 279-282.

357. R. Steinberg: Subgroups of SU_2 and Dynkin diagrams, (manuscript).

358. R. J. Stroeker (1981): On the diophantive equation $(2y^2 - 3)^2 = x^2(3x^2 - 2)$ in connection with the existence of nontrivial tight 4-designs, Indag. Math. 43, 353-358.

359. M. Suzuki (1961): On a finite group with a partition, Arch. Math. 12, 241-254.

360. M. Suzuki (1977, 1978): 群論 (上・下) (Group Theory I, II), Iwarami, 1974, Tokyo (in Japanese), English translation, Springer-Verlag (1982).

361. G. Szegő: Orthogonal polynomials, Amer. Math. Soc. Colloq. Publications 23, Amer. Math. Soc., Providence, R.I., 4th edition 1975, (1st edition in 1939).

362. G. Szegő (1982): Collected Papers, 3 vols, Birkhäuser.

363. M. Takeuchi (1975): 現代の球関数 (Modern spherical function theory) Iwanami, Tokyo.

364. O. Tamaschke (1963a): Zur Theorie der Permutationsgruppen mit regulärer Untergruppe, I; Math. Z. 80, 328-354.

365, O. Tamaschke (1963b): Zur Theorie der Permutationsgruppen mit regulärer Untergruppe, II, Math. Z. 80, 443-465.

366. O. Tamaschke (1964): S-rings and the irreducible representations of finite groups, J. Algebra 1, 215-232.

367. O. Tamaschke (1970): Schur-Ringe, BI-Hochschulskripten 735 a* Bibliographisches Institut, Mannheim, (also see the references at the end of this book).

368. T. Tannaka (1938): Über die Dualitätsatz der nichtkommutativen topologischen Gruppen, Tohoku Math. J. 45, 1-12.

369. D. E. Taylor (1977): Regular two graphs, Proc. London Math. Soc. 35, 257-274.

370. D. E. Taylor and R. Levingston (1978): Distance-regular graphs, in Combinatorial Mathematics (D. A. Holton and J. Seberry, eds.), Springer Lecture Note Series 686, 313-323.

371. P. Terwilliger (1982a): The diameter of bipartite distance-regular graphs, J. of Combinatorial Theory (B), 32, 182-188.

372. P. Terwilliger (1982b): Eigenvalue multiplicities of highly symmetric graphs, Discrete Math. 41, 295-302.

373. P. Terwilliger (1983): Distance-regular graphs and (s,c,a,k)-graphs, J. of Combinatorial Theory (B), 34, 151-164.

374. P. Terwilliger (to appear): Distance-regular graphs with girth 3 or 4, I and II, (preprint).

375. J. G. Thompson (1970): Bounds for orders of maximal subgroups, J. Algebra, 14, 135-138.

376. A. Tietäväinen (1973): On the non-existence of perfect codes over finite fields, SIAM J. Appl. Math. 24, 88-96. (See also, A. Tietäväinen and A. Perko, Ann. Univ. Turku. Ser. AI, 148.)

377. A. Tietäväinen (1977): Non-existence of nontrivial perfect codes in case $q = p_1^s p_2^t$, $e \geq 3$, Discrete Math. 17, 199-205.

378. H. C. A. van Tilborg: Uniformly packed codes, Ph.D. thesis, Tech. Univ. of Eindhoven, June 1976.

379. J. Tits (1955): Sur certaines classes d'espaces homogenes de groupes de Lie, Memoires Acad. Roy. Belg. Cl. Sci. 29, fasc. 3.

380. J. Tits (1974): Buildings of spherical types and finite BN-pairs, Springer Lecture Notes, 386.

381. J. Tits (1981): A local approach to buildings, in Geometric Vein, 519-547, Springer-Verlag.

382. T. Tsuzuku (1961): On multiple transitivity of permutation groups, Nagoya Math. J. 18, 93-109.

383. W. T. Tutte (1947): A family of cubical graphs, Proc. Cambridge Philos. Soc. 43, 459-474.

384. N. J. Vilenkin (1965): Special functions and the theory of group representations (in Russian), Moscow, Izdat. "Nauka" 1965, English translation published by Amer. Math. Soc., Providence, R.I., 1968.

385. E. de Vries and A. J. van Zanten (1970): On the number of roots of certain equations in finite simply reducible groups, Physica 49, 536-548.

386. H. C. Wang (1952): Two point homogeneous spaces, Annals Math. 55, 177-191.

387. H. N. Ward (1974): Quadratic residue codes and symplectic groups, J. of Algebra, 29, 150-171.

388. R. Weiss (1976): Über lokal-s-reguläre Graphen, J. of Combinatorial Theory (13), 20, 124-217.

389. R. Weiss (1979): Groups with a (B,N)-pair and locally transitive graphs, Nagoya Math. J. 74, 1-21.

390. R. Weiss (1981): The non-existence of 8-transitive graphs, Combinatorica (Hungary) 1, 309-311.

391. H. Wielandt (1935): Zur Theorie der einfach transitiven Permutationsgruppen, Math. Zeit 40, 582-587.

392. H. Wielandt (1949): Zur Theorie der einfach transitiven Permutationsgruppen, II, Math. Zeit 52, 384-393.

393. H. Wielandt (1956): Primitive Permutations gruppen vom Grad 2p, Math. Zeit, 63, 478-485.

394. H. Wielandt (1964): Finite permutation groups, Academic Press, N.Y.

395. E. P. Wigner (1959): Group Theory and Its Applicatior to Quantum Mechanics of Atomic Spectra, Academic Press, N.Y.

396. E. Witt (1938): Über Steinersche Systeme, Abh. Math. Sem. Univ. Hamburg 12, 265-275.

397. E. Witt (1941): Spiegelungsgruppen und Aufzählung halbeinfacher Liescher Ringe, Abh. Math. Sem. Hamburg 14, 289-322.

398. J. A. Wilson (1978): Hypergeometric series recurrence relations and some new orthogonal functions, Ph.D. thesis, University of Wisconsin, Madison, 1978.

399. J. A. Wilson (1980): Some hypergeometric orthogonal polynomials, SIAM J. Math. Anal. 11, 690-701.

400. R. M. Wilson (1972): An existance theory for pairwise balanced designs, I, II, III, J. of Combinatorial Theory (A), 13, 220-245, 246-273, and J. of Combinatorial Theory (A), 18 (1975), 71-79.

401. J. A. Wolf (1974): Spaces of constant curvature; Publish or Perish, Boston MA, 3rd edition 1974.

402. W. J. Wong (1967): Determination of a class of primitive permutation groups, Math. Z. 99, 235-246.

403. S. Yamamoto, Y. Fujii and N. Hamada (1965): Composition of some series of association algebras, J. Sci. Hiroshima Univ. Ser A-I 29, 181-215.

404. K. Yamamoto (1965): A necessary condition for the existence of partially balanced block designs with an m-subset association scheme, the Memoirs of the Faculty of Science, Kyushu University, Series A, Math., 76-98.

INDEX

IMPORTANT NOTATION